T0093969

Digital Forensics for Handheld Devices

Digital Forensics for Handheld Devices

Eamon P. Doherty

CRC Press is an imprint of the
Taylor & Francis Group, an informa business

CRC Press
Taylor & Francis Group
6000 Broken Sound Parkway NW, Suite 300
Boca Raton, FL 33487-2742

Version Date: 20120626

International Standard Book Number: 978-1-4398-9877-2 (Hardback)

Visit the Taylor & Francis Web site at
http://www.taylorandfrancis.com

and the CRC Press Web site at
http://www.crcpress.com

This book is dedicated to my lovely wife Ester, my mom, my sister-in-law Elly,
and the memory of my dad, Edward T. Doherty

Contents

Preface

This book was written to teach someone about all the areas of mobile device forensics, which include topics from the legal, technical, academic, and social aspects of the discipline. It is hoped that the reader will now have an idea how to use a variety of digital forensic tools to examine flash drives, cell phones, PDAs, digital cameras, and netbooks. It is also hoped that the reader will understand the differences between a corporate investigation and a criminal investigation as well as some of the issues regarding privacy and the Fourth Amendment. The book ends with a discussion of the education and certifications that one needs for many possible careers in mobile device forensics.

Disclaimer

The views expressed in this book are Dr. Doherty's own views and ideas and do not necessarily represent the ideas and viewpoints of his employer or any organization he is associated with. This is an academic book that explores ideas, mobile device forensic techniques, and tools. This book is not a forensic manual or a legal manual. Dr. Doherty is not endorsing any products and was not paid to endorse any products. One should consult their organization's general counsel and follow the policies of the organization before attempting anything from the book.

Author

Eamon P. Doherty, PhD, CCE, SSCP, CPP, is an associate professor and the Cybercrime Training Lab director at Fairleigh Dickinson University (FDU), New Jersey. Dr. Doherty is a member of the High Tech Crimes Investigative Association, ASIS International, the FBI Infraguard, the American College of Forensic Examiners Institute, the FDU Digital Forensics Club, the IACSP, and the American Society of Digital Forensics & eDiscovery. Dr. Doherty has also assisted with some law enforcement cell phone investigations and is the chairman of the New Jersey Regional Homeland Security Technology Committee. Dr. Doherty previously worked for Morris County Government in their M.I.S./I.S.D. section. Presently, Dr. Doherty has developed and taught many continuing education classes for FDU on the subjects of cell phone forensics, PDA forensics, and digital camera forensics.

CHAPTER 1

The Cell Phone

THE CELL PHONE IS INVENTED

The telephone is the ancestor of the cell phone, so it is advisable to first learn something about the history of the telephone. The telephone was invented in 1876 by Alexander Graham Bell. In 1877, President Rutherford B. Hayes had a telephone installed in the White House. By 1878, there was a commercial telephone exchange in New Haven, Connecticut with 21 subscribers and 8 landlines [1]. People could then make and receive calls with the limited number of people who had a telephone. The telephone then grew in popularity as millions of people worldwide became subscribers. One serious complaint about the telephone was that one had to be within earshot of the bell ringer to hear and answer a call.

People then told the telephone company that they wanted a phone that they could take with them and not be tethered to a wire. One example of a profession of people who needed wireless telephones was the private detectives, known as the P.I. The P.I. often needed to interview people in the field as well as collect documents from the local court house. They also needed to be in touch with people for frequent updates about leads in existing investigations. Having to be at prespecified times and locations to receive calls made their work inefficient and lowered productivity. As other professions voiced a need for a portable telephone system, a wireless phone then became a research and development goal of the telephone companies. Bell Research Labs and Motorola both had research and development initiatives to invent a wireless telephone which could connect to the existing telephone networks. In 1973, Dr. Martin Cooper of Motorola invented the first modern wireless telephone. The patent number for the telephone was 3,906,166 [2]. The first wireless phone call is reported as being from Dr. Cooper to his rival Joel Engel at Bell Research Labs [3]. This wireless telephone system used analog signaling just as the landline telephone did, but Dr. Copper is still generally considered the father of the modern cell phone. Dr. Cooper's first phone was the analog Motorola Dyna-Tac. This phone weighed approximately two and a half pounds and sold for about US$4000.

Five years after Dr. Cooper filed for his patent, there were many customers in Chicago. The early analog cell phone networks that existed soon reached maximum caller capacity, so capacity was increased by switching to digital networks [3]. Analog signals have a large range of values while digital signals are either a one or a zero. Digital signals are simple because a value is either one or zero and then error correcting codes can be embedded in the signal thus increasing the accuracy of the transmission and sound quality. An example of an analog protocol used on early wireless telephones is known as frequency distributed multiplexing (FDM) where the signal is distributed along a wide bandwidth

[3]. Two examples of digital protocols commonly found on older cell phones are the collision detection multiple access (CDMA) and the time division multiple access (TDMA). The 800 MHz band was used for the original cell phone services. However, some old cell phones with TDMA and CDMA protocols have an analog mode which makes them compatible with old networks in certain parts of the world where digital modes are not available.

Another motivation for learning such material about digital protocols can also be made to appear more credible in court. If an investigator ever testifies as an expert witness, also known as an expert, he or she may be asked technical details about the first calls and the protocols used. This can happen during a process known as "Voir Dire" where an expert is questioned to determine his or her ability to be an expert [4]. If an expert does not know who made the first cell phone call and when it took place, the opposing counsel might use that simple omission of knowledge to decrease the credibility of the expert witness. It would also be advisable to understand a few of the common protocols and memorize their definitions for the same reason.

In 1999, the United Nations Secretary General Kofi Annan said to the International Telecommunication Union that "half the world's people have never made or received a telephone call" [5]. Fast forward in time, 11 years to 2011, and see how that has changed.

"The planet has 6.9 billion people alive, from babies to great grandparents. Now there is an active mobile phone subscription for 75% of them" [6]. That means that they rent or own a cell phone. That means that three out of four people on the planet use a cell phone as part of their activities of daily living, and some of these people commit crimes, which means a significant number of cell phones may hold some evidence related to a crime. Two examples of such evidence might be some people who called accomplices before a robbery or others who used camera phones to take pictures of a contracted killing to give proof to a client. It is important to consider that a tremendous number of people worldwide have cell phones and law enforcement may need to examine them as part of a lawful investigation.

In 2011, there are some unofficial estimates that perhaps two thirds of humanity may have a cell phone. Many of these modern cell phones also support the use of the following: email, web surfing, picture taking, video creation, short message service (SMS) text messaging, and phoning people. In this chapter, we will learn about the history of the cell phone, some examples of cell operating systems, some criminal activities that the cell phone may be party to, and then learn how to examine the cell phone. There are a variety of tools that can be used to examine cell phones from around the world. In this chapter, we will survey some cell phone forensic tools, discuss an examination machine, and learn about some precautions to take for high-profile investigations.

CELL PHONE MODELS AND CELL PHONE MUSEUMS

There are a number of museums and associations where one can see both old and new cell phones. These places also contain numerous resources to learn about the various models that exist. It is important to know about old cell phones because some criminals may specifically choose and utilize them with the rationale that investigators will be unfamiliar with old technology. There is a museum in South Korea that has a very extensive collection of both antique and modern telephones, including cell phones from all over the world. The privately owned museum and its director Lee Byeong-Cheol was featured in the *Korean Times* [7].

The Infoage Science/History Learning Center in Wall, New Jersey has many exhibits that enable the visitor to peruse the history and exhibits of the telephone and other branches of telecommunication. Some of the volunteer tour guides there have retired from the telecommunication industry and can discuss some technical aspects of the traditional telephone networks as well as the newer cell phone networks. Some academics say that it is important to understand the history of the telephone since the cell phone is an offshoot of the land-based telephone.

If one cannot make a visit to Wall, New Jersey, there is also the virtual telecommunication museum which is online and has various topics related to telephony [8]. This museum is convenient because it is online and does not have to be visited in person. There are also references on the website which is important for validating certain facts and also allows the viewer to seek additional knowledge. Al's Cell Phone Museum is also another online museum that offers close-up pictures of many models [9].

There is also another value for cell phone museums which may be less obvious to the reader. Suppose there is an old model of a rare cell phone that is seized from a suspect. The cell phone may be damaged by the suspect upon surrender. Cell phone forensic tools may not be usable on the phone in its present condition. It may be necessary to contact the museum and ask for another phone of the same type. Then internal storage areas may be swapped so that the device can be powered on and examined. This is not an ideal situation because the examiner wishes to preserve evidence as much as possible. However, if the situation and motivations for such actions are documented, then it should most likely be acceptable for investigation purposes.

I once assisted the Morris County Prosecutor's Office with a cell phone investigation. The cell phone was damaged; it appeared that the suspect had taken a tool such as a screwdriver and damaged the port where the USB cable connects between the cell phone and the computer. I suggested that it might be necessary to contact someone with a cell phone museum to swap parts of the cell phone with another phone of the same model so that an acquisition could be done. The detective told me that some suspects might have a tool such as a small screw driver in the front seat of the car and may damage the USB port before surrendering a cell phone to law enforcement. The detective said that the suspect had a pail of water on the front seat of the car and some phones were dropped in it. Water damages electronics.

Many of the museums may also be used as a teaching tool to show students the progression of technologies such as the cell phone and to demonstrate concepts such as Moore's law. Moore's law states that the number of transistors in a chip doubles every 18 months. This trend allows significant gains in processing power which in turn enables cell phones to do many of the things that were previously reserved for computers such as email, web surfing, and video conferencing. The student who studies cell phone forensics and views the museums should easily discover for themselves that lifelong learning is unavoidable.

The Spy Museum in Washington, DC, is not a cell phone museum but does have a large collection of devices on display that have secondary purposes. There is an umbrella with a dart gun in it and some innocuous-looking devices with small-caliber hand guns in them. It is also good to know that some devices appear to be cell phones but covertly hold a taser—stun gun. Other cell phones have a small-caliber gun in them. There is a video on YouTube showing a cell phone that is also a four-shot gun [10]. Since cell phones are owned by 3/5 of humanity and do not create any case for alarm, they are often used to hold a number of other items which may include cigarettes, explosives, drugs, and many other items. This is why people have to have their cell phone x-rayed at the airport before getting on a plane or visiting many federal buildings, including the White House.

CELL PHONE PROTOCOLS AND OPERATING SYSTEMS

The cell phone originally was a simple telecommunication device that was only used to make calls. The operating system was originally an embedded system with a simple interface that allowed people to conveniently operate the hardware, namely the phone's keypad, to make or receive calls. From the time the cell phone was invented until the early 1990s, it was only for making and receiving calls. Phones such as the Motorola D520 that was announced in 1998 were part of a widely accepted trend of new cell phones that allowed for SMS messages. The operating systems had to become more complex as consumers now wished to synchronize their computers and their cell phones to update large phone books with points of contacts for both calling and sending SMS messages.

In the early 2000s, there were many sophisticated cell phone operating systems for phones such as the Microsoft CE versions, Palm OS, Android, Symbian, Apple's iOS, and others. These phones acted like small computers and each their own following of people who used software development kits (SDKs) to create applications. By 2011, some forensic tools such as Mobil Edit divided smartphone operating systems into the following groups: Apple, Blackberry, Android, and Windows 6.5. However, this is not a complete list because there is the Symbian operating system which is used on smartphones all over the world. Other cell phones might have the Palm operating system which has been around since 1996 when it was first introduced on personal digital assistants (PDAs).

Many cell phone forensic practitioners and students have noticed that the cell phone has become a small handheld computer with the ability to send and receive email, surf the web, watch movies, and do many activities that they could only do previously on their desktop computer. There is an operating systems concept called "feature migration" which means that the capabilities of large computers move to mid-sized computers over time and then eventually to handheld devices [11]. If one wishes to know what trends will eventually evolve on handheld devices, one needs to look at the operating systems and activities of large powerful desktop systems. The sixth edition of *Operating Systems* by Siberschatz, Galvin, and Gagne is an excellent source for gaining more understanding of feature migration as well as the theoretical aspects of how small-scale and large-scale operating systems work.

If one reads the user agreements that accompany mobile devices, it appears that legally replacing operating systems with newer ones is not possible. This could be the reason why so many people frequently purchase new cell phones and mobile devices with newer operating systems. It is also generally well known that older operating systems offer less protection than new ones because hackers and others learn and publish online the vulnerabilities of older operating systems. Preston Gralla, a writer of a book on malware, says that keeping your operating system, applications, and security software updated increases security [12]. Preston Gralla mentions one cell phone virus in his book, but in 2011, it is generally well known and highly publicized that there are over 1000 viruses and/or strains on cell phones and smartphones. Viruses were once something only computer users worried about, but now they have migrated to cell phones.

CELL PHONE OPERATING SYSTEMS: FINDING THE ESN AND IMEI

If one navigates through the cell phone operating system, there is always a subsection about system hardware models, device serial numbers, electronic serial numbers (ESNs),

IMEI numbers, and the version of the operating system one is using. In Windows Mobile 6.0, it is easy to get to the name, address, phone, and email of the owner. Just go to settings, owner information, and the details appear. If an examiner wishes to get the device serial number, go to settings, system, identity, and advanced details. It is very important for anyone interested in cell phone forensics to learn about the ESN. The ESN is a unique identifier that the manufacturer of the cell phone embeds in it. Each time a person makes a call, a signal is transmitted and is picked up by a cell phone tower that provides support for that phone. A mobile switching office can check the ESN and any other data embedded in the signal associated with that phone number against a network database to see if it is valid and part of the network. If the data is correct, then the call is allowed to proceed through the telecommunication network and also logged for billing. The call may also be routed through other networks in other countries if it is an international call where someone may answer it.

The IMEI is the international mobile electronic identifier. It is 15 digits long and numeric. It includes numbers 0–9 and does not use the hexadecimal number system. Some dual-SIM cell phones such as the SciPhone have two different IMEI numbers in the phone. The IMEI is found in GSM phones and is used worldwide to help prevent cell phone fraud and theft of service to the cell phone service provider. The IMEI also has a bar code next to the 15-digit number. If a person scratches out the IMEI and bar code to avoid identification, it is possible that enough of a bar code is left to scan. If one does not have a bar code scanner, an investigator may bring the cell phone to the library and ask the librarian to scan it where they scan library cards. Most modern supermarkets also have a bar code reader at the checkout counter where one should be able to see the results.

The IMEI is found by removing the cell phone's plastic back cover and battery. In New Jersey, some municipal law enforcement officers who are continuing education students have told me in an academic setting that if they wish to obtain the IMEI number without the consent of the suspect, a search warrant is needed to remove the back plastic cover from the cell phone. This was because the IMEI is not in plain view and the phone is considered a closed container. The Fourth Amendment protects U.S. citizens from unreasonable search and seizure. The same is true about a SIM card inside a phone. However, since I am not a lawyer or legal authority, it is best to check with your state's Attorney General's Office.

The SIM is an acronym for the subscriber information module found within the cell phone. It contains the information about the phone subscriber, the cell phone number, and a code that lets them use the cell phone network. SIMs can often be taken out of one cell phone and put in another. The second cell phone can often be powered on and used almost immediately. I performed an experiment with a vintage Blackberry model 6230 that had an LCD monochrome screen and a newer Blackberry model 7290 with a color screen. I took the SIM out of the first phone, placed it in the second phone, and then successfully placed a call to a colleague. The SIM was needed to make the call and the IMEI or the ESN was sent to the network too. It may be possible for a law enforcement officer with a subpoena to get the information about the call that was placed. It may also be possible for the law enforcement officer to know which phone was used to place a call if the telecommunication provider keeps information about the ESN that is embedded in the signals.

There are logs that the service providers keep about calls which may prove important in certain cases. In the United States, telecommunication service providers are required to assist law enforcement officers in lawful investigations where a search warrant was issued. The United States Federal Legislation is known as CALEA and became effective in 1994.

It was passed for matters of national security as codified at 47 United States Code (U.S.C.) §§ 1001-1002.

Bryan Miller wrote an article for the *New York Times* on July 20, 1995 about phone cloning and being vulnerable at chokepoints such as the airports and near the George Washington Bridge in New Jersey. One possible way that the phone cloning can happen is if people would park a car on the side of a road near crowded chokepoints such as a place where highways would converge to one wide area with toll booths. The George Washington Bridge is a crowded place at any time, day or night, as cars pass to or from New York City. Phone cloners are people who may work as part of a ring. One person might sit in a car with a wireless sniffer and collect mobile identifier numbers (MINs), ESNs, and possibly subscriber names. This person might get so many pieces of data with the sniffer that a laptop is needed to organize and create records in a spreadsheet for each caller that was eavesdropped on.

Policemen have told me that when phone cloning goes on, the person with the spreadsheet would later sell the spreadsheet to another person. That second party would create SIM cards with the stolen information and configure the phones to have the same ESN. These phones were then sold to people to make expensive out-of-country calls. The owner of the real original SIM card was the one who was billed. This activity was illegal and once common in New Jersey until the New Jersey State Police ran a successful campaign to stop it.

There is another interesting discussion of an illegal cell phone exchange racket in India [13]. Phones were being cloned for making calls between India and Dubai [13]. Cloning is when the real cell phone number of a person and the cell phone's ESN or the ESN is put on another phone and then the real owner gets the bills for the calls [13].

CELL PHONE OPERATING SYSTEMS AND PROTOCOLS: SYNCHRONIZATION

When people speak of cell phone synchronization, they could mean two things. The first meaning could be to make sure that the files in the desktop computer are the same as the ones in the cell phone. Many people add contacts to their address book and wish the latest copy of the address book to be the same on both the desktop and the cell phone. Many people synchronize their cell phone with their desktop at the end of a workday and put all the latest pictures, address books, and documents on the desktop with all the changes. The second type of synchronization has to do with updating the time and date of the cell phone with the present time zone one is in. An example of this type of synchronization would be if someone drove through Eastern Indiana to Western Indiana where the time zone changes from Eastern Standard Time (EST) to Central Time (CT). The following paragraph will discuss a real example of this time and date synchronization.

Some cell phones such as the Blackberry Bold 9700 came preconfigured to synchronize with the time and date of the locality where it is. The date and time is checked against the network and if there is a difference between the phone and the network, the operating system loads the date and time from the local telecommunication network. Consider this example: if one flew from Japan to New York and the flight only took 11 hours due to a tailwind, the cell phone will go back in time 1 hour as soon as the cell phone is powered back on in New York.

Some cell phones such as the Blackberry Curve will often need to be manually reconfigured when time zones are crossed because they do not update automatically.

Name	Date	Type	Size	Date created	Date modified
Roebling	8/19/2010 1:38 AM	File folder		8/19/2010 1:38 AM	8/19/2010 1:38 AM
checksums	10/12/2010 9:30 PM	Message D...	1 KB	10/12/2010 9:30 PM	10/12/2010 9:30 PM
IMG00001-20100827...	8/27/2010 1:28 PM	JPG File	272 KB	8/30/2010 9:41 PM	8/27/2010 11:28 AM
IMG00002-20100827...	8/27/2010 1:28 PM	JPG File	272 KB	8/30/2010 9:41 PM	8/27/2010 11:28 AM
IMG00028-20100317...	3/17/2010 1:39 PM	JPG File	357 KB	8/19/2010 1:38 AM	3/17/2010 11:39 AM
IMG00029-20100317...	3/17/2010 4:39 PM	JPG File	336 KB	8/19/2010 1:38 AM	3/17/2010 2:39 PM
IMG00030-20100317...	3/17/2010 4:39 PM	JPG File	150 KB	8/19/2010 1:38 AM	3/17/2010 2:41 PM
IMG00031-20100317...	3/17/2010 4:41 PM	JPG File	191 KB	8/19/2010 1:38 AM	4/7/2010 10:50 PM
IMG00033-20100321...	3/21/2010 4:53 PM	JPG File	532 KB	8/19/2010 1:38 AM	3/21/2010 2:54 PM
IMG00034-20100321...	3/21/2010 4:54 PM	JPG File	500 KB	8/19/2010 1:38 AM	3/21/2010 2:54 PM
IMG00035-20100327...	3/27/2010 4:27 PM	JPG File	437 KB	8/19/2010 1:38 AM	3/27/2010 2:27 PM
IMG00036-20100402...	4/2/2010 1:34 PM	JPG File	378 KB	8/19/2010 1:38 AM	4/2/2010 12:34 PM
IMG00037-20100402...	4/2/2010 5:24 PM	JPG File	218 KB	8/19/2010 1:38 AM	4/2/2010 4:25 PM
IMG00038-20100402...	4/2/2010 5:25 PM	JPG File	218 KB	8/19/2010 1:38 AM	4/2/2010 4:25 PM
IMG00039-20100403...	4/2/2010 5:25 PM	JPG File	187 KB	8/19/2010 1:38 AM	4/3/2010 9:20 AM
IMG00040-20100403...	4/3/2010 2:43 PM	JPG File	314 KB	8/19/2010 1:38 AM	4/3/2010 1:50 PM
IMG00041-20100403...	4/3/2010 2:53 PM	JPG File	340 KB	8/19/2010 1:38 AM	4/3/2010 1:53 PM
IMG00043-20100403...	4/3/2010 3:01 PM	JPG File	376 KB	8/19/2010 1:38 AM	4/3/2010 2:02 PM
IMG00044-20100403...	4/3/2010 3:02 PM	JPG File	365 KB	8/19/2010 1:38 AM	4/3/2010 2:03 PM

FIGURE 1.1 The date and time stamps on Blackberry Bold 9700 cell phone pictures.

The question of automatic date and time updating becomes important to the investigator because files and logs have time stamps. Files have a creation time and date as well as a modification time and date (see Figure 1.1). If someone is a suspect in a harassment investigation, does the time and date of certain activities coordinate with the events on the phone? Fortunately many cell phone investigative tools give the current date time of the acquisition and the date and time on the cell phone. Many of the cell phone tools such as Susteen Secure view show the MD5 hash values for the files in addition to the date and time. If one seized the system files from the cell phone and compared the MD5 hash files with the known correct MD5 hash values of the operating system for that phone, one could find possible tampered files from malware.

CELL PHONE DIFFERENCES WORLDWIDE

It is generally known that European cell phones are approximately 6 months ahead of American cell phones in terms of storage capacity, features, and in the number of megapixels on the digital camera. The cell phones in South Korea, Japan, and Hong Kong are generally considered to be 1 year ahead of those in the United States. To illustrate an example of this advancement concept, please consider this example that William Bulkeley, a writer for the *Wall Street Journal*, gave on February 8, 2007 [14]. Ten megapixel cameras were for sale in Europe and Asia while only 3.2 megapixel cameras were available in the United States. The problem with cell phones with advanced hardware is that they often contain newer versions of operating systems than American cell phones. Newer foreign phones most often cannot be examined by most of the state of the art cell phone forensics software available in the United States.

I addressed this concern with cell phone forensic tool vendors and law enforcement officials at the September 2010, High Tech Crimes Investigation Association Conference

(HTCIA) in Atlanta, Georgia. People whom I spoke to from the law enforcement community expressed concern that their present cell phone investigative tools could not allow them to investigate new models of foreign phones. This could be a problem if there is a need to search a cell phone in a timely manner. The phone would need to be sent to a Regional Computer Forensic Laboratory, known as an RCFL. A case manager would then interview the officer about the case and assign it a priority. If there were other many high-priority cases, then the phone could not be examined in a timely manner, thus allowing the chance for the recovery of digital evidence to be lost.

Mobil Edit was one tool that was displayed at the 2010 HTCIA conference in Atlanta that provided support for a range of foreign phones. Many digital device investigators said that they would like to see their present forensic software vendors also include support for foreign phones. They told me that the support would be best provided in an upgrade of their present forensic software because limited budgets often prevent them from purchasing additional tools.

Many of the foreign phones with larger-capacity storage take longer periods of time to examine because there is more material to examine. With more processing power, higher-resolution screens, increased storage, and more available bandwidth, people are surfing the net, creating and sending video, making phone calls, and sending as well as receiving email. This volume of digital media becomes an increased burden on the examiner. Since the certified computer examiner's (CCE) time is expensive, many law offices just ask the CCE to image the phone and pass the image of the cell phone to a paralegal to examine and then create a report on it.

Imaging is a term that means to copy all allocated and unallocated parts of storage. Solomon, Barett, and Broom, three authorities on digital forensics, also refer to the image as a forensic duplicate [15]. The forensic duplicate starts at byte zero and goes to the last byte of the storage device. The duplicate or image contains parts of files, parts of sessions, and many pieces of important forensic evidence that can help the CCE get a clearer understanding of the case and create a report. An image is different than a backup because a backup only includes files that are presently listed in the file allocation table. A backup does not include existing files that were marked for deletion.

Many law offices that were using paralegals rather than CCEs to save expenses also found that the volume of material to sift through was still too voluminous. This also made the use of paralegals too expensive. Now, many law firms have resorted to hiring document readers who may have little legal training. This becomes a problem because they may miss some pertinent evidence because of their limited training and knowledge of law. However, who can manually read all these documents, emails, watch the videos, and completely report on all relevant items in the phone? Now there is expensive e-discovery software that organizes the contents of an acquisition, produces charts of associations, and helps sift through the enormous amount of material. E-discovery tools can help save time and labor but since e-discovery tools are very expensive, there are some who feel that the number of law firms or private investigators that can perform large-scale quality digital forensics investigations will dwindle to a few large companies with deep pockets.

CELL PHONE DIFFERENCES WORLDWIDE: VARIOUS BANDS

Some people use the term "tri-band" with regard to various types of cell phones. Tri-band means that the cell phone uses three very different sets of frequencies. The reason that every country does not use the same frequency is due to different parts of the world using different frequencies for television, radio, aircraft, and cell phones. A band plan for a country may show what frequencies are used for aircraft, cell phones, and other broadcasts. The band plan may also show acronyms such as GHz or MHz so it is important to first obtain the electromagnetic spectrum to learn what signal frequencies are microwaves, VHF, AM, FM, shortwave, radar, x-ray, infrared, and ultraviolet light. The electromagnetic spectrum gives one an idea what type of signals exist and how they are classified. Shortley and Williams have an easy-to-read electromagnetic spectrum chart in their book *Principles of College Physics*.

The cell phone networks use not only different frequencies but also different protocols such as GSM or CDMA. It is advisable for the student of cell phone forensics to start his or her education by looking at some cell phones that are in use in his or her locality and learning what frequencies and protocols are used there. Some computer science students who attended Fairleigh Dickinson University in 2003 were from Hong Kong and had what was known as tri-band phones. These cell phones could be used in Hong Kong, the United States, and in the United Kingdom. These phones were a necessity for students who often flew to these countries to visit relatives, go to school, or assist in their family's international business.

This paragraph will illustrate a real example of a quad-band phone that utilizes four bands. In 2010, the quad-band cell phone such as the SciPhone mode i68++ became popular on eBay. It is made in China and sells for approximately US$50. It uses a GSM quad-band protocol which includes the 850, 900, 1800, and 1900 MHz frequencies. The interesting concept about frequencies is that the higher the frequency, the shorter the wavelength. Low frequencies have longer wavelengths. The shorter waves do not travel as far as the long waves. Wave propagation is the term used with how far a wave travels. Jerry Wilson, a writer of a physics book, describes the mathematical equation that explains frequency and wavelength as [16].

$$\text{Speed of light/frequency} = \text{wavelength}$$

Cell phones that utilize higher frequencies must use more power than cell phones that operate on lower frequencies, or there must be cell phone towers and repeaters in closer proximity to retransmit the signal. Shannon's law of information basically states that the higher the frequency, the more the bandwidth, and the more information that can be sent. Higher frequencies are often preferred for cell phones with high-resolution digital cameras where high-bandwidth video needs to be sent. *Newton's Telecom Dictionary* says that Shannon's law is "a theorem defining the theoretical maximum at which error free transmission can be transmitted over a bandwidth limited channel in the presence of noise" [17]. *Newton's Telecom Dictionary* states that the theorem works out to approximately 10 bits per hertz in analog circuits [17]. However, in 2011 there are many ways of getting more information on a signal such as using digital circuits and quadra angle signaling. For more information on signaling, frequency, and bandwidth, Roger Freeman's *Fundamentals of Telecommunication* would be a good source of further information [18].

A cell phone examiner should be able to Google search the brand and model of a phone to see the protocols of that cell phone and learn which geographic regions utilize that protocol. Can a cell phone that utilizes certain protocols and frequencies make a call in the area where the alleged threat took place? If a person has a vintage monoband American cell phone that uses TDMA, it probably should be ruled out as a possible phone that was used to make a threatening call in the United Kingdom in 2011.

The terms "MHz," "KHz," and "GHz" are often used without much explanation. A hertz or cycle is a sine wave. A series of 1000 cycles is a kilocycle or now what is commonly referred to as a kilohertz. It is based on 10 multiplied by 10 multiplied by 10. It is different than a kilobyte which is 2 to the 10th power which is 1024. A MHz is also based on the power of 10 and not 2. A million hertz or MHz is a million cycles whereas a megabyte (MB) is more than one million bytes. An MB is 1,048,576 bytes or 2 to the 20th power. Shortley and Williams give advanced explanations of signaling and may be useful for the cell phone investigator who needs to understand the physics behind the technology [19].

Many cell phones have very high-quality miniature microphones. The digital signal also allows a high fidelity of audio to be transmitted across the phone networks as compared to analog circuits. That is why the cell phone is being used as an electronic listening device by stalkers. There are numerous news stories of cell phone stalking victims that are available on the Internet, which are worth watching to learn about the extent these devices can be used to frighten people. For those unfamiliar with the case, certain family members would often receive calls on the cell phone commenting on clothes they wore or activities that they were involved in. The specificity of the comments showed they were being watched.

The mechanics of this stalking will be explained now for the purpose of educating cell phone forensic examiners. A person may download spytones for the cell phone because they do not ring. Then he or she would set the phone to autoanswer and hands-free calling. This allows the phone to be called and used as an electronic eavesdropping device to covertly collect video and audio. A cell phone investigator should also consider the time, date, call logs, frequencies used, and the nature of the case to determine if the phone was also unknowingly used for stalking someone or spying on them.

CELL PHONE INTERNAL AND EXTERNAL STORAGE

Cell phones have a variety of internal and external storage configurations. Some Blackberry Bold models sold in 2010 in the United States for example have a 2 GB internal storage card as well as a considerable internal storage capacity. The Blackberry Bold 9900, sold in India in 2011, is one example of a device that was reported to have 8 GB of internal storage [20]. The increased storage capacity is the result of research and development that increases the number of transistors within a chip. Here, we see an example of both Moore's law and the availability of advanced cell phones outside the United States.

It is also a good idea to review the units of storage if one is going to examine cell phones. The unit of storage encountered in digital devices is the byte. Early examples of cell phones used kilobytes of RAM. The kilobyte (KB) is actually 1024 bytes but some people will round it off to be 1000. Megabyte (MB) is often rounded off to one million bytes but is actually 1024 multiplied by 1024 bytes. Gigabytes (GB) are usually rounded off by people to mean one billion bytes but are actually 1024 multiplied by 1024 multiplied by 1024 bytes. These little differences in theory and practice are good to know if one has to appear in court and must answer questions from the opposing counsel.

A cell phone will have random access memory known as RAM. This RAM is volatile and means that if the power is lost, the contents of RAM are lost. However, some operating systems may use a swap file and the contents of RAM may be found in a swap file. This is why knowing the model of phone and some details about the operating system becomes so important in an investigation. There are often a variety of external cards such

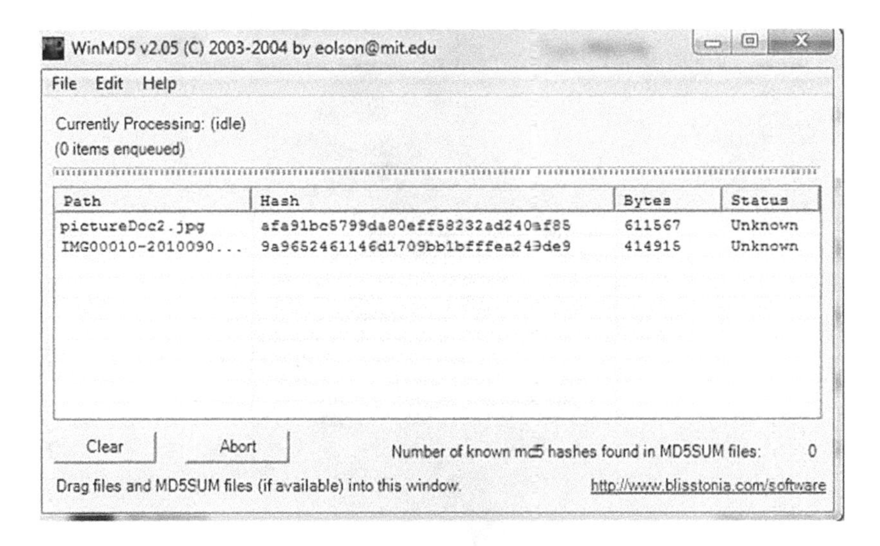

FIGURE 1.2 Different MD5 hashes of identical-looking pictures.

as micro SD cards that can be found in cell phones or nearby in the possession of a suspect. External cards are important because they may hold important evidence that could be known as exculpatory data, or data that proves someone innocent. Investigators need to look for both types of evidence that can prove someone guilty or innocent [21]. In Figure 1.2, one can see the MD5 hashes of some files of simulated exculpatory evidence showing that the identical-looking pictures were different.

If one performs an investigation of a cell phone, he or she should make sure that the e-forensic tool can be operated to perform a seizure of the internal memory, internal storage, external cards, and a SIM card. A SIM card is mainly for identifying a caller initiating a call on a telecommunications network but its secondary purpose is its ability to be a storage device for a person's address book. The address book or books of the cell phone owner are not necessarily saved in the SIM but it may be optionally saved there. Because it is optional, it is also important to check it for data too. The acronym SIM means subscriber identity module. Since the capacity of SIM cards has increased, a new tertiary purpose is to use it as a storage area for saved text messages. It is important to remember that the SIM card holds the phone number and identifier of the subscriber at the minimum, but could also possibly store address books and text messages. The SIM card has traditionally worked with GSM digital networks but recently there are SIM cards in China that can also work with CDMA protocol phones [22].

In Figure 1.3, we see a phone on a camera stand waiting to be examined. The dual-SIM card phone means the cell phone simultaneously holds two different SIM cards. Each SIM card has its own identity and telephone number. It is not uncommon for people to have more than one SIM card for legitimate purposes. A person may wish to have an identity for work and an identity for his or her personal life, in order to keep one's business life and personal life separate. Each card may be in a different name and even have a different service provider such as Verizon and T-Mobile. Some people have two SIMs because of the lack of quality coverage by any one company where they live. Cell phone service provider companies provide varying qualities of service for the same geographical areas because some service providers may have standard antennas on a pole along a highway while others may have a high gain antenna on top of a water tower. Some people may also have a SIM card with a number for the United States and one for

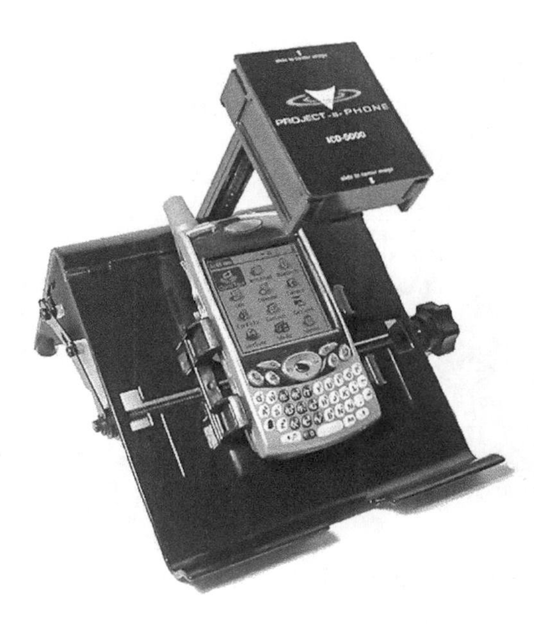

FIGURE 1.3 A cell phone being examined using a camera and stand.

another country such as England. When people go to another country, they often prefer to have a local phone number and a card from the service provider of that country. The calls are much cheaper when it is considered a local call. There are also some models of cell phones that hold three SIM cards such as the Intex IN5030 that is available in the Far East. This type of cell phone can be used to hold one personal identity and perhaps two business identities.

The important consideration when doing an examination of a cell phone is to make sure that one uses a separate research machine that goes online and looks up the specifications of that model of cell phone. The examination machine stays offline to avoid malware and connects directly to the cell phone to collect and collate the evidence. The research machine is connected to the Internet and can be used for purposes such as surfing the web to learn if a certain cell phone model supports multiple SIM cards and external memory cards. The examiner should systematically check the internal memory, external cards, and all the SIM cards for possible evidence. Lastly, it is important to consider that a person may have an illegitimate business that is done with a completely different identity on a removable SIM. The examiner should look for evidence of wear in the other SIM card slots and also examine the call logs of the phone and operating system files to see what identities were used in that phone.

INTERNAL CARDS: SIM CARDS/LOCKED AND UNLOCKING

Many times people have SIM cards which bind that SIM card and GSM phone to a certain network. The Sony Ericsson T610 is one example of a cell phone that will give an error message to put the original authorized SIM card back if someone replaces it with an unauthorized SIM card. This appears to be a technical security measure built into certain cell phones that is done to make sure that a person who received an expensive cell phone as part of a promotion cannot dissolve their business relationship and go to a cell phone service provider who charges less money. Many people have complained about the locked

SIM cards. A group of programmers created programs that can be found online that "unlock" your phone and remove the restriction of the cell phone to only accept a particular SIM. This allows a person to go to another cell phone service provider.

Many people use the term "unlocking a phone' synonymously with the term "jail breaking a phone." However, the two terms are very different. A cell phone investigator may hear the term "jail broke phone" from someone in a chat room or as part of an investigation. The term is often confused with unlocked. Unlocked means that the security feature within a GSM cell phone was reconfigured to accept another SIM card other than the one originally sold with the device and provided by the cell phone service provider. The term "jail broke phone" means that the cell phone was reconfigured by a special application to allow programs that are unapproved by the cell phone manufacturer for that phone.

If one has a jail broke Apple IPhone, for example, he or she may be able to download and run applications that may easily allow the cell phone camera to zoom in on a subject as a picture is being taken. Another example of an unapproved application available online for jail broke cell phones may conveniently allow the user to choose fake GPS coordinates and embed them in the phone. Suppose a person is at a casino in Atlantic City but wants to tell the boss at his or her company that they are in New York City at a conference. The person may take a picture of the Atlantic City conference room but use the unapproved application to embed the GPS coordinates of the conference where they are supposed to be in New York City.

THE NEED FOR A FARADAY BAG

Cell phones are devices that connect to telecommunication networks through wireless signals. Some cell phones can also connect to Bluetooth devices or wireless networks. If the security features are not enabled on the cell phone, then it may be possible to connect to the phone and alter, delete, or add digital evidence to the phone. It is very important that the digital evidence be preserved from the time of seizure until it is presented as evidence in court. If evidence is suspected of being tampered with, it could be ruled as inadmissible in court. Therefore, it is important for CCEs to preserve digital evidence by using a Faraday bag and noting its usage on the chain of evidence form. Smith and Bace, two authors of a forensic testimony book, discuss the importance of preserving evidence and protecting the integrity of digital evidence [23].

Faraday bags look very similar to antistatic bags. The difference is that the antistatic bag prevents damage to the device from small electrical charges that have built up and are discharged from static electricity, but it does not protect the device from outside connectivity. Static bags are obtained commonly when purchasing electronic equipment such as a wireless weather station, computer memory chips, or an EZPass transponder for the car that makes it convenient to pay tolls. The Faraday bag is based on the concept of a Faraday cage. The Faraday cage is an enclosure that prevents outside signals from penetrating the cell phone or examination equipment [24]. The Faraday bag is made with materials that block wireless signals from entering the bag, thus protecting the integrity of the device in the bag from outside influences.

The Faraday bag will not prevent the device from internal data alteration by items such as logic bombs. A logic bomb is set to go off if certain conditions are met. If a person was supposed to simultaneously press a set of keys daily to keep a destructive program from running on the cell phone, this would be one example of a logic bomb. The phone that was seized from someone may be protected from outside control of hackers with the use of a Faraday bag, but the phone may be victim to a logic bomb if certain conditions are not met while the phone is in possession of the CCE.

I was once teaching a class about cell phone forensics to a class of visiting cybercrime students from Kyungnam University from South Korea [25]. One of the students asked if another type of metallic bag such as an aluminum foil bag could be used in an emergency situation if no Faraday bag was available. To demonstrate an answer to the student's question, I placed various cell phones in aluminum foil bags and asked students to call the phone. The signal was blocked. The phones also did not appear as wireless device icons on the student's laptop display. The lesson learned was that aluminum foil bags give some protection from connectivity but a proper Faraday bag is best. I explained that the effectiveness of the aluminum foil bags was not known and that could offer an unwanted line of questioning in court. Each tool and methodology used in the collection, preservation, and examination of digital evidence should be able to withstand the Frye Test [26]. The Frye Test helps ensure that the tools and methodologies used to gather, process, and examine evidence in an investigation are accepted as general practice by authorities in that field.

SURVEY OF TOOLS TO INVESTIGATE A CELL PHONE

Since at least half of humanity has been reported to have a cell phone, there will exist some occasions when some of those phones will need to be examined. Cell phone examiners, like most other investigative professions, will need to have properly trained people to examine the cell phones before they can use these forensic tools. Training has not yet been standardized in this new discipline so there are many options that one can choose from. For some, training might consist of a 5-day certification class, while for others with less resources, it may only consist of a half-day continuing education class. Training is important.

In this section of the chapter, a brief survey of training will be discussed. It is not possible to create an exhaustive list of all the training that exists due to the numerous classes that are becoming available on a daily basis as the need for cell phone forensics and the tools for examining these devices increases. There are also a variety of classes with different focuses appearing on the Internet daily. Some examples showing the variety of training options for tools and techniques of cell phone forensics will be addressed in this chapter. Some of these entities provide certificates of completion for the class, while others provide more comprehensive testing options and certification. The cell phone examiner should consider getting as much training as possible in as many tools as one can obtain. Many teachers of cell phone forensics are originally from the technical side or information technology security side of private industry, while others are from a law enforcement background. The same material could be taught with an emphasis on policy and legal issues, while other instructors may emphasize the cell phone operating system details and the acquisition of data from flash memory or SD cards. Each instructor brings a different perspective and may focus on different aspects of cell phone forensics. All the training and tools only widen one's knowledge and give the student the opportunity to try to collect the evidence from the cell phone and process it for a report.

Some students will sometimes make their requests known to the instructor so that the training that they receive is more beneficial and customized for their work needs. Here is an example. I taught a class on cell phone forensics in 2009 to a group of law enforcement officers and private investigators. A corporate leader asked me to focus on certain organizational cell phone policy topics concerning incident response for him and the technical details of how to use the tools for his assistant.

Many people consider the smartphone to be a small handheld computer with the ability to make and receive phone calls. Therefore, the cell phone examiner would be well advised to obtain a CCE certification. This is available from the International Society of Forensic Computer Examiners. The website for them is www.isfce.com. Many corporate and law enforcement cell phone examiners also become EnCase certified and purchase the EnCase Smart Phone Examiner. Guidance Software, makers of EnCase, give a certification after the successful results of one or more tests. Others may elect to enroll in a class from Susteen, makers of Secure View, and get a certificate. Students who need both hands-on training and theory may take a half-day continuing education class called "Cell Phone Forensics" taught by me at Fairleigh Dickinson University. Others may take a Cellebrite class from a place such as http://www.sumuri.com/.

The Paraben Corporation offers a certificate in cell phone forensics. The class includes elements of both theoretical and practical training. Many classes conclude with both testing and a certificate. Smart Phone Forensics is a company that offers training and certificates and uses guidelines from the National Institute of Standards and Technology. The SANS Institute offers training and certificates in SIM card examination, a specialty of cell phone forensics. The High Tech Crime Investigation Association occasionally sponsors classes on cell phone forensics. The Public Agency Training Council offers training and a certificate in cell phone forensics training in Las Vegas, Nevada. All these options for training and the places they are located can be easily searched on the Internet.

Cell phone forensic examiners can be found in a diverse group of places throughout the world. The CCE list server often shows written dialog concerning cell phone seizure and examination in places such as Iraq, Somalia, Morocco, the United Kingdom, and the United States. Many examiners may not have the resources or opportunity to fly to the United States, stay at a hotel, and take an expensive class. I wrote a paper called "Teaching Cell Phone Forensics and E-Learning" that discusses the need for online learning, test taking, and certification in all parts of the globe. The paper discusses how theoretical and practical training could be given at low cost using online learning environments and the students' own phones [27].

The Mobile Forensic Certified Examiner (MFCE) can be a very good credential too because it requires that a person take a series of practical examinations and theoretical examinations. This is very good because the investigative agencies that hire the people holding the MFCE can be confident that a person has a certain common body of knowledge (CBK) and certain practical skills about how to collect or acquire, preserve, examine, analyze, and report on digital evidence. The link for the group is the following URL: http://www.mfce.us/. Many academics and employers feel confident about having students or employees with multiple certifications because it shows that they can repeat the process of demonstrating various proficiencies in both practical and theoretical skills with regard to seizing and analyzing digital data from cell phones.

EXAMINING CELL PHONES WITH OPERATING SYSTEMS UNSUPPORTED BY YOUR TOOLS

Suppose an examiner has a cell phone with a classic operating such as Windows Pocket PC 2002. Perhaps the examiner's present tools do not support this version of a classic operating system. This is a possible situation because Windows Pocket PC 2002 supports

GSM smartphones and someone may still be using it. This is important because GSM phones are used in over 120 countries [28]. Many blog writers said that this version of operating system looks similar to Windows XP. Making an operating system interface for phones that appeared to some like a slimmed-down version of the Windows XP desktop operating system seems to be another example of feature migration. This similarity in operating system appearance could also save time to a cell phone investigator who was unfamiliar with Windows Pocket PC 2002 and had to examine it manually. Sometimes, cell phone forensic tools occasionally do not work with certain phones and operating systems and it becomes necessary to manually navigate the phone and take pictures of the cell phone display with a high-resolution camera. If there were certain custom applications on the cell phone that may contain evidence to an investigation, manually going through the application and taking pictures of screens may be the only way to collect the evidence within a limited amount of time.

Examiners do not like to manually examine cell phones and take pictures of the screens because the process can change certain files with logs and time stamps or possibly create additional temp files. However, sometimes it is necessary because one may seize a new foreign phone from a suspect and there is no forensic tool available anywhere that supports that phone. A phone must be on the market a few months before a forensic tool can be available for that phone.

Cell phone providers do not give forensic examiners or forensic tool makers advanced release of any phone. For such situations, tools such as the Paraben that has a Project-A-Phone ICD-5200 are a must. This tool holds the cell phone still and connects to the USB drive of a computer. The device also contains a camera that allows the examiner to take pictures with a resolution of 5.2 megapixels. Figure 1.3 shows the phone in the ICD-5200. This tool allows manual examination of custom applications and viewing files as well as capturing each screen.

Windows Mobile 2003 was an operating system that was issued standard with many Pocket PCs which are also known as PDAs. The version for PDAs had many standard features such as Internet Explorer. It also contained games such as Jawbreakers, but this will be addressed in a future chapter on PDAs. There was also the Windows Mobile 2003 Smartphone edition that became available that same year that included applications that supported activities for calling. However, there was a software development kit known as an SDK that allowed developers to create customized software applications for cell phone users with Microsoft Windows 2003 for smartphones. This becomes important because investigators may find nonstandard applications that may hold important evidence related to an investigation.

Windows Phone 7 is an operating system that one may encounter on phones in 2011. There are the usual applications such as email, video creation and sending, SMS messaging, using social media such as Facebook, and websurfing. However, there is a new feature in Windows Phone 7 that cell phone examiners with top secret clearance who do work that touches national security issues should be aware of. Such examiners could include people who do contract work for the Department of Defense or work for one of the United States' 16 spy agencies. This feature is the new connectivity with Xbox Live. This is another example of feature migration. It also means that people can play games and perhaps communicate in multiuser environments. If a person wanted to communicate with others but leave a less obvious trail, one might elect to meet others in a virtual environment such as Second Life or in a multiuser game. In 2008, *Wired Magazine* reported that the United States Congress was worried about terrorists meeting and exchanging funds in the virtual world known as "Second Life" [29]. In the United

Kingdom, *The Sunday Times Online* had an article titled "Virtual Jihad Hits Second Life Website" [30]. Though most cell phone examiners may never encounter terrorists, it is good to be aware of national security issues because one may examine the phone of someone who turns out to be part of a sleeper cell.

There are many tools that are available to the cell phone forensic examiner. Some are hardware based such as the Logic Cube Cell Dek and the Cellebrite Universal Forensic Extraction Device (UFED) Physical Pro. These tools are great for those who wish to have a self-contained unit that one easily connect a cell phone to and collect the contents for saving on a piece of sterile media. The advantage of using both pieces of equipment are that both have devices that utilize embedded operating systems that appear to be straightforward to use, and do not require the burden of a computer in order to collect the information. Both units are housed in sturdy Pelican cases that protect all the cables and equipment and allow it to be easily ported to an incident response vehicle.

Some CCEs prefer a lower-cost solution to cell phone forensics that requires the use of licensed software, a cables kit, and an examination machine. Paraben's Device Seizure and Susteen Secure View have easy-to-use software interfaces that work well with many laptops. The cables kits have a number of cables that work on multiple cell phones and simplify the need to have hundreds of cables. Susteen Secure View support also sends cables and software updates to user at no extra cost during the time that the licensed software is active. This is important as new cell phones and operating systems are constantly being introduced to the marketplace. Paraben's Device Seizure is also good about providing new updates for phones as they become available on the market.

Some cell phone examiners with limited resources will find open-source tools such as BitPim an excellent solution for them. The software is free and open source which means that one can download it, use it, and modify it without licensing issues. However, there is no warranty on the software that may be available on software you purchase. The BitPim website also has links for downloading the source code for the software which allows investigators and their support staff to program in Python and add phone support for new foreign cell phones or American ones.

Some cell phone examiners who use Guidance Software's EnCase, which is great for laptops and desktops, may just purchase an add-on such as SmartPhone Examiner for a relatively inexpensive fee. Some may choose this option because they may be highly invested in training and certifications for EnCase and this add-on allows digital evidence examiners to add on cell phone examination capability as an extension to their present tools and training. It may be difficult for some departments to justify a set of completely different cell phone examination tools and training sessions in times of fiscal restraints. An additional application for use on an existing product that is presently invested in may be politically easier to purchase in the public sector.

CELL PHONE FORENSIC TOOLS: MOBIL EDIT

Many cell phone examiners have told me that they think Mobil Edit is a good tool because it offers forensic examination capability of foreign cell phones. The software can be purchased and then many new popular phone drivers can be downloaded from the website so that the examiner's software stays current. The Mobil Edit software enables examiners to acquire the contents of four major operating system families of cell phones. The first family is the Blackberry operating system family of phones that include the 8120 Pearl, 8310 Curve, 8520 Curve, 8900 Curve, 9000 Bold, 9500 Storm, 9530 Storm,

9700 Bold, and the 9800 Torch. The second family of cell phones supported has to do with the Android operating system. Some examples of phones in this family are the Dell Streak, Vadaphone 845, T-Mobile G2, Samsung Vibrant, Samsung Galaxy Tab, Samsung Epic 4G, HTC Magic, Motorola Droid, and the Google Nexus One. The third family of operating systems supported by Mobil Edit is the Apple OS. Some examples of Apple operating system phones are the IPhone, IPad, IPhone 4, and iPod Touch. The fourth family of operating systems is the Windows Mobile 6.5 which includes phones such as the Samsung i600, TMobile Wing, TMobile MD Vario V, TMobile MDA III, Sony Ericsson Aspen, Sony Ericsson XPERIA X2, Qtek A9100, O2 XDA Orbit II, Acer 960, and the i-mate X-JAM.

Mobil Edit also offers a cables kit that provides cable or IRDA infrared connection for its supported phones. It also has hardware for reading SIM cards which may hold SMS messages, phone books, and other data in addition to the mandatory basics to connect to the phone network.

Many cell phone examiners that I spoke to at the 2010 HTCIA Conference in Atlanta said that Mobil Edit was affordable and provided support for foreign phones. Mobil Edit is also popular because it has a newsletter which promotes a community of users and discusses concerns and solutions. The newsletter also discusses ways in which technically inclined examiners may get the software development kit and create new solutions to acquiring more data on newly released phones. Mobil Edit also allows people to become beta testers of new operating systems which may be a solution for corporate users with limited security budgets.

SURVEY OF CELL PHONE FORENSIC TOOLS: PARABEN DEVICE SEIZURE

The Paraben Corporation is the creator of a tool called Device Seizure. Many CCEs say that they like this product because it has some capability to collect information from cell phones, PDAs, and some GPS devices for use in the car. Companies that wish to save money often look for one product that can do more than one purpose because sometimes it is difficult to obtain both justifications and funding for two products and two sets of training. Device Seizure was the response to consumer demand that said, "Why don't you combine PDA Seizure and Cell Seizure?"

Paraben also sells a cable kit for Device Seizure that allows mobile device forensic examiner personnel the opportunity to find many of the cables that they will need to seize the data from a phone and examine it. Paraben also gives training and certificates which is very important because cell phone examiners need to show some type of audit for the knowledge that they have in order to withstand the scrutiny of court. The latest software for Device Seizure can be downloaded and used immediately as long as the user has a valid license key. The license key is good for a period of time between 1 and 2 years. By having a valid license key, it also keeps people from copying the software and having others illegally using it. If the investigators used an illegal copy of the software, the results of the investigation will be dismissed by a legal concept known as "Fruit of the Poisonous Tree." The valid software license key is kept secure by the license holder and is actually a safety for everyone who might be pressured by others to keep costs down and use an unlicensed copy. Device Seizure is very interesting to digital forensic examiners because it has a hex viewer and one can view the file structure of the seized device as well as the content of the file at the bits and bytes level. One can see metadata in a document that would not otherwise be viewable if one opened up the file in a word processing program such as Microsoft Word. The hex viewer also enables the examiner to view the metadata about the camera type within a digital picture that otherwise would not be seen in a general digital picture viewing program. Some metadata gives clue to the make, model, or possibly the serial number of the camera or scanner that created the picture.

Acquiring the data from the cell phone is not difficult to do with Device Seizure. One starts the program, clicks on an icon of a mobile device, and then follows the prompts on the screen. The prompts will ask about the operating system of the cell phone being seized and the port of the examination that connects with the cell phone. There is also a report wizard that asks the examiner about his or her name, agency, address, and information about the case. There is also an option to add a jpg or gif file of the logo of the examining agency. Then, the materials are captured from the phone and encapsulated in a report.

SUSTEEN SECURE VIEW

This product costs approximately $1500 and consists of the forensic license dongle, the software, and the cables kit. The product is supported for a year and a half, the time that the license dongle is good for. The company frequently sends cables to registered users for new phones that appear on the United States market. Susteen also sends frequent links for new versions of forensic software that include support for new phones as well as dynamic new features such as the one that allows examiners to extract GPS coordinates from the phone and display them on Google Maps. Some CCEs say it is very straightforward to use because it is menu driven from the start.

Secure View provides support for a large number of cell phones. I visited their software development lab in October 2009 and saw a team of developers creating the code to get all the data from each phone. Each cell phone that they support had to be bought by the company and they must spend considerable time creating the code to acquire the data from the phone and then validate the results. People who criticize the developers for the price of their forensics products should visit the company and will probably quickly understand all the development costs that go into the product. If one considers that they must pay a large staff, rent a facility, buy many phones, market their product, and go to conferences, the price seems reasonable.

One item that distinguishes Secure View from the other tools is that it has a component called SVProbe. There is a feature that takes the call file and maps out all the names, phone numbers, and frequency of contact, with all parties who are in communication with the owner of the cell phone. In 2007 and 2008, I and some cybercrime students visited the wire room of a county prosecutor's office in New Jersey where calls are forwarded to when part of a lawful investigation. Certain investigations had the suspect's phone calls mapped on a large paper poster with a directed graph to each party they speak with. It became obvious that tools such as SVProbe save the law enforcement agencies significant time by automating the process of graphing the number of calls between the suspect and other parties they speak with. An example of a graph can be seen in Figure 1.4.

SVProbe also has the "Gallery" feature which lets a viewer see all the photos within the seized phone. There is also a button for GPS extraction and linking to Google Maps which allows the photos that contain GPS metadata to be shown on a map with street names. This is important for investigators who need to see the origin of where digital pictures were taken. It is also important that some sophisticated suspects might have cell

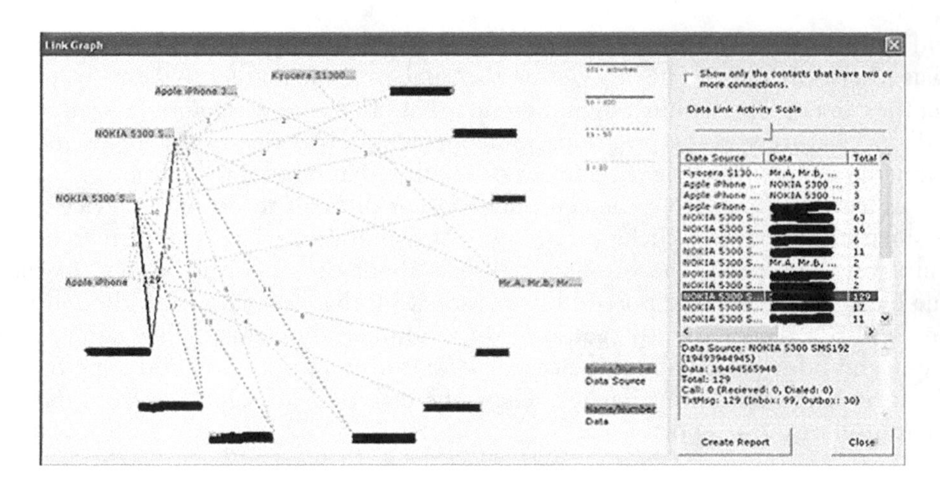

FIGURE 1.4 SVPRobe directed graph of callers.

phone pictures with spoofed GPS data in them to throw off investigators. Good follow-up police work still needs to be done with any tool.

Susteen Secure View also has a feature called SVPin which can read and provide the personal identification number (PIN) on a CDMA locked phone. This is important because some people use a pin to lock the cell phone so that others, including cell phone examiners, cannot get in there to view evidence. This feature does not work on smartphones but works on phones that use the Qualcomm chipset.

Now that we have discussed Susteen Secure View, let us discuss how it can be used as an examination tool in a sexting case. Sexting is often in the news, so here is a hypothetical example of sexting and how an investigation might proceed. Perhaps a man in an apartment building was fascinated with a tenant and got her phone number from her garbage. Then he might have called the number to find out if it was a landline or cell phone. One cannot perform sexting to a landline. Then he found out that it was a cell phone. Perhaps the man performed sexting by taking a picture of his crotch and used the multimedia messaging service (MMS) to send it to the neighbor. The woman was then appalled and called the police.

The police would then arrive and interview the woman and make a report. They would most likely ask for her cell phone so that they could examine it. Since she is the victim and wants justice, she would consent to surrendering the phone to the police. The police may then take the phone and fill out a chain of custody form. It would be assigned a case number and there would be a file associated with it. The officers would also put it in a Faraday bag so the evidence would not get tainted. It would also be locked up so others could taint the evidence. The locked box would also be in a temperature-controlled area because heat and cold both destroy digital media storage as well as electronic devices. The police chief might assign it a high priority depending on the caseload of crimes and their severity.

Then the detective might take the phone and bring it to the examination room where the examination computer is kept. Then he or she would verify that there are no connections by wireless, Ethernet, infrared, Bluetooth, or a modem which could compromise the investigation. The detective would run antivirus on the machine and sterile media where the results of the investigation would go. The detective would put the license dongle in the examination computer and start up a tool such as Susteen

Secure View. Then the person would go through the menu options and collect all the data from the phone. The data would include phone call logs with date and times. Any pictures that were sent could possibly be captured too. The call log might show the telephone number and crotch shot picture. This is important evidence. A reverse lookup can be done at a site such as www.spokeo.com and this might show the name and address of the person who pays the account and has that cell phone.

The examiner might go to SVProbe and select "Gallery" so that all the pictures are shown. The detective who is the examiner in this case might find the crotch shot and use the mapping feature to extract the GPS data if it exists, and show the origin on the map. Such a picture could have been taken at home or a place such as a public bathroom nearby the suspect's home or workplace. The detective would then have some probable cause and sign an affidavit in front of a judge. The judge would probably issue a search warrant.

The detective and another policeman might knock on the door of the suspect and then enter after being invited in. The suspect would be asked to surrender the cell phone and come downtown for questioning. The phone would go in a Faraday bag and a chain of custody form would be filled out. The cell phone would also be locked up. They would probably go downtown and the suspect might be in a room where the detective interviews him. Others may watch through a two-way mirror. If it is a real mirror, there is no gap between your thumbnail and the image of a thumbnail in the mirror. In a two way mirror, there is a gap between the fingernail and the image of a fingernail if a person puts his fingernail right up to the glass.

The detective might get the person to confess. Simultaneous with the interview, another police technician may do a cell phone examination with one or more tools such as Paraben's Device Seizure, Susteen Secure View, Encase Smartphone Examiner, or the Cellebrite UFED. In cases where people do not surrender a phone, the police must have to subpoena the records of the telephone service provider to get the records of the call. A report with all the evidence may get the person to confess and plea bargain. In cases where the person is arrested and does not confess, it is the Grand Jury's duty to decide if there is enough evidence to go to trial.

CHINESE CELL PHONE EXAMINATION TOOLS

It is common knowledge that the People's Republic of China has approximately one quarter of planet earth's population. The CIA Factbook says that 1.337 billion people live in China—Hong Kong and the mainland. MobiThinking reported that there are "859 million mobile phone subscribers" as of December 2010 [31]. Some unofficial estimates on the Internet say that India is slightly ahead in 2011 with 865 million subscribers. However, the statistics show that there is a massive number of Chinese cell phones and a certain percent will probably need to be examined for corporate investigations, intelligence agency, and law enforcement investigations. A recent email blast from Cellebrite as well as an article from *Newsblaze* stated that Cellebrite in Glen Rock, New Jersey, created a tool called Cellebrite UFED CHINEX which gives the examiner the ability to support thousands of Chinese phones [32].

UFED CHINEX works with UFED ULTIMATE and supports logical extraction and physical extraction of phones. The tools are reported to be able to recover deleted email, SMS messaging, contacts, and multimedia content which may include pictures as well as video. Cellebrite reports that they are number one in the world with 12,000 devices in use in over 60 countries. The UFED CHINEX could greatly expand that market which could make any certifications they offer more valuable since the devices are so prevalent.

Another tool was invented for analyzing Chinese cell phones. It is called the Taruntula Chinese Cell Phone Analysis by eDec Digital Forensics. The tool includes a hard case with software and 29 cables that can be used to seize the deleted phone books, call records, and SMS messages [33]. This device is interesting because it also works with what is commonly called knock-off phones. The knock-off phone is a phone that is made to look and perform the same as a name brand phone but has nothing to do with the real phone it is trying to imitate such as the IPhone. This tool also allows the examiner to create an image file of the memory which is exportable to another piece of media.

Many cell phone examiners are very concerned about changing evidence on the phone and often go to great efforts to preserve cell phone evidence. Examiners may perform actions such as using a USB write blocker or changing the registry on the examination machine so that the USB port is read-only. The Tarantula Chinese Cell Phone Analyzer is reported not to write to or harm evidence on the cell phone. The tool also identifies the MTK chipsets and collects the phone identification information. This is important because some information can be spoofed at the operating system but the real original identification data should be obtainable from the chipsets.

Report generation is also very important. The results of the call logs, contact lists, and SMS messages are output in an HTML format. This is important because this format takes up less room than some proprietary formats. Most tools create a standardized report for the examiner with all the data included in it because it is easy to forget something. Standardized formats also save everyone time because both the prosecution and the defense get used to knowing where to look to get the important data.

It is also important to be able to power up a phone in order to examine it. The tool kit comes with a universal battery charger to power up and recharges a multitude of phones. The cell phone investigators will often tell academics, officers of the court, and others that the cables and recharging units are often not available at the time of the seizure. This means that the examiner needs to have both a large group of cables and a universal charger. It is possible that a low-priority case may not be investigated for months. This would mean that the battery would most likely be dead at the time of examination. Charging the phone in a controlled environment with a Faraday bag may be a good idea if the phone will not be examined for a long time.

TRANSLATION OF CHINESE CALLING CODES AND DOCUMENTS WITHIN A PHONE

It is quite possible that law enforcement, intelligence agencies, and corporations in Western countries at some point in time may have a need to examine a Chinese cell phone. It seems that the Cellebrite tools would be a natural choice in such circumstances. However, when it comes to translating Chinese documents and text, a translator may not be available. I and my wife used Google Translate to translate some documents posted on the Fu Jen Catholic University website as part of a test of Google Translate. The documents were written in traditional Chinese characters and not the simplified Chinese found in mainland China. Google Translate appeared to work well for about two pages of text at a time. Larger selections of texts and translations did not appear to work in a timely fashion. Some professional agencies in New Jersey may charge as much as eight cents a word for translation and one may have to wait a significant period of time. Google Translate will give an approximation of what the document says and this may be enough to give justification for hiring someone to do a more accurate translation.

If a person only does a physical acquisition of the Chinese cell phone by themselves by starting at byte 0 and going to the last byte, they will collect data but it may not be useful. There are certain symbols that appear unintelligible to the examiner so it is necessary to use a tool that decodes the data into something meaningful. This is not changing evidence but applying code tables to it and creating the intelligible information that refers to calls, contacts, and messaging. If I look at a digital photograph, I want to see what it represents in a form that I understand. I do not want someone to show me a page of ones and zeroes and tell me that this is a picture.

VALIDATING YOUR CHINESE CELL PHONE EXAMINATION TOOLS

It is important to be able to validate your tools and do a comparison of what the tools locate and how they interpret it. One could start by taking a Chinese IPhone knock off and using it to call four people. Then, one could send four SMS messages, four emails, and use the camera to take and send four pictures. Perhaps one email and one picture could be deleted. Then, one could use the Cellebrite tool and the eDec device to see what is recoverable and compare the results. This of course is a simple test but is one that could get you started on the path to data validation. One could also note the identity in the chip set and see if both tools accurately discovered both identities.

PARABEN'S DEPLOYABLE DEVICE SEIZURE

There is a tool called the "Deployable Device Seizure" (DDS) which is made by Paraben and is considered fabulous by some digital evidence examiners for quickly seizing the data from certain cell phones. Paraben gives the digital evidence examiner the opportunity to purchase the DDS on a PC tablet that can be held in one hand and operated by touch screen with the other hand. The device appears to be rugged, portable, and useful in the field examinations of cell phones, certain digital cameras, certain PDAs, PDA phones, and SIM cards. It would be advisable to upgrade this item to include the field kit and get the SIM card reader and porn stick detection device. The DDS and tablet is also good from a usability perspective because it is preloaded with device drivers, software, and is simple to operate since it is menu driven. The device comes with a cable kit so that it allows one to connect the phone and then use the tablet to both collect and examine the data. Since the device is also in a tablet form and can be operated with a touch screen, it could also probably be operated by someone in a Hazmat suit with a stylus in hand. This might make it possible to go to areas where chemical or biological WMDs may have been used and collect evidence from cell phones left behind.

The tool could also be used by probation officers to examine the cell phones of released sex offenders to make sure they are not violating their parole by previewing the SMS messages, call logs, and any pictures in the device. The DDS lets the digital evidence investigator quickly preview cell phones and other devices with a minimum of training since the device could be considered highly intuitive. However, it is probably advisable to get the Paraben's Certified Mobile Examiner (PCFE) certification so that one could better answer questions about the tool in court. The examiner could use the DDS to do a logical acquisition of the phone to see active content or do a full physical acquisition to take back to the office for further analysis. DDS has some data carving capability. Some examiners may wish to do further analysis with a group of data carving tools back in the office. If one has a physical image, a number of tools such as FTK, Encase, Data Lifter 2, and many other programs could be used to recover pictures and deleted attachments.

The DDS Interface is very important in that it collects all the evidence, adds MD5 hashes, and then allows one to create a report that can be given to others. The interface also allows for the insertion of the agency logo, the case number, the address, the examiner's name, as well as any notes about the case. It is very important to be able to incorporate SMS messages, received call information, missed calls, dialed numbers, recovered pictures, call logs, and other data in this one computer-generated report. Some of the important data concerning calls that may be captured with this tool is known as "call duration." This could be important to investigators who may be interviewing a suspect who may say those conversation were meaningless and were only wrong numbers or short calls. The calendar and datebook with all its entries can be put in the report too. There could be important evidence here linking the crime with the activity in the calendar. This report could also be incorporated in Susteen Secure View's report if one chooses to use other tools later.

One of the noteworthy items is that approximately 4000 mobile devices can be seized with the DDS. What can be seized will vary with different devices. Some tools such as Susteen Secure View will tell the user by notification on the wizard what data is possible to obtain when the seizure process is being implemented. The DDS can also obtain the Blackberry PIN messages. This PIN number is given by the service provider at the time of configuration. Many people share their PIN number with friends and so that no-cost text messages can be shared, even across countries. This PIN is different than the password number one uses to lock the device.

Another important consideration is that people often back up their Blackberry. The backup is done with a synchronization tool made available from RIM. This tool allows one to copy some or all of the contents such as phonebooks, pictures, email, Internet favorites, videos, and other items to a desktop, netbook, or laptop computer. People may also delete much of this content once they transfer it and/or back it up from the Blackberry to the other computer. This is an important consideration because suspects may try to keep evidence off their Blackberry, yet still have it stored at home. Keeping mobile devices clean of important evidence may be an important consideration for some suspects because it is easy to have a Blackberry pick-pocketed or just lose it somewhere. Paraben's DDS, tablet, and cables allow one to get information on the desktop backup record from the Blackberry. This type of information may give the examiner in a criminal investigation the hard evidence that is needed along with an affidavit to obtain a search warrant for the suspect's home computer or netbook.

The hardware and software for this investigative tool has an annual fee of $580. The first year is free so there is no extra expense for updates with the software, cables, or technical support by phone or email. The renewal costs of the second year helps pay for research, development, and continued support so that many of the new mobile devices that appear on the market can be investigated with the DDS. Paraben often has to add new software updates since there are sometimes very different phones appearing which use protocols such as TDMA, CDMA, or GSM. The DDS is not supported for use with VM Ware. This is important because many digital evidence examiners wish to run tools and images of seized systems in a virtual window on a desktop. The DDS also comes with a 12 V car adapter. This is important because if batteries die and no other power sources are available, the car adaptor can be put in a cigarette lighter and then the DDS can be powered or recharged. Many digital evidence examiners bring uninterruptible power supplies, gel cells, extra batteries, and other adaptors so that they can finish the examination in spite of all types of power problems. The DDS also comes with a 15 V alternating current adapter.

The DDS software that runs on the tablet needs at least 1 GB of RAM and a central processing unit (CPU) with at least a speed of 1.4 GHz to run properly. The higher the speed of the CPU, the quicker the results can be obtained. The tablet or other device that runs DDS needs at least 200 MB of storage free to hold this program. The operating system for DDS should be Windows 2000, Windows XP, Vista, and Windows 7.

PARABEN'S SIM CARD SEIZURE VERSION 3.1

There is a program from Paraben called SIM Card Seizure. There is a demonstration version that lets digital evidence examiners try it for a limit of 23 times over a period of 30 days. The price is listed as $129 [34]. The program is very intuitive and easy to use. First, one has to have permission to have the SIM card and search it. Then, one has to take the back cover off and remove the battery. Then the SIM can be removed. The next step is to carefully place the SIM card in a SIM card reader so that the gold contacts touch the inner circuit board. The reader is then plugged into the USB port. SIM Card Seizure is started up. The program will run properly and not in demo mode if the person paid for the software, registered it, and downloaded the license key to the prescribed directory. Once the program is run, it is menu driven. There are a series of questions that it asks about the SIM card and about the examiner, his or her organization, the case number, and the address where the examination takes place. All this will be used later in an html report that can be read on any web browser.

Once the acquisition is done, there is a tree on the left side. If one used a Motorola Razor cell phone, some of the operating system files may be displayed in the tree. Then, another area may show incoming and outgoing calls with the telephone numbers of both the caller and the person called. Another branch may show the telephone number of both incoming and outgoing SMS messages on GSM networks. However, it is not always possible because some phones are configured to put SMS activity on the phone and not the SIM card [34]. It is possible that deleted SMS messages may be possible to retrieve depending on the activity of the phone. There is also ample support for the product with email, online help, and a telephone number to call. The support is a very important feature that cannot be overlooked. It is possible that Paraben may try to increase the number of SIMs it supports to help law enforcement. If one has already purchased Device Seizure, this product would not be needed because its functionality is included in Device Seizure. Paraben also frequently updates versions of its software in order to improve the software and increase the number of SIMs it supports. The software works nicely on Windows 98, Windows XP, Windows 2003, and Windows 7. This program needs a processor (CPU) with a speed of 1.4 GHz or higher and at least 200 MB of hard drive space for installation. The program also supports other languages such as Arabic, English, and Russian. This is because of its inclusion of Unicode support. This could be important to defense contractors who are assigned to places such as Iraq. The latest version also supports CDMA SIM card acquisition.

The tool has a 1 year subscription. Many programs only last for a limited amount of time because the development costs are so high and the business model could almost be considered a leasing agreement as one leases a car. The program is very good at hashing the evidence in a MD5 or SHA-1 algorithm. This is very important for showing that evidence was not tampered with and such information should be kept with the chain of custody. There is also an option for bookmarking evidence. This may be important for rerunning reports or knowing where important evidence is so it can be visited later.

REPLACING THE USB MINI PORT ON A CELL PHONE

Sometimes the mini USB port on a cell phone becomes damaged. This can happen from plugging and unplugging the cables very often, and pushing the connector too hard. Sometimes the result is that a pin becomes damaged. The phone will charge but not be able to transfer data because the power and data use different pins. There are also times that a suspect will put a tool in the mini USB port on a cell phone before surrendering it. The motivation can be to damage the port in order to avoid examination. If an examination is needed immediately, there are some options. The first is to see if there are other connectivity options such as infrared or Bluetooth. Regardless of the speed, it may be possible to get some information from the cell phone using alternative wireless connectivity methods. Many cell phone forensic tool kits have options for this. The other option is to replace the mini USB port yourself or to have it professionally repaired.

It is best to make sure that the diagnosis of a problematic mini USB port is correct. The first step is try the phone and cable on another examination machine with a good USB port. This may solve the problem because it could have been a problem port on the examination machine. If it is not the answer, it may be a software program problem. One should go to the Device Manager in Windows and then uninstall and reinstall the USB drivers [35]. Then, after the device drivers are changed, it is advisable to restart the computer and the problem might be fixed. If it is not the problem, it could be with the cable. It is good to get a replacement cable from the manufacturer or a forensic cable kit such as Paraben's Device Seizure Cable kit or Susteen Secure View's kit. Some dollar stores also sell cables for a multitude of cell phones since people are often losing them or damaging them. There are places such as CablesToGo online too. Priority shipping might be worth the money.

Then, there are a number of places that fix cell phones. One can also fix the mini USB connector on the cell phone by replacing it. There are a number of places that discuss how to do this such as Hack a day website [36]. Will O'Brien wrote an informative article on the subject that was also high documented with pictures. The article also gave a list of what tools were needed and some of the tools were pictured. It is important to order the correct replacement part. The mini USB connecter which goes inside the phone was given by the article on the Hack a day website as Mouser electronics (Part #538-67503-1020). The cost on the website was under two U.S. dollars plus shipping. There needed to be a soldering iron, screw drivers, pliers, and a rotary tool to remove remnants of parts that would inhibit the correct seating of the new mini USB port. A desoldering braid and a device with alligator clips to hold the printed circuit board of the phone was also needed.

There is a website called www.pdarw.com that has a repair service for mini USB receptacles on cell phones. They claim a 24 hour turnaround from the time that they receive it until sending it out. The cost is approximately between sixty and eighty U.S. dollars. This is a good option for a nontechnical person that might have a problem unsoldering the part, using a rotary device on what could not be removed, and then putting the new part in. The pressure not to make a mistake on the installation of a part might be too much for some people, so sending the phone to a professional might be money well spent. If it is a corporate investigation, then just bill the client. The examiner really needs to do a cost–benefit and risk analysis before making a decision. It might be better to discuss with one's legal counsel to get a second opinion. Such matters should not be taken lightly because people are quick to sue.

One of the services that www.pdarw.com does that is very interesting is that they will perform a free diagnostic and send it back. The only cost is shipping. If one decides to use their service, then it is already there. Suppose that the Blackberry that someone had fell into liquid or became damaged in the rain. Then it would be possible to send that cell phone to them and get an estimate. If a suspect broke a screen or broke the cover, it would also be possible to get that repaired too. In an important case, it would first have to be asked if it could be sent to a third party for repair. The shipping method would also have to be agreed upon so that the chain of custody would still be maintained. These are very important issues. The defense lawyer or prosecutor may not want a Blackberry sent to a phone repair shop where people who are not law enforcement can work on it. Both sides may not like the fact that it may be necessary to power up the phone multiple times to test the port and data transfer capability. The powering on and off of the phone would change various date and time stamps of some files. However, it might be OK; some type of ruling and agreement needs to be made ahead of time.

The same company also does a dusting and cleaning on a device. It may be necessary to use the dusting and cleaning service if the device was in an unsanitary place, because the examiners and jury may have to handle the device. In any case there needs to be a discussion with the correct legal people to see what can be done without jeopardizing the case. I once used a screen replacement company in Texas to purchase a replacement screen on a Blackberry 6230. They provided prompt service and sent a special screw-driver tool to open the unit and replace the screen. They provided support on how to perform the replacement by phone. I successfully replaced the screen and seized my own phone for a demonstration in class. Many people are not aware that the data can be seized with a broken screen. The only disadvantage is that one cannot see activity on the screen. It is also possible to make and receive calls too with a broken screen.

It might be worth it for one investigator in a corporate setting or in a law enforcement lab to learn how to solder, use hand tools, and a rotary stripping device in order to do their own light repairs on cell phones that come in. Tools such as a Weller 15 W soldering gun, a screwdriver and wrench set, and solder are rather inexpensive. Many county colleges have electronic technology programs that teach people to build and repair electronic devices. Video instruction may be a low-cost option to learn to solder and replace items if classes are at inconvenient times and too costly.

Work that requires significant electronic rebuilds and a clean room should still be outsourced to larger labs that do that for a living and specialize in that. One might consult with people at Kroll Ontrack for an opinion as well as members of ASIS International and the High Tech Crimes Investigative Association. If one goes to a chapter meeting of one of these organizations, there are always people with business cards who can be contacted further and asked about other computer forensic service vendors that they had good experiences with. Many people really enjoy talking and having someone genuinely listen to their stories and experiences. There is much to learn and many times these organizations such as ASIS International have monthly periodicals with contacts for vendors for all types of security parts and services. Sometimes when one talks to them, they will provide a reference to someone who will really help.

The power of face-to-face networking combined with searching websites and paper periodicals can point one in the correct direction.

PARABEN'S LINK2 V2.5

This tool works with Device Seizure by Paraben and allows directed graphs to be created that help investigators see relationships that exist between the suspect and other callers. The call log contains the information about who calls the suspect and how often. The call log also shows how often the suspect calls other people. It is then easy to determine who are the most important influencers and key associates of the suspect. The program is similar in functionality to the Link Graph feature in Susteen Secure View's SvProbe. Link2 is free and is available anytime from the website. Link2 works with one or more of the PDS case files that are stored in Device Seizure. Link2 is good in that it can create graphs from multiple files from numerous devices.

An examiner can link the calls, attachments, and SMS communications between devices. There are also numerous filter options, so custom queries can be created and graphed.

The Link2 program can create reports in an HTML format. The program works with many file formats for export. This is good for law offices and examiners who prefer to use a certain format such as HTML, PDF, or SVG. Link2 uses significant resources to filter data from multiple files for complex queries that create graphic results. This takes a CPU with significant processing power with 1.4 GHz or higher and at least 1 GB of RAM. The unused hard drive space needs to be at least 200 MB for the program to reside on. The operating system requirements are Microsoft 2000, XP, 2003, Vista, or Windows 7.

ELCOMSOFT PHONE PASSWORD BREAKER

This is an interesting program because it allows the digital investigator to recover the password from a backup of certain smartphones such as the Apple iOS and RIM Blackberry platforms. The backup would probably be on the netbook, laptop, desktop, or perhaps all three machines that belong to the suspect. The vendor says, "The password recovery tool supports all Blackberry smartphones as well as Apple devices running iOS including iPhone, iPad and iPod Touch devices of all generations released to date, including the iPhone 4S and iOS 5" [37].

The backup password might be the same one used for the phone. If one cannot get the password from the user for the Blackberry, this might be another method to try to get it. Many Blackberries are so set up that if you get the password wrong 10 times, then the data is made unavailable forever to the investigator. Sometimes people will not give the password even when the judge issues a compel order because the evidence on the phone may get them more time in jail than disobeying a compel order. This tool may solve the problem of getting the backup password in its plaintext form. The backups can contain the same data on the phone at a different point back in time. If the password is obtained and the new phone is imaged, it would be interesting to compare the differential in the data. The backup may contain call logs of incoming and outgoing calls. It could also have the address book which can contain name, work and home addresses, phone numbers, email addresses, and personal notes. If the person uses SMS messaging, the important SMS archived messages will be there.

Then, the backup should have all the bookmarked favorites, web browsing history, and cache. The Internet sites visited and bookmarked favorites could have things related to a drug case such as how to make a meth lab. There should also be snapshots. If the GPS was enabled, then the location of that picture as well as the date and time should be obtainable, thus putting a person at a certain time and place. There could also be saved voicemail messages, but unlisted to voice mail would need an additional warrant in New Jersey known as a CDW or Communication Data Warrant. The calendar could also be a clue to the activities of the suspect and perhaps help put a person at a certain place and time. If the calendar and GPS embedded photos put a person at the crime scene, then that is helpful evidence in getting a conviction.

If a person wishes to purchase the tool, it is available on the Elcomsoft website for $79 for the home edition and $199 for the professional edition. The examination machine should have any one of the following operating systems installed on it: Windows Server 2003, Windows Server 2008, Windows XP, Windows 7 (32 bit architecture or 64 bit architecture), or Windows Vista. It is also good to have a machine with a multicore processor and a NVIDIA graphics card, so GPU acceleration can be taken advantage of. The iPod Touch, iPad2, iPhone 4, and iPhone 3G, are just some of the smartphones that the software package works with. It is best to check the website for the latest list of supported phones. The website also posts data about performance times. One fine example listed on the website includes the AMD Radeon HD 5970 graphics card installed on a machine with a multicore processor. Then, it is possible to use the tool and the previously listed hardware to check 20,000 passwords per second on a Blackberry 6.x [37]. If one checks the graph, the type of graphics card used greatly impacts the number of passwords per second checked.

INVESTIGATIVE COMPUTER AND PRECAUTIONS TO TAKE

It has already been stated in this book that it is very important to preserve the integrity of the digital evidence obtained from the telephone from the time it was seized until it is presented in court. It is important for many digital evidence incident response team members to use a checklist so that they know that they did not forget anything [38]. Once the evidence is obtained, a chain of custody form should be filled out. Each time the evidence is copied, processed, or transported, it should be documented on the chain of custody form. If others receive a copy of the evidence for prosecution or defense purposes, they too should sign for it.

When the examiner is ready to investigate the phone, he or she may have a checklist to make sure that the examination machine is ready. This computer that is known as the examination computer can be a laptop or a desktop. The main requirement is that it has at least a Pentium 90 for processing speed and enough RAM to operate the cell phone forensic software. There must also be enough available storage for the contents of the seized phone. This may be a challenge with an older computer but fortunately old computers with Windows 98 had USB ports and this allows USB external storage devices to be used. The examination computer should have a current, properly licensed copy of the examination software used for the phone.

Since this examination computer is used to download the data from the phone and then prepare it in a readable format for examination, it is best to check in the operating system that the USB port is working properly. In Windows 7, one can go to Control Panel,

Device Manager, Universal Serial Control Bus, and then double click. It would also be a good idea to close all unnecessary programs.

This examination computer should also be protected from unwanted outside connectivity. That means that the wireless ports, infrared ports, Bluetooth ports, modem port, and Ethernet port should all be disabled. This examination computer may also be checked for viruses by running a current version of a properly licensed antivirus program to remove the possibility of a virus altering the data. A properly licensed program for antispyware should also be run. A scrubbing program such as CCleaner that deletes all unallocated hard drive space and clears temp files should be run too.

The examiner may also wish to prepare a Faraday cage or a stronghold tent such as the one from the Paraben Corporation so that no signals will escape or penetrate the area where the cell phone examination is being conducted [39]. The examiner may elect to sit in the stronghold tent when conducting the examination so that the evidence is not tainted by outside signals from hackers or people who wish to compromise the examination. Each person that enters and leaves the examination facility should sign in and sign out. If an examiner retires, transfers, or quits, the cylinders for the locks should be rekeyed. It is recommended that each examiner have a badge with their name, organization, weight, and height. If a person has a dramatic change in appearance due to weight or facial hair change, then a new picture should be placed on the badge. Cameras should also be placed at the entrances and exits so that visitors, examiners, and illegal entries are documented on film. If one contacts ASIS International, they have a bookstore that sells the Certified Protection Professional (CPP) set of study guides that address these physical security best practices.

PRECAUTIONS: EXAMINING PHONE—HIGH-PROFILE CASE

If it is a high-profile case, then it may be important to make sure that the examination room is in a Tempest facility with no possibility of eavesdropping. This means that only filtered power is used so that the wiring for the outlets cannot be used for transmitting data from the examination room [40]. Heating and air conditioning ducts should have a grate in them so that someone cannot climb in the facility and eavesdrop on the investigation or compromise the data. The Spy Museum in Washington, DC, has an air conditioning duct that the public can crawl through to get a real example of this method of spying. The walls of the examination facility should be like the ones at the New Jersey RCFL that contain a mesh wire system so that a criminal cannot easily breach the security of a wall. There should also be no false ceiling or raised floors where people can easily hide eavesdropping equipment or crawl into. The walls should have copper so that signals cannot escape. A guard and close-circuit television are also necessary. It is also suggested that the examination of the digital device should not be conducted in a room where there is a flat roof above. This is because criminals can use a Sawzall or reciprocating saw to cut a hole in the roof after hours and get to the evidence. The NIST has a set of guidelines that discuss the security needed for Tempest facilities [40].

The digital examiner should also make sure that his or her credentials are current and not out of date before performing the examination on the cell phone. If he or she took online instruction, attended a workshop, or got any continuing education, it should be noted on the resume in case the examiner's credentials are questioned in court. Before the examination begins, the examiner should question all possible weaknesses or conditions that could cause doubt with a jury if the case should ever go to court.

PRECAUTIONS: PROTECTING EQUIPMENT FROM STATIC ELECTRICITY

Cell phones are electromechanical devices with operating systems and can be affected by low batteries, humidity, temperature, and other complex environmental factors. Sometimes, examination machines have intermittent hardware failures on USB ports and hard drives. The operating systems may also get corrupted. It is therefore important that cell phone examiners have more than one forensic tool and examination station because of all the complex variables that must work together in order for a cell phone to be seized and the electronic evidence to be collated by the examination software. I have demonstrated the same cell phone on the same examination machine numerous times with varying results. The cell phone may be acquired on the first time but many times it takes three or four attempts before the process is successful and complete. It is not anyone's fault but there are so many factors concerning the environmental factors. The discharge of static electricity in the winter months is a big concern. That is why it is important to wear a grounding strap, use an antistatic mat, or use a static potential equalizer known as a "Static Buster" as pictured in Figure 1.5. If none of these option are possible, at least touch a metal object before touching the phone or computer.

CCEs often tell me that it can take one or many times to acquire the evidence on the cell phone because of the correct parameters on the variables of the phone and examination machine that must be within acceptable boundaries for the process to work. Some cell phone examiners that use powerful desktops to run the forensic cell phone software will connect the desktop to an uninterruptible power supply because the power often intermittently goes out in the summer due to storms. Brown outs from too many air conditioners running on hot days is another reason for intermittent power losses. Even a short bit of power loss will cause the desktop computer to reset and reboot, thus ruining the data acquisition from a cell phone with high-capacity storage in it. Therefore, many cell phone examiners will have two examination machines with uninterruptible power supplies, and a licensed copy of BitPim, Device Seizure, Susteen Secure View, and Mobil Edit. Others will have a Cellebrite and a small portable storage device for incident

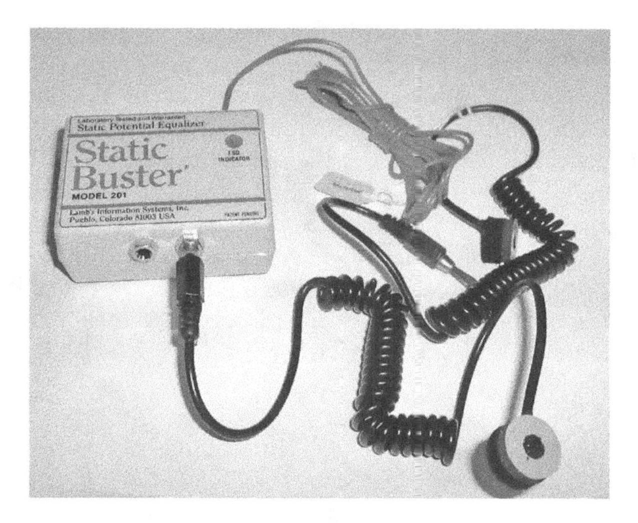

FIGURE 1.5 Static potential equalizer—Static Buster.

response in the field. Because of the number of things that must be right, having a wide range of tools is considered a must.

Susteen Secure View can now incorporate the data of its e-forensic competitors in its reports so now juries and judges do not have to read a multitude of reports for each device. Cloke, Goldsmith, and Bennis, experts in workplace conflict and resolution, give numerous examples that one might consider parables, which appear to teach that more can be gained from cooperation than competition [41]. This lesson has appeared to be internalized by Susteen, the maker of Secure View. Many forensic examiners have said that they have also bought Secure View so that they can incorporate the data from other forensic tools such as Device Seizure of the Cellebrite UFED products in their main report.

GPS CAMERA PHONES

It is good to start with the discussion of the acronym GPS before discussing the GPS camera phone. Harry Newton defines GPS as an acronym for "global positioning system" in his telecom dictionary [17]. There are both military- and civilian-grade GPS receivers that give an approximate to precise location of where they are located. Many cell phones have small internal GPS systems that allow the location of the cell phone to be known to the user and perhaps others. Before discussing the modern GPS camera phone, it is good to review the history of camera phones. The camera phone is a cell phone that has a digital camera built into it. Scott Calonico, a writer for *Know Your Cell* said that the first commercially available camera phone for the consumer was the Kyocera VP-210 "Visual Phone." This phone became available in 1999 [42]. The Motorola v710 is an example of a camera cell phone that has a fish eye lens on the front. It may be used for taking digital pictures and the Motorola v710 also had GPS. Cell phones such as the Blackberry Bold 9700 contain a digital camera that is capable of taking video clips or still pictures. There is also an option for advanced settings that allows the GPS services to be turned on or off. If the GPS feature is enabled, the GPS data is also embedded in the picture so that when one takes a picture and sends it to someone, the location of where the picture was taken can easily be extracted and mapped. The data is not visible to the eye but can be extracted with a tool such as the online website www.gpsvisualizer.com or by utilizing the gallery feature in SVProbe in a tool such as Susteen Secure View.

It is unfortunate that many camera phones have been used for criminal purposes. Many people have used camera phones to take unauthorized pictures of others in locker rooms, showers, changing rooms, and tanning salons and then uploaded these pictures to a website without consent. If the subject of the picture is naked and under 18 years old, a charge of creating and distributing child pornography may apply. If the picture is mailed to others with negative comments about the subject's weight or private parts, additional cyberbullying charges may apply. The Video Voyeurism Law of 2004 was a Federal law that was created to protect people from having their picture taken in a place where a reasonable expectation of privacy exists. Changing rooms, locker rooms, showers, and tanning salons are some examples of places where one has an expectation of privacy [43].

GPS DATA IN PICTURE

The GPS data that is in the picture is an example of metadata because it is data about the picture. This data is easily viewed in Windows 7 or in Windows XP by right clicking on

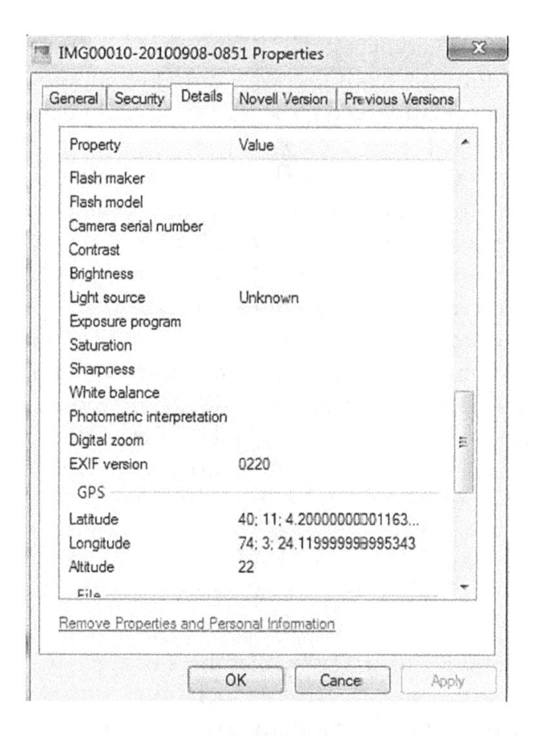

FIGURE 1.6 GPS metadata displayed in Windows 7.

the picture and selecting properties. There is a tab called details which leads to a section that says latitude, longitude, and altitude. Other details about the digital camera that took the picture are viewable too. Sometimes the make, model, and the aperture of the lens of the camera are displayed too.

All or part of the metadata can sometimes be scrubbed by right clicking on the photo, selecting properties, and then further selecting a link titled "Remove Properties and Personal Information." One can select all or choose from a long list of items to choose from. A person can choose up to three items to remove in the section entitled GPS. The first item is latitude, the second item is longitude, and the third item is altitude (see Figure 1.6).

GPS DATA AND CRIMES

A group of detectives at the 2010 HTCIA conference in Atlanta told me that in the past, some young men or women in various parts of the United States were reported to have taken pictures of themselves while drunk and posted them on social networking sites. It might have a caption such as "guess where I am." In the past, there were occasions where people were found and assaulted. In order to safeguard people from their bad judgment, Facebook removes the GPS metadata from the uploaded pictures in order to help shield them from people who may wish to locate and harm them.

A New Jersey detective took a digital camera forensics class that I taught in the Fall of 2010. The detective said that he heard of story of someone in another state who was assigned to a missing person/possible homicide case. The questioning of various people and good police investigative work led to a prime suspect. The suspect's phone was seized after a search warrant was issued by a judge. The contents of the phone were downloaded from the seized phone unto an examination computer. Photos of the missing person appeared. The missing person appeared to be a victim of a murder and the dead body and blood was photographed.

Another photo showed the victim in a shallow grave in the woods. The GPS data was taken from the phone and a team of law enforcement people were dispatched to the location of the grave. A body was found and the case proceeded. It was said to be a contract murder and the pictures provided proof that the job was done. Other evidence was uncovered and the detective said it was an open-and-shut case.

President Ronald Reagan, also known as the great communicator, once said in a speech to the American public that regardless of age, background, or religion, one thing that we have in common is watching old movies. In the movie *Road to Perdition* with Tom Hanks, we see that Tom's character is killed by a contract killer who takes a picture of him after shooting him. This was to show the bosses that the job was completed. Some people then ask if life imitates art or does art imitate life?

GPS ACCURACY AND VARIABLES

There was a discussion of GPS accuracy by members of the law enforcement community at the 2010 High Tech Crime Investigative Association Conference in Atlanta, Georgia. One detective asked the people in the room to do an experiment for themselves to test the accuracy of the data. The experiment was to take a picture in the underground floor of the hotel where the conference was held on a cloudy morning. It was important to note where one stood when one took the picture. The conference hotel was also surrounded by other large buildings and one did not have a clear view of the sky. He then said that the second part of the experiment was to take a short ride on the metro to a football field on a sunny day and note the location where the picture was taken.

Then he said to use a website such as www.gpsvisualiser.com to link the files on a Google Maps. The metadata is then extracted from the picture and mapped with Google Maps. We were guaranteed that the picture in the football field would have a more accurate location than in the hotel in Atlanta, Georgia. There was a discussion among some of the attendees who said that there are a number of factors that affect the accuracy of the GPS data in a picture. The factors of accuracy are influenced by the weather conditions and how many GPS satellites in the sky you can you reach with your present location. Roger Freeman, an authority on telecommunications, states that in any location on the earth's surface, three or four satellites should be in view [44]. Harry Newton, an authority in telecommunications, also says that canyons, urban environments, and being under canopies of trees can degrade the accuracy of GPS [45]. In Figure 1.7, we can see an example of using the www.gpsvisualizer.com and the link to Google Earth to show the location of where the GPS coordinates are displayed.

The detectives said that they only use the data in the phone to get an approximation of where the picture was taken because of the known factors of inaccuracy. They look for landmarks in the pictures near those GPS coordinates. They also said that there are a number of tools where false GPS coordinates can be put in the phone so that the evidence of the digital camera will lead investigators on a false trail. GPS spoofing tools are getting more popular and evidence of such applications should also be searched for within the seized cell phone of the suspect. The GPS coordinates can be a good indicator of an

FIGURE 1.7 GPS coordinates mapped with www.gpsvisualizer.com.

approximate location of where the picture was taken if the weather was good, there were no obstructions in the sky, and spoofing tools were not used.

The more satellites that can be reached, the more accurate the position of the GPS receiver or phone can be calculated. The latitude, longitude, and height should be most accurate on a sunny day in a clear field with no obstructions to the GPS satellites in the sky. There are two types of GPS positioning systems, one is the standard positioning service (SPS) system that is used for civilians. It is reported to be as accurate as 1 m in ideal conditions. The other type of GPS system used by the U.S. Department of Defense (DOD) is a military version that is very precise called the precise positioning service (PPS). It is accurate to within centimeters [45].

If there were terrorists known to be using GPS cell phone for some type of missile system, the accuracy of the GPS can purposely be degraded by the U.S. military to be inaccurate to as much as 100 m or about 300 ft [45]. If the reader wishes to know more about GPS signaling and synchronization, then it would be a good idea to seek further knowledge in Roger Freeman's *Fundamentals of Telecommunication* that discusses more of the technical data.

METADATA: LINKING PICTURE TO GOOGLE MAPS

One can first obtain the GPS data from the properties area of a picture and then copy and paste the latitude and longitude data in an application to calculate the location of the coordinates. However, the application may or may not be accurate. It is better to use a tool such as Google's Picasa 3 which links to Google Earth so that an accurate location can be obtained.

It is important to use trustable sources. In the intelligence field, there is what is known as strong and weak intelligence. Weak intelligence is from a source that is not reliable and has limited credibility. It is better to obtain strong intelligence which is reliable and from credible sources. If a CCE is performing an important investigation on a phone, then he or she should use the most trusted applications possible because the work may be scrutinized in court. Google Earth, for example, is a much more trusted application than something written by an unknown computer geek in a basement and uploaded to a website. The unknown application could also have malware embedded in it that leaves a trap door for the writer to enter. Trusted applications are important for both security and credibility reasons.

One may start Picasa 3, a picture editing program made by Google, and then choose a Microsoft Windows directory of pictures that one wishes to investigate. If one then clicks on the Geo-Tag button, the pictures located in that directory may appear on the map as small thumbnails. One can see an array of thumbnails on the map and learn precisely where the pictures may have been taken. This is a good tool for investigators.

FAKING GPS DATA USING PICASA 3 AND GOOGLE EARTH

Picasa 3 is a software tool for viewing and editing pictures. It also gives the metadata about the picture and one can learn about the type of camera that took the picture. The aperture information is also given. The date and time that the picture was taken can be seen too. The file size, picture resolution, and file format can be determined too. Picasa 3 will also show the focal length, information about the exposure, and will show the file's digital signature. The digitization date and modification dates are also given. There is a wealth of additional information provided that may be of use to a photography expert.

Some pictures that a person takes may not necessarily have the GPS coordinates in them and someone might forget where they are taken. There is an option called Geo-Tag that lets one embed new GPS coordinates in a picture. The Geo-Tag button starts up Google Earth and then one can select a location on earth to go down to. It is possible to take the camera down to a level where a person may geocode the GPS coordinates of a particular swimming pool at a particular hotel in a digital picture. This is a good feature for those who wish to put in truthful data in a picture. However, it could also be misused to create an alibi by making it look as though someone might have been at a different place at a particular time and date. It is also important to consider the sophistication of the suspect using the digital camera and computer. If they can barely operate the camera and using a computer browser is challenging work, chances are that they did not change GPS data in the pictures. However, it is also important to check the email and MSS traffic to see if an accomplice could have altered some digital pictures for them too.

PUTTING IT ALL TOGETHER: CELL PHONE HARDWARE

We reviewed many aspects of the cell phone in this chapter and will look at a dissected one in Figure 1.8. There is the case front and back case along with a separate latch for the battery. There is also a coiled-up antenna. In the 1990s, many phones had telescoping metal antennas. There should also be a speaker that allows the person to hear the conversation. There is also a screen which may be LCD on 1990s phones or a color display on twenty-first century phones. There may be holes that allow combination headphones and microphone sets to be plugged in. There is a full keyboard on modern cell phones but older 1990s cell phones have a dual tone multifrequency (DTMF) board which only allows number calling, on/off, and dial number. All cell phones have a

FIGURE 1.8 The internals of the cell phone.

speaker or built-in microphone for speaking into. There is also a printed circuit board which holds all the components and has a bus system on it for data travel. Many people are familiar with large printed circuit boards in computers which are called "motherboards." There is usually a metallic hinge that holds a SIM card. The SIM can be used for storage and also has phone subscriber information that is used for identification and billing purposes. There may also be internal storage built in the cell phone which could be volatile but probably is not. There should also be a large microprocessor which acts like a CPU in a computer. Then there is a radio frequency and power section. Some college students have simplified the concept of cell phone hardware and say that it is half radio/transmitter and half computer phone. Newer phones may also have connectors for external cards such as SD cards and receptacles for cables to connect with computers. It is important to understand the basic hardware so that one knows where to look for evidence. There is the internal chipset with the phone's identity, internal cards, external cards, the SIM, internal memory, and perhaps other storage or computing devices that are connected to the phone.

REFERENCES

1. Doherty, E. et al. (2007). *Eforensics and Investigations for Everyone*, p. 3, Authorhouse, Indiana, ISBN 978-1-4343-1614-1.
2. Cooper, M. et al. (October 17, 1973). *Radio Telephone System*, US Patent number 3,906,166; Filing date October 17, 1973; Issue date September 1, 1975.
3. Moultan, P., Moutlan, J. (2001). *The Telecommunication Survival Guide*, pp. 514–517, Prentice Hall, Upper Saddle River, NJ, ISBN 0-13-028136-0.
4. Smith, F., Bace, R. (2003). *A Guide to Forensic Testimony, The Art and Practice of Presenting Testimony as an Expert Technical Witness*, pp. 8, 9, 276, Addison-Wesley, Boston, MA, ISBN 0-201-75279-4.
5. Annan, K. (1999). Secretary General of the United Nations, Kofi Annan's Speech to the International Telecommunication Union on October 9, 1999 at the Opening Ceremony. Telecom. URL accessed August 27, 2011. Retrieved from http://www.itu.

int/telecom-wt99/press_service/information_for_the_press/press_kit/speeches/annan_ceremony.html.

6. Ahonen, T. T. (February 17, 2011). All the numbers, All the facts on mobile the trillion-dollar industry. Why is Google saying: Put your best people on mobile? Communities Dominate Brands from TomiAhonen Almanac 2011. URL accessed February 27, 2012. Retrieved from http://communities-dominate.blogs.com/brands/2011/02/all-the-numbers-all-the-facts-on-mobile-the-trillion-dollar-industry-why-is-google-saying-put-your-b.html.
7. Shim, C. (2009). The world's first phone museum and gallery. URL accessed August 27, 2011. Retrieved from http://www.koreatimes.co.kr/www/news/art/2009/04/153_43336.html.
8. Telecommunication virtual museum. URL accessed August 28, 2011. Retrieved from http://www.telcomhistory.org/vm/scienceCellPhones.shtml.
9. URL accessed August 28, 2011. Retrieved from http://www.forrestfotos.com/OtherGalleries/ALs-Cell-Phone-Museum/10383782_83FGVm/742594990_5wo3U#!i=742594990&k=5wo3U.
10. *Youtube*. Cellphone gun. URL accessed November 15, 2011. Retrieved from http://www.youtube.com/watch?v=Sxq14T0JElQ.
11. Silberschatz, A. et al. (2003). *Operating Systems Concepts*, pp. 20–21, Wiley and Sons, USA, ISBN 0-471-25060-0.
12. Gralla, P. (2005). *PC Pest Control*, pp. 102–103, O'Reilly Media, Sebastopol, CA, ISBN 0-596-00927-7.
13. yourdeadin. (January 16, 2005). Cell phone cloning, antionline.com. URL accessed February 24, 2012. Retrieved from http://www.antionline.com/archive/index.php/t-262738.
14. Bulkeley, W. (February 8, 2007). Finally, cell phone photos worth sharing. *Wall Street Journal*. pp. D1, D3.
15. Solomon, M., Barrett, D., Broom, N. (2005). *Computer Forensics Jump Start*, p. 104, Wiley Publishing, San Francisco, CA, ISBN 0-7821-4375-X.
16. Wilson, J. (1982). *Technical College Physics*, p. 329, CBS Publishing, New York, NY, ISBN 0-03-057912-0.
17. Newton, H. (2002). *Newton's Telecom Dictionary*, p. 663, CMP Books, New York, NY, ISBN 1-57820-104-7.
18. Freeman, R. (2005). *Fundamentals of Telecommunications*, 2nd edition, IEEE Press, Hoboken, NJ, ISBN 0-471-71045-8.
19. Shortley, G., Williams, D. (1959). *Principles of College Physics*, p. 600, Prentice Hall Inc, Engelwood Cliffs, NJ, Library of Congress Catalog Card Number 59-11031.
20. Lal, A. (November 8, 2011). RIM launches BlackBerry Bold 9900 Smartphone in India at Rs. 32,490. thinkdigit. URL accessed December 28, 2011. Retrieved from http://www.thinkdigit.com/Mobiles-PDAs/RIM-launches-BlackBerry-Bold-9900-smartphone-in_7319.html.
21. Nelson, Phillips, A., Enfinger, F., Steuart, C. (2005). *Guide to Computer Forensics and Investigations*, pp. 24, 64, Thompson Course Technology, Boston, MA, ISBN 0-619-13120-9.
22. *Wisegeek*. What is a sim card. URL accessed August 28, 2011. Retrieved from http://www.wisegeek.com/what-is-a-sim-card.htm.
23. Smith, F., Bace, R. (2003). *A Guide to Forensic Testimony, The Art and Practice of Presenting Testimony as an Expert Technical Witness*, p. 121, Addison-Wesley, Boston, MA, ISBN 0-201-75279-4.

24. Vacca, J. (2005). *Computer Forensics, Computer Crime Scene Investigation*, 2nd edition, p. 516, Charles River Media, Boston, MA, ISBN 978-1-58450-389-7.
25. Doherty, E. (2005). Investigating cell phones in the international arena. *Eighth New Jersey Universities Homeland Security Research Consortium Symposium Homeland Security*, Friday, December 5, 2008 at Princeton University. URL accessed September 5, 2011. Retrieved from http://dimacs.rutgers.edu/NJHSConsortium/Princeton Abstracts.html.
26. Smith, F., Bace, R. (2003). *A Guide to Forensic Testimony, The Art and Practice of Presenting Testimony as an Expert Technical Witness*, p. 407, Addison-Wesley, Boston, MA, ISBN 0-201-75279-4.
27. Doherty, E. (2011). Teaching cell phone forensics and e-Learning. Worldcomp 2011, Las Vegas Nevada, July 18–21, 2011, *EEE'11 International Conference e-Learning, e-Business, EIS, and e-Government, Conference Proceedings*, pp. 161–164, ISBN 1-60132-176-7.
28. Harte, L., Levine, R., Livingston, G. (1999). *GSM Superphones*, p. 1, McGraw Hill Publishing, New York, NY, ISBN 0-07-038177-1
29. Weinberger, S. (April 4, 2008). Congress freaks out over Second Life terrorism. *Wired Magazine*. URL accessed August 28, 2011. Retrieved from http://www.wired.com/dangerroom/2008/04/second-life/.
30. Gourlay, C., Taher, A. (August 5, 2007). Virtual jihad hits Second Life website. *The Sunday Times Online*. URL accessed September 4, 2011. Retrieved from http://www.timesonline.co.uk/tol/news/world/middle_east/article2199193.ece.
31. Mobithinking. URL accessed November 15, 2011. Retrieved from http://mobithinking.com/mobile-marketing-tools/latest-mobile-stats.
32. Cellebrite launches UFED Chinex (October 24, 2011). News Blaze. URL accessed November 15, 2011. Retrieved from http://newsblaze.com/story/2011102407360200002.pnw/topstory.html.
33. Tarantula Chinese Chipset Analysis tool. eDec Digital Forensics. URL accessed November 15, 2011. Retrieved from http://www.edecdigitalforensics.com/tarantula-chinese-chipset-analysis-tool.
34. Paraben Sim Card Seizure v3.1. Paraben Corporation. URL accessed January 22, 2012. Retrieved from http://www.paraben.com/sim-card-seizure.html.
35. Ty, A. (nd). How to repair a laptop USB port. Ehow. URL accessed February 15, 2012. Retrieved from http://www.ehow.com/how_5030220_repair-laptop-usb-port.html.
36. O'Brien, W. (November 8, 2007). How to replace a mini USB port on your cellphone. Hack a day. URL accessed February 15, 2012. Retrieved from http://hackaday.com/2007/11/08/how-to-replace-a-mini-usb-port-on-your-cellphone/.
37. Elcomsoft Phone Password Breaker. Elcomsoft. URL accessed February 4, 2012. Retrieved from http://www.elcomsoft.com/eppb.html.
38. Mandia, K., Prosise, C., Pepe, M. (2003). *Incident Response & Computer Forensics*, pp. 78–80, McGraw Hill/Osborne Publishing, Emerysvile, CA, ISBN 978-07-222696-6.
39. Doherty, E. et al. (2007). *Eforensics and Investigations for Everyone*, p. 9, Authorhouse, Indiana, ISBN 978-1-4343-1614-1.
40. Nelson, Phillips, A., Enfinger, F., Steuart, C. (2005). *Guide to Computer Forensics and Investigations*, pp. 24, 64, Thompson Course Technology, Boston, MA, ISBN 0-619-13120-9.
41. Cloke, K., Goldsmith, J., Bennis, W. (2005). *Resolving Conflicts in the Workplace*, Jossey-Bass, USA, ISBN 0787980242.

42. Calonico, S. (2011). A retro salute to the Kyocera VP-210. Know your cell. URL accessed August 31, 2011. Retrieved from http://www.knowyourcell.com/culture/cellhumor/1022860/a_retro_salute_to_the_kyocera_vp210.html.
43. Doherty, E. et al. (2007). *Eforensics and Investigations for Everyone*, p. 6, Authorhouse, Indiana, ISBN 978-1-4343-1614-1.
44. Freeman, R. L. (2005). *Fundamentals of Telecommunication*, 2nd edition, p. 140, Wiley-IEEE Press, Hoboken, NJ, ISBN 978-0-471-71045-5.
45. Newton, H. (2002). *Newton's Telecom Dictionary*, p. 331, CMP Books, New York, NY, ISBN 1-57820-104-7.

CHAPTER 2

Digital Camera Forensics

THE HISTORY OF THE DIGITAL CAMERA

It is difficult to find an exact starting point for the digital camera, so this chapter will start with the first mention of it in a recognized publication. Eugene Lally was a NASA scientist who wrote a theoretical paper in 1961 that described digital photography using a mosaic array of photo detector circuits, analog to digital converters, and a place to send the resulting pictures [1]. Others will sometimes start the history of the digital camera with a story about Texas Instrument's initiatives on a filmless camera in 1971.

The history of the digital camera differs because there are many types of digital cameras and opinions about the history and whose contributions are significant. This is why definitions are important. If you consider a device that used a traditional movie camera lens, a coupled charged device (CCD) image sensor, digital circuits, and a process that takes 23 s to save an image on a cassette to be a digital camera, then Steve Sasson is the inventor of the first digital camera [2]. The U.S. Patent Office would also agree with that because Steve Sasson and Lloyd Gareth were awarded patent number 4131919 on December 26, 1978 for developing the digital camera [3]. The patent expired in 1997 which meant that many restrictions on the manufacture and sales of digital cameras after that date would be lifted.

The mass-produced fully digital cameras that we are used to today and that can save images quickly to electronic media did not happen until the 1980s. The student of digital camera forensics would most likely be interested in Steve Sasson's contribution to the field of digital cameras but would be more interested in digital cameras that are mass produced, used in crimes or policy infractions, and must be examined as part of an investigation. The history of the modern mass-produced handheld consumer device that is known as the handheld digital camera, or the one that is built into the cell phone, is the result of many innovations. The history could be divided into some major components, which are the conversion of analog pictures to digital signals, the storage of signals on magnetic media, the transmission of digital images, Kodak's 1.4 Megapixel Sensor, and the Apple Quick Take 100 digital camera in 1995. This section of the chapter will discuss each component.

DIGITAL CAMERA HISTORY: CONVERSION OF ANALOG PICTURES TO DIGITAL SIGNALS

Digitizing pictures is said to have come out of the television industry in the 1950s. Charles Ginsburg led a team that developed the videotape recorder (VTR) in 1951. The VTR converted analog television signals into digital signals that were captured on magnetic tape. Charles Ginsburg's VTR was sold in 1956 to the public by the Ampex Corporation [4]. The changing of analog signals to digital signals is the main function of our codecs today. Digitizing signals was a necessary hurdle to pass in order to later invent a handheld digital camera.

DIGITAL CAMERA HISTORY: THE TRANSMISSION OF DIGITAL IMAGES

Many great technologies have been improved due to by-products of the research and development of the U.S. Space Program started by former U.S. President John F. Kennedy. It is common knowledge that on July 20, 1969, the astronauts of Apollo 11 landed on the moon and the digitized pictures of the moon were transmitted back to earth and seen by an estimated five hundred million people worldwide on television. The transmission of digital signals by a short wire and ultimately the long distance to earth wirelessly could be a process that is considered an ancestor of our ability to take pictures from our modern digital cameras and connect them to our computers via cable or wirelessly.

DIGITAL CAMERA HISTORY: KODAK'S 1.4 MEGAPIXEL SENSOR

The digital camera that Steve Sasson invented was something that could be considered to be an expensive prototype that proved the concepts of a digital camera in the labs of Kodak in the 1970s. Then, in 1986, the Kodak Corporation created a 1.4 megapixel sensor that allowed pictures to be taken and digitized with the use of the sensor [4]. This was the last component that needed to be invented so that someone could take all the achievements of the past and create a digital camera. This 1.4 megapixel sensor allowed for the capturing of an image to digital media and thus could be immediately sent to a printer. Kodak would have probably been the innovator and lead digital camera maker except that a patent dispute over instant pictures with Polaroid was decided in favor of Polaroid.

DIGITAL CAMERA HISTORY: THE APPLE QUICK TAKE 100

The Apple Quick Take 100 is considered by many to be one of the first truly digital cameras that were affordable, mass produced for the consumer, and interfaced with the home computer [4]. It was available in the United States in 1995. Some of the early cameras in the 1980s such as the Sony Marvica could arguably be traditional cameras that used a type of codec to save the picture to digital media. That is not a fully digital camera but is perhaps a hybrid digital camera. The Sony Digital Mavica MVC-FD5 that was introduced in 1997 was a fully digital camera. The Quick Take 100 used the Kodak megapixel sensor and had a cable that allowed one to download picture using a RS232 port and connect to a personal computer.

DIGITAL CAMERA HISTORY: THE WEBCAM

Haag, Cummings, and Phillips are authors of a management information systems book that divides digital cameras into three categories [5]. They describe the *webcam*, the *digital still camera*, and the *digital video camera*. Most of this chapter will focus on the digital still camera but the webcam is worth addressing since it is connected to the laptop, notebook, netbook, desktop computer, and the Xbox. Xbox forensics is a new area of digital forensics and there is now a webcam for it called the Xbox 360 Live Vision Camera.

The webcam can be used to take a still camera shot or video. It is a digital camera. Webcams can be separate pieces of hardware that sit atop computer monitors and plug in a USB port. Webcams can also be small devices embedded in PDAs, cell phones, or PDA cell phones. Some of the same video conferencing applications made for desktops and laptops such as Skype now appear on mobile devices. The webcam still shot photo can be exchanged with another person through email or by Skype phone or by a video conferencing program such as Microsoft Messenger. America Online (AOL) Instant messenger is also a program that lets people send text messages and still picture file attachments.

Sometimes, some sex offenders misuse communication programs and then get caught contacting minors to meet for sex. The majority of the sex offenders are adult men looking to “hook up” with teenage girls. This has been the subject of a television show called *To Catch a Predator*. There are two ways that the sex offender can be caught. The first is to report the offender while engaged in a conversation with the minor. In this situation, the minor can tell his or her mom or dad and they can call the police. If the police can get there while the conversation is still active, then law enforcement can get the IP address of the offender and use a reverse look up to find out where that IP address is and to whose account that is registered to.

The second way is for the minor to engage in the conversation with the offender and set up a meeting for the minor and the offender. Of course the police will be present to arrest the offender when he shows up. The police can also use a screen capture program to collect all the evidence of the inappropriate chat and videoconference while it takes place. The screen capture might even collect stills or videos of the offender exposing him or herself to the minor. There are also logging features in certain instant messaging and videoconferencing programs that can be enabled even before law enforcement arrives.

Once the sex offender is arrested, he or she will go to trial or plea bargain for a lesser jail sentence and parole. The sex offender’s name and address will be added to the sex offender database because of the Megan’s Law legislation. There are over five hundred thousand registered sex offenders in the United States [6]. Once these offenders are back in the community and on parole, a parole officer will visit their residence on a weekly basis and question the offender about his or her activities. The parole officer will also check the offender’s computer to see if he or she has been violating parole by going into chatrooms and engaging in conversations or sending still camera pictures or video with minors.

The offender’s computer can easily be checked for inappropriate video and chat by using covert programs that are installed in the background of the offender’s machine and take snapshots of activity that the user is engaged in. Email, video, chats, and web surfing can all be easily checked by reviewing the results of one of these capture programs such as Spectorsoft’s Spector Pro 2011 [7]. When a person agrees to parole, he or she also agrees to visits to the home, questioning about activities, and monitoring of the home computer with products such as Spectorsoft’s Spector Pro 2011. Since the sex offender agreed to be monitored, there is no breaking of laws or invasion of privacy. It is all legal and agreed to as a condition for parole.

DIGITAL CAMERA HISTORY: dSLR

In 2012, there are often a number of digital cameras on the market with the designation dSLR. It might be asked what the difference between an SLR and a dSLR camera might be. The fact that they are both SLR tells us that they are both of the same genus, metaphorically speaking. The main difference between SLR and dSLR is that the dSLR uses digital media instead of film to store the image. Film needs to go to a processing lab to be processed while images on a XD, SD, miniSD, or CF2 card could be printed on color photo paper but need not be printed. The digital image could just be kept on a hard drive or distributed to others by email, social media, or website. There are often places on the Internet where there are discussions of prosumer dSLR cameras and consumer dSLRs. Prosumer is a word that combines consumer and professional. Some digital camera experts will say that prosumer is a term for electronic products with a standard that is generally thought to be somewhere between what is found at large outlets such as Walmart and what is the top-of-the-line professional standard. It most likely has to be with standards referring to ruggedness, resolution, and the shutter speed variability.

In 2012, one might read an ad for a dSLR camera such as the following, "Canon EOS 7D 18 MP CMOS Digital SLR Camera with 3 inch LCD and 28–135 mm f/3.5–5.6 IS USM Standard Zoom Lens." Most of the technical information about this camera should be embedded in the picture it takes. This is known as EXIF metadata and follows format standards as described by Kevin Cohen in the *Small Scale Digital Forensics Journal* [8]. It is important to understand how to read the acronyms and understand what they mean. Canon is a brand. There are many brands of digital cameras in 2012 such as Canon, Nikon, Olympus, Sony, Panasonic, Kodak, Fujifilm, Pentax, Casio, GE, HP, Leica, Samsung, and Sigma. Sometimes, one can see the term ISO. This has to do with sensitivity to light and ranges from 100 to 3200. A large range is good because it allows for various lighting situations. The number and term 18 MP means eighteen megapixel. This means it can take very-high-resolution pictures.

The term EOS is an electrical optical system and SLR is an acronym meaning single lens reflex, meaning that there is a mirror and prism so what we see is what we shoot with the camera. The term "CMOS" is complementary metal–oxide–semiconductor and this type of chip is used in desktop computers to store the time and date. The CMOS is famous for not wasting power on heat so the battery can last longer. The 3 inch LCD means that it has a 3 inch liquid crystal display viewer that shows the picture before putting it on digital media. The term "IS" means image stabilization and helps prevent blurry pictures. The number associated with f is the aperture and f/3.5–5.6 has to do with how much light comes in. The USM is an acronym for ultrasonic motor and is part of a system for a digital zoom. The EOS 7D has to do with the optical system and there are firmware upgrades sometimes issued by Canon to improve the optical system. The SLR has come a long way since one of the first ones was seen in East Germany in 1949.

GETTING TO UNDERSTAND DIGITAL CAMERA HARDWARE

One of the ways that people who wish to become future digital camera forensics experts can learn about the hardware is to ask friends and relatives for old digital cameras and documentation. If the cameras did not work and they do not care about them, one can take a screw driver and pry the molded plastic frame apart. Then one can Google search the make and model of the camera and any associated hardware specification sheets. Then, one can look at the printed circuit board, the lens, the sensors, and look for chips that appear to be memory. One can look at the flat gold-plated paths known as the bus systems that take data around the system board. Online documentation that is detailed will explain the advanced features of the camera and often tell something about the hardware too. Just as people dissected a fetal pig or frog in high school, a person may also learn from dissecting a digital camera. One can also visit the library and take out some books on digital cameras to learn how they work. I once bought a digital camera from a local camera store from a salesman named James. Later, I went to purchase another camera and asked for the same person. The store owner said he was sad that the knowledgeable man had left. The man only stayed on the job to learn about digital cameras and moved on to learn about other devices at another store. The man was going to do something in computer science or forensics and he was learning about the equipment on the job before moving forward in that career. This is actually more common than one might think. I have met people in cell phone forensics who started their career by first working for a cell phone store. The cell phone examiner first started with sales and configuring SIM cards. Other examiners discussed how they once had jobs in the cell phone store at the mall and first learned about collecting the contents of a phone by using a Cellebrite to transfer data from the old phone to the new one.

HISTORY OF THE DIGITAL CAMERA: CRIMINAL ACTIVITY

Once computers and digital cameras became popular and could connect to one another, people were misusing digital cameras. Computer forensic examiners are sometimes called by computer technicians at stores such as Best Buy when customers leave computers for service. The technicians call when they find child pornography. Once child pornography is found, law enforcement is called and investigation starts. The law enforcement investigator would sign an affidavit, get a search warrant, and then question the suspect. The investigator would most likely write the search warrant broadly enough so that any digital cameras and media at the suspect's home could be seized too. The pornography that was found on the computer could have been downloaded from the Internet. The other possibility is that the pornography was created with the computer owner's digital camera and transferred to the computer by a cable.

According to Steven Yagielowicz, "there has been an explosion of online child pornography" [9]. Steven says this is because of digital cameras getting in the wrong hands [9]. Shinder and Tittel say that when people create child pornography, it is considered a violent crime [10]. Distributing it is another crime [10]. Some adjunct professors who are detectives have told their classes that they have investigated people in their full-time police jobs where the people said that the child pornography got on their machine because they might have clicked on a link for washing machines and were bounced to a porn site that downloaded some images to their computer. The police said they may believe that for a few images but not when there are hundred on the computer. Steven also said, "A surfer seeking legal, amateur erotica could easily download illegal content without ever intending to seek out or download child pornography, but it could be up to a jury decide his intentions—and all after a life-destroying arrest and expensive trial" [9].

DIGITAL CAMERA OPERATING SYSTEMS

Many of the digital cameras use an embedded operating system that has a limited number of functions. The important basic functions are setting the date, time, and time zone. All the file creation dates and times are inaccurate if this function is off. One should check this setting when examining a camera because it impacts all the evidence. All differentials should be noted. Embedded systems also provide the interface to select file size, resolution, and exposure time. The embedded systems usually have a feature to let the user select saving the pictures to internal memory or an XD or SD card or some type of external media. If the seized camera has an option set for storage on external media and the card is missing, then evidence needs to be found. The embedded operating system often uses a FAT 32 file system for storing and retrieving and deleting photos.

It is advisable to take a class on file systems because there are so many file formats that one may encounter with regard to digital images. BMP is an acronym for bitmap because the bits of the digital image are said to be stored in an organization that was mapped out. Information is uncompressed and there may be up to eight bytes of information with regard to each pixel. Other common formats are TIFF, JPG, enhanced JPEG, and GIF. Digital images are files that may have been taken with the camera where they reside or may have been placed there from other sources when the camera was in mass storage mode. It is important to learn about headers, footers, and how the file is saved across clusters and chained together. It is also important to learn about data carving and how a file is reassembled.

The file system standard that the digital camera investigator should be familiar with is the JEITA—CP-3461 standard. This standard is produced by the Standards of Japan Electronics Information Technologies Association [11]. This format is used in digital cameras and one can recognize it when we see a root, followed by a DCIM directory. DCIM means digital camera images. Underneath the DCIM directory, there may be up to 900 subdirectories. Each subdirectory starts with three numbers, followed by five letters. This standard discusses the use of a format such as FAT 12, FAT 16, FAT 32, or exFAT on the use of Digital Still Camera Image File Format Standard (EXIF) standards on Design Rule for Digital Camera File Systems (DCF) media. Pictures files may have the extensions JPG while thumbnails have a file extension of THM for thumbnail. Each file type has its own data structure. It is also stated that thumbnails be 120 × 160 pixels. The document on standards can really help the digital examiner understand the hardware and file system which is crucial to recovering digital images.

DIGITAL CAMERA FILE SYSTEMS: FAT

There are many versions of the Microsoft File Allocation Table (FAT) file system. Perhaps you remember FAT 12 being used in the early 1980s with 5.25 inch floppy diskettes. Twelve meant 12 bits of addressing and that meant a 16 MB limit. Then, there was FAT 16, FAT 32, and VFAT, virtual (FAT). If you are a video game player, you might have an Xbox which has FATX. Some digital cameras use a FAT file system so it is worth studying.

The FAT file system was a result of the inspirations and the actuations of those thoughts by Bill Gates. The FAT file system has a file name, a starting cluster, file size, and information both about the creation date of the file and its most recent modification. With a set of tools such as those created by Norton Utilities, it is also possible to see the chaining of clusters for a file until the end of file (EOF) marker. There is often not enough contiguous room or one continuous space to store the file, so the chaining of clusters is used. Part of a file may be at cluster 20, and then continue at cluster 50, 51, 52, and end at cluster 64. When a person deletes a file, the first letter of the file name in the file allocation table changes to E5(hex) indicating that the space reserved for that file is now free for the operating system to use and write over.

There are also two copies of the FAT table on a piece of digital media in case one is damaged. Many digital forensic tools and data recovery tools will examine the FAT tables and restore the first character with another letter so it becomes a valid entry. If those clusters were not written over and the chain is intact the file or digital picture can easily be restored.

EXTERNAL MEDIA

The Ampex Corporation's storage of these new digital signals of video on a magnetic media was a major success. A form of flexible magnetic media was a key component of floppy disks in the late 1970s. Later in the 1980s, some pictures from the analog camera family of Sony Marvica cameras were stored on these magnetic disks and later 3.5 inch floppy disks. Though these cameras were not digital, the saving of data on digital media was a significant step toward the truly digital camera. One could argue that the digital camera would not be a reality without inexpensive digital media to store data. Because of Moore's law, the number of transistors stored in a semiconductor device doubles every 18 months. This has allowed for the affordable storage media known as SD cards to be the media *du jour* of digital cameras in 2011.

One may find a variety of external media cards with digital cameras. The CF card or compact flash storage card may often have 16 megabytes of memory. The card is approximately one square and has two rows of holes that connect in a camera. The card has to be kept out of direct sunlight, heat, humidity, and away from water and shock. This CF card fits in the universal card reader and can be used with many of the forensic tools surveyed in this chapter.

SURVEY OF TOOLS TO INVESTIGATE A DIGITAL CAMERA

A fact-finding corporate committee that may oversee an investigation headed by the authorized requestor (AR) may ask, "What is the difference between data recovery and digital forensics and can the same tools be used?" The data recovery process is used to recover files or data that was lost. However, digital camera forensics uses data recovery techniques to recover lost files, namely digital pictures, as part of an investigation. That investigation could be a policy infraction investigation in the private sector or as part of a criminal investigation in the public sector. The investigator should look for digital evidence that exonerates the suspect from wrongdoing. This is called exculpatory evidence. The investigator should also look for evidence to help prove the suspect of the investigation as guilty.

It is necessary to power on a camera in order to collect the pictures from the resident memory of a digital camera. Once the camera power is on, the device appears as a mass storage device when it is connected to an examination machine. The certified computer examiner (CCE) then uses a tool such as Guidance Software's Encase, Recover

My Files, Access Data's FTK, or Paraben's Device Seizure. The camera must be powered on before using these tools and sometimes cameras having been sitting in an evidence locker for some time will have batteries that had died. The problem with just putting any batteries conveniently nearby in the digital camera and powering it on is that the batteries may not be the correct type or the investigator may not install them correctly. Some cameras are known for using up batteries, so start with the correct new ones so that one does not have to replace them and create additional temp files and further change evidence. It is important to have exact replacement batteries and to put them in exactly as it shows. The bump or plus should go in exactly as the picture directs. Some cameras even have the polarity marked with etchings inside the case for the batteries. It is important to also read the documentation and put in the correct type of battery. There are lithium, zinc oxide, and alkaline. Some cameras also specify not to use rechargeable batteries.

It is better to follow the directions for the battery exactly as the directions for the camera specify and to note that in the examination documentation. If the digital evidence in an investigation cannot be recovered and it was discovered that the CCE was negligent, due to ignoring battery selection and installation, then a lawsuit or sanction may follow. Some digital camera store owners will also tell you that sometimes the mechanisms for the batteries go bad and it is best to use an external power source to power the camera on.

When it comes to external power supplies for digital cameras, it is absolutely critical to observe the polarity. The inside may be plus or minus. A connector may fit and if one has the wrong polarity, it may damage the digital camera. It is also important to have the correct type of voltage such as 3.3, 4, or 6 V. It is also important to check that the external power source is giving direct current or D.C. voltage and not A.C. or alternating current. One time I accidentally connected a 6 V A.C. external power supply to a router that needed 6 V D.C. voltage. The result was that the router became extremely hot, stopped functioning, and began to smell like burning electrical components. It is therefore important to double or triple check the power requirements for the device and to check the equipment before powering it on.

It is also very important to note the amperage needs of the camera and what the device specifications call for. The external power supplies often say they supply a certain voltage, a polarity on the connector, amperes (amps) or milliamps, D.C. or A.C., and a certain frequency that is measured in hertz. Having too much amperage draw on an external power supply with thin wiring can be a fire hazard. Once the device is powered up correctly, the following devices may be used below.

SURVEY OF TOOLS: FORENSIC PROFESSIONAL KIOSK CARD READER EX-3U READ ONLY

There have been some emails from some CCEs that mentioned the "Forensic Professional Kiosk Card Reader EX-3U" sold on eBay. The device was made by Atech Flash and marked as brand new. It was priced at $65 with a label of buy it now. That means that it is not an item that has to be bid on, but can be purchased immediately. There were also more than 10 available. This meant that it was not just a good deal but that it was one of a kind. The shipping charge was $10. The tool at that price is affordable for academics, law enforcement, and corporate investigators. The device was for sale from California which meant that it did not have to clear customs and could be obtained in a reasonable amount of time. One could purchase it with PayPal or a “Bill Me Later” card and there was an option for eBay buyer protection.

The piece of equipment had a metal housing. This was a very good feature in its design because it could fit in a 3.5 inch bay of a workstation or PC and be accessible to anyone who was in front of the computer. That size bay is smaller than the 5.25 inch bays many people are used to for 5.25 inch floppy disks or IDE hard drives. That means that this device goes in the forensic workstation and will not get lost as universal card readers often do. Examiners could easily misplace external card readers which are often the size of an old-fashioned cigarette lighter and connect by a USB cable. The device dimensions are 102 × 25 × 125 mm, that is, the width, height, and depth. The device also includes USB 2.0 Internal Motherboard Cables. It also featured color-coded bezels which many people find easy for identifying the correct card slot. Many digital media cards have similar sizes and it is easy for someone to jam the media card in the wrong slot and damage something if it is not color coded. Many people may not realize that performing an investigation is a stressful act and it is easy to make a mistake under duress. Anything such as color coding simplifies the process, helps reduce stress, and may reduce errors and prevent card damage.

The fact that it is labeled forensic and professional and read-only may significantly help it become accepted by both the legal community and the digital camera forensic examiner community. Since the device is read-only, that means that one does not have to change the registry key of the examination machine to make the USB port read-only. The device has read-only firmware to prevent data from being corrupted. Firmware is programmed instructions on a chip that are activated when the device is going to transfer data. If one studied computer architecture, one might have encountered the term “firmware” with regard to read-only memory (ROM) or erasable programmable read-only memory (EPROM) chips. The transfer rate is rated up to 480 Mb/s with USB 2.0. The forensic examiner workstation also has to be able to support USB 2.0 and an operating system of Windows 2000/XP/Vista must be used. The device is also good with Mac OS X and later versions. It was also able to work with Linux systems and it might be a good experiment to see if it works with UBUNTU. This is important since many corporate examiners are no longer working exclusively with Windows and Intel processors.

The supported media cards are XD—Picture Card Type H, M, M+. The XD M+ card often comes in 2 GB capacities and can be used in many Olympus and Fuji digital cameras. The XD card with H designation is often found with capacities of 512 MB and works with Olympus cameras. The XD M card is often encountered in both 512 MB and 2 GB sizes. If one examines the previously discussed speeds of USB transfer and the card capacity, the transfer of data should take far less than ten seconds in the worst case. That seems quite acceptable. When data transfer takes excessive amounts of time, corporate examiners will sit there and supervise the task while multitasking. This allows the person to maximize their workload in a day. The other cards supported are Memory Stick, Memory Stick Duo, and Memory Stick Micro. Memory Stick Duo cards are for many Sony cameras. The device also supports MultiMediaCard are often called MMC cards by students. They come in capacities as low as 16 MB. The device also supports CompactFlash often called CF cards that are often made by Promaster for SLR cameras and have capacities that range from 16 MB to 16 GB. This read-only device also supports the CompactFlash UDMA standard. There CF UDMA 6 cards often found in capacities as large as 32 GB and have transfer speeds of 90 MB/s. The transfer of pictures on a card that is full could take quite a few minutes. This type of card can be used on a wide range of cameras. The device also supports MicroDrive cards which are often found in 340 MB capacities, such as “IBM 340 MB Microdrive CF+ Type II Micro Drive 340 MB Digital Memory Compact Card.” The device also supports the Secure Digital Card, often called an SD card. These are often made by ScanDisk and may be found in a wide range of capacities from 32 to 16 GB. The device also supports miniSD cards which are often found in digital cameras and cell phones. It also supports microSD cards which are often found in capacities of 2–16 GB and made by ScanDisk. The device also supports the Secure Digital eXtreme Capacity SDXC card which can be found with a capacity of 64 GB.

It is also interesting that this device is also built for the professional photo kiosk machines. Such devices may be found at the local pharmacy or department store where one goes shopping. Such stores would want to avoid lawsuits from damaging people’s photos so a forensic card reader makes sense for them. The desire not to destroy or alter data is shared by both digital forensic examiners and photo developers. Hardware that is solely developed for the digital forensic market is often expensive because the market is small and high development costs must be recovered. If the digital examiner can identify synergies such as the photo kiosk market and the digital forensic market, then one should take advantage of the lower pricing for a larger market.

This device is labeled as read-only, but it is still good to do a validation test to try to write to the card when it is in the device. If one was doing an examination on a 16 MB and a 16 GB CompactFlash card, it would be good to get two other cards of the same type and run a test on them. The test could be as simple as putting the blank 16 MB card in the forensic professional kiosk reader and trying to write to it five times. Then, the same could be done with the 16 GB blank card. The results should be put in a notebook and then filed with the case. This type of validation is not a set of new methodologies but an adaption of the same type of validation that was done previously with other types of media. Consider this classic example: examiners would test the read-only notch on 3.5 inch floppy disks by trying to write to the disk when the window was closed. Before that, examiners would test the read-only notch on the 5.25 inch floppy by trying to write to the blank disk when a piece of tape was put on the notch on the side of the disk. Classic methodologies should be applied to new technologies.

The fact that the device also has a USB 2.0 port on it also means that the digital examiner can connect any USB cable to it. That means that the digital forensic examiner can connect a USB cable and digital camera to it to seize the internal pictures. Most digital cameras include an internal memory with a limited number of pictures and a space for external storage cards. It would be advisable to open up a session of VMware with Windows XP running and then run Access Data’s Forensic Toolkit. The next step is to use this device in conjunction with a cable and digital camera to carve out all the deleted pictures and thumbnails from the internal memory of the camera. It is important to realize that a person using a digital camera could easily have taken some important pictures that resided in internal memory and then deleted them. Data carving is an important task and also helps show that one did due diligence to uncover exculpatory evidence.

NEW 3.5″ ALL-IN-ONE INTERNAL CARD READER USB FLASH MEMORY METAL SILVER

This device is available on eBay for less than five U.S. dollars. This device is not labeled as read only. This means that one would have to change the registry so that the device becomes read-only. Then one would need to verify that it is read-only. This device works with TF Trans Flash SD cards and many of the same cards as the previously discussed device. This device works with Windows 98 and Windows 98 Special Edition (SE). This is important because many CCEs have DOS-based examination software that is very stable and runs on the Windows 98 SE platform. If one has an examination machine with certified forensic tools, one is reluctant to upgrade the platform to a newer operating system. That may mean that the old tools may not be supported by the newer platform and it means buying all new tools, software, and the computer. People may opt to stay on the older examination machine and look for inexpensive tools such as this one. Then they do actions such as change the registry to make sure the device becomes read-only.

The other option is to have one workstation with virtual machines (VM) running different operating systems such as Windows 98 SE and Windows XP and having various pieces of hardware for each VM. This option may be better than having large amounts of space at home for different computers with different operating systems and pieces of hardware. However, if one computer fails, then one has another computer to possibly use. If one has one workstation with various VMs on it, then one cannot do anything if that has a hardware failure or the machine experiences damage from a power surge. The digital forensic examiner really needs to sit and think about what type of setup they want and the type of work philosophy that they have.

The device just discussed was advertised as only shipping from Brooklyn, New York, and no other information is available about the brand. If one is going to do examinations and present the evidence in court, one may wish to buy a more expensive piece of equipment from a more famous brand such as LogicCube which makes equipment for the forensic market. The more famous company might have more information available about testing, quality control, and its use in court cases. Both devices might be great, but it is up to the digital evidence examiner to determine what he or she wishes to pay and what type of equipment to use. It all comes down to what you choose.

THE 26-IN-1 USB CARD READER BY DZ-TECH

This is another good option for the digital camera forensic examiner who needs to examine a variety of external media cards such as the Compact Flash 1 and 2 cards, Mini SD, Micro SD, XD card, MMC Mobile, and the MMC reduced card. This device is low cost, but not read-only and not forensic, so the registry would have to be changed so the device was read-only. Then the verification process of proving that it was read-only should be done. MMC Mobile cards are found on digital camera devices that record sound and often use an MP3 format. The digital forensic examiner needs to think about the digital media card that he or she needs to examine and then make sure the reader supports it. The next step is to make sure that the registry can be changed so evidence is not altered.

DIGITAL IMAGE RECOVERY

Zero Assumption Recovery (ZAR) makes a program called Digital Image Recovery. The demo version is free but the complete version cost $59.95. The vendor reports that the program works best if the lost pictures were on a memory card. The Windows-based file systems supported are FAT 16, FAT 32, and NTFS [12]. The recovery tool also works with Linux file systems such as ext2, ext3, ext4, and XFS. Digital Image Recovery runs on Windows NT, 2000, XP, 2003, Vista, and Windows 7 [12]. The tool supports recovery from many formats, including the popular GIF, JPEG, and TIFF formats. It also supports MOV and AVI movie formats. The latest version of this software also supports the Canon Raw data known as CRW and the newer CR2, Canon RAW Format. The ORF Olympus Raw format is also supported and the website has a list of over 40 models of cameras supported.

The website for the tool discusses some of the reasons why pictures are lost. There may be a camera failure or there was an accidental formatting of the digital media card. Loss of pictures can also occur when the process of copying pictures from a digital camera to a personal computers fail. The more image recovery tools that one has, the better. One cannot have too many tools in one's incident response toolkit. The fact that this tool appears to specialize in raw formats for Canon and Olympus cameras is a good thing for those who investigate those types of cameras. Lastly, the tool has an easy-to-use interface that gives a graphic about the fragmentation of files on the storage media. It is good for the examiner to know something about the disposition of the media.

ZERO ASSUMPTION RECOVERY TOOLKIT

The ZAR Toolkit includes a tool named Zlon which can be used to make a forensic image of a drive. Zmeil is another item in the toolkit that is used for recovering corrupted mailboxes. The demo version allows for the recovery of nine messages. The Zmeil tool by itself cost $39.95. Mailboxes can often be corrupted by voltage spikes, accidental deletion, or by people who add too many items and never delete them. That causes the file size to exceed its limit. It is also possible that software problems, possibly by bad clusters that hold code, can cause the database to become corrupt. Microsoft Outlook and Microsoft Outlook Express use the .pst and .dbx files. Both email programs can often get a benefit from Zmeil. Other mail programs such as Thunderbird and Eudora can often be helped by Zmeil.

DISK SPACE VISUALIZER VERSION 1.2

This is a program has a graph that displays what programs use up large amounts of space. This tool would be of more use to the investigator on his or her machine because it may be possible to identify software that is no longer used, or log files or cases that should have been removed but were not. It could be considered a forensic tool but it could also be a diagnostic tool that could be used to identify unneeded items.

ZERO ASSUMPTION RECOVERY TUTORIALS

ZAR has some online tutorials that discuss how data can become irretrievable and some methods to try to restore it. The tutorials are on a variety of subjects and could be of use to the computer forensic professional or information technology security professional. The tutorials and products have multilingual information that is in German, English, and

Russian. They also have a website with thousands of threads where questions are asked and answered.

EXIF DATA STANDARDS

Another place to learn about digital cameras and metadata is from the EXIF standards. The Digital Still Camera Image File Format Standard V2.2 discusses everything from the layout of diodes which translate into bytes until the coding of metadata. When someone takes a picture, it is scanned from left to right and from top to bottom. This type of scanning methodology was also used to populate the analog television picture tubes of yesteryear. When someone takes a picture, it is a one-to-one relationship with each square pixel being part of that image. Each pixel color is made of some combination of red, green, and blue.

EXIF FIELD TYPES

If one is going to study metadata, then one must learn about some of vocabulary related to file storage on digital media. A byte is an eight bit unsigned integer. ASCII uses a byte of storage that contains the seven bit ASCII code and ends with a null character. Sometimes the term "short" is used. Short is an unsigned integer of 16 bits or two bytes. Unsigned means that there is no bit assigned to be an indicator of negative or positive. This allows a longer number to be stored since there is no place required for the sign bit. Long means that there are four bytes or 32 bits that are allocated for storage. Nothing is signed. A rational is equal to two longs. Slong is a format four bytes like long except that it is in a signed two's complement notation. Signed means that there is a prefix bit before the byte that means that the number is either positive or negative. Then we take a number and code it to binary, flip all the bits, add one, and that is two's complement. A book written by James Saxon is a seminal work on data processing mathematics from 1972 that explains all the bits, bytes, and two's complement forms in an easy-to-read fashion. The ISBN is 978-0130589095. When one reads the EXIF standards, there are many formats described as byte, ASCII, long, short, slong, rational, and undefined. It is good to look at Jim Saxon's book first to learn the basics of binary, decimal, hexadecimal, signed numbers, unsigned numbers, and then two's complement. Once these basics are mastered, then the EXIF standards document is straightforward to read. If one is going to understand metadata, then one first needs to understand how information is formatted and where it is stored and what numbers mean what. There are also tags that exist in the TIFF metadata standards that are reserved for artist and copyright. In an intellectual property dispute concerning images, an examiner first looks into the EXIF standards for the location of the hexadecimal address of the tag for copyright and the location of the tag for artist. Then the examiner can go into the seized evidence which is a picture. Going into the bits and bytes of a picture is done with a file viewer such as WinHex or with a tool such as Thumber. Then the digital evidence examiner must proceed to the location of that tag that is reserved for the artist and copyright. It is there that one looks for ownership of the picture and if it is copyrighted. That picture could hold important evidence. However, that data could be overwritten by the suspect. Consider this example. If the "date modified" on the stolen picture file is February 10, and the picture was hacked and reported as stolen on February 9, that would indicate tampering.

GPS EXIF DATA

There is an amazing amount of EXIF data in the picture concerning global positioning system (GPS). Most people only consider latitude and longitude but that is only the beginning. There is also a field that indicates the number of satellites reached. This could give some clue to the accuracy of the picture too. If three or four satellites are in view, the picture should be relatively accurate. There is a metadata field that tells the examiner what version of GPS it is. This data is stored in a byte. There is the field labeled as GPS altitude which tells the altitude. There is also the GPS time which uses the atomic clock. There is also a GPS speed tag which gives the speed of a moving GPS receiver. If the GPS receiver is in a Humvee going 50 mph, the tag will collect that speed at that instant. There is also a GPS precision field indicating some accuracy. There is a direction of movement field. There is a GPS date. There are 31 fields that get populated with GPS data that may or may not all be filled in a picture. If a picture is spoofed, then generally there is going to be some inconsistency in those fields.

EXIF READER

If one is interested in looking at these GPS tags, then one should download a free program called EXIF Reader. It is a freeware program that is written in Visual Basic. The program is very simple to operate and shows the tag, its position, and the value. An expert on GPS may want to look at the data from the 31 fields since some are very complex and have to do with the position in relation to true north and magnetic north.

PixelZap AND PZapGui

PixelZap and PZapGui are two programs that the investigator can use together. They allow the digital evidence examiner the ability to save the picture without any loss. The program is a shareware program that was written in Visual Basic and can work on a Windows 95, 98, and the Windows XP system. As digital cameras age and diode arrays fail, cameras can produce consistent dead spots where there is no image. The same hardware problems can cause some areas to create bright spots too. PixelZap and PZapGui can fix both types of spots. It might be good to print a picture its original state and then print one that was fixed with the two programs. Then a digital evidence examiner could use a variable spectral comparator and a microscope to compare both to locate the uniquely flawed locations. A questioned document examiner might also be a good person to give a second opinion on the printed pictures.

THUMBER

The thumber program was once freeware but is now considered shareware. The shareware option is good because it means that if you try the program and find it useful, then the cost is only $18. It is better to pay the cost so that one gets the full operational value of the software without any type of limits on the number of images that can be processed. Thumber is written in Visual Basic. It would be advisable to get Visual Basic and install it on a machine so that one has all the necessary dynamic link libraries and other files. This will reduce the chance of run time errors or having necessary files be missing. It should also help with any other forensic tools that one might have that are also written in Visual Basic.

It may not be evident to digital evidence examiners that when they download pictures from a camera and then use a viewer to rotate the image, they are losing data. Thumber has a special feature built in it that is called "LossLess Image Manipulation." The feature was made because the JPEG committee permitted the creator of Thumber some access to the source code [13]. The loss of bits can mean the loss of color as well as defects that might help identify that picture with a certain camera. Anytime the loss of data can be minimized, it should be. Some information about certain pixels are lost when a document is embedded in a picture with a steganography tool [14]. If one compares a picture with a document embedded in it and the same picture with no document in it, then the latter is much larger in bytes and the colors are slightly different. The file modification date and file last accessed date operating system (OS) metadata associated with the file in the OS are different too.

From reading the website, it appears that one can also email the person who has a custodial relationship of Thumber and get an answer to technical questions [15]. Thumber seems to be a very useful tool because it allows people to organize pictures into thumbnails. Each thumbnail has all the information about the jpg formatted data contained in that picture. Thumber extracts the metadata if it exists. Some metadata can be scrubbed by people using various tools so it may have been removed before the examiner looks at it. If there is metadata, Thumber may be able to obtain such useful information, including the camera model, camera make, and information about the light source. The light source could be a tungsten light, daylight, or flashing device. Thumber does a lot of translation on a picture's metadata and presents something that is more understandable to humans. However, the problem is that sometimes the camera manufacturer codes some data in a format that Thumber does not understand and cannot be displayed [15]. It then becomes necessary for the investigator to get the EXIF standards book and contact the camera manufacturer to get some information about how they code the data that is embedded in the picture.

The Thumber website has a tool to repair deadspots and bright spots that are caused by some type of defect in the digital camera [13]. However, from a forensics perspective, that would not be a good feature to use because it is those unique defects on those pictures that associate it with that particular camera. A rifle or handgun that fires a bullet leaves a certain ballistic fingerprint on the bullet. The defects in a camera's diode array cause unique defects in the picture which may be obvious or only detectable by a skilled eye using a microscope.

FRAMER

This tool appears to be very useful for the examination of digital camera pictures that were done in multiframe mode. Sometimes people who take pictures of young women modeling will use a camera such as the Epson PhotoPC 600 to take a succession of pictures in a few seconds. Framer is a tool that can be used to take the succession of pictures and make them a movie. Framer was also written in Visual Basic.

FILE LISTER

This is a Visual Basic freeware tool that is used to create a text file with all the files in the folder. Then one can also select what sublevels and sub-sublevels and all their text files that should be listed on screen. The choices are one, two, three levels or all sublevels. File Lister can also be used to search for a list of certain types of files such as images. Custom lists can be created and these can be saved and then imported into a report.

MinUpTime

This is a Visual Basic tool that lets the investigator know exactly how long he or she has been working on the system during that session. It is a useful freeware tool for logging, auditing, and report writing in computer forensics and mobile device forensics. MinUpTime could be useful for running in a VMware window with the suspect's forensic image. Then the exact amount of time investigating the computer image could be billed to the client and documented in a report. A lengthy time factor might help show that due diligence was done in looking for both exculpatory evidence and evidence used for the prosecution.

HideWin

This is a Visual Basic program that is freeware and can be downloaded from the TawbaWare website. HideWin could be used by an investigator to limit the number of windows currently viewed on the Windows Desktop. The URL is http://www.tawbaware.com/hidewin.htm and the website has a place to get in contact via email with the custodian of the program. If an investigator is investigating a pop-up window flurry, then one would not want to kill all the windows but rather hide some for investigation later. The program allows one to select which ones to hide and this allows the examiner to focus on which ones to keep on the Windows desktop. HideWin can be used for both investigation purposes and nuisance control. The current version is 1.0 and it was released in 2002.

FileMonitor

FileMonitor is a Visual Basic program that is distributed as freeware from TawbaWare. The current version is version 1.1 and was last updated on August 5, 2003. The program allows one to select up to four files in a window and concurrently display the file size, last modification date, and last modification time. This could be very useful in an investigation of four identical-looking images files for comparison in a steganography case.

CamWork

This is a very classic program that runs on IDOS 3.3 or higher. It can also run in Windows 3.1 and Windows 95. The program may be useful for determining the battery level and the total number of images in the camera. It is available for download on TawbaWare's website. The program requires 640 kilobytes of memory and runs on an 8086 Intel processor or higher. This software is also useful for downloading pictures from certain digital cameras from the Olympus, Epson, and Agfa family of cameras [16]. The program claims to support four download speeds and four COM ports.

CRead

This is a freeware program that can run on Windows 95, Windows 98, and NT. It is used to read embedded comments within the picture since a number of tools exist that enable embedded commenting.

TOOLS FOR THE CAMERA INVESTIGATOR: X-WAYS "FORENSIC SOFTWARE"

This is a very good tool for examining the contents of the memory or a drive. It also has an option to examine files or unallocated space. In devices with RAM, it can be easily used to examine RAM and collect all the contents of RAM to save in a file. One can also acquire an image of the disk storage and burn it to a CD. The software is reasonably priced and comes with good support. The interface is also what people call intuitive so it is not that difficult to learn. There is also a feature to search for key words. There is also a built-in documentation file which is important since books get lost. The software works well with FAT-based file systems.

TOOLS FOR THE CAMERA INVESTIGATOR: RECOVER MY FILES

The Recover My Files software can be reasonably obtained from the Get Data Software Development Company. The license is good indefinitely with the version that one purchases but it would behoove the investigator to purchase a newer version every so many years to get a nicer interface or more features. The software is easy to use and the interface is intuitive meaning that it takes little time to get proficient with. The program works well with FAT-based file systems. There is also a free trial download where one can see if it can possibly recover many files, then the license key can be purchased and the full recovery of files can be completed. The newer versions of the software work with Windows 98/ME/2000/2003/XP/Vista/Windows 7 and works with FAT 12, FAT 16, FAT 32, NTFS, and NTFS5 file systems.

TOOLS FOR THE CAMERA INVESTIGATOR: ProDiscover BASIC

The ProDiscover Basic Tool has a nice Windows interface and lets students image a disk or storage device. Then they can save those results and use the built-in viewer to analyze it. This tool is useful for both computer forensics and digital camera forensics. The imagining part is good because it allows the collection of both free and allocated space. One could also use another data carving tool such as Data Lifter 2 to collect fragmented pictures. ProDiscover Basic is great because it is freeware. It gives the investigator a break to get something for free. Some of the components also enable one to create a report on the evidence. There are also workshops at the High Tech Crime Investigative Association conferences and meetings where one can learn more about this tool.

TOOLS FOR THE DIGITAL CAMERA PHONE INVESTIGATOR: SUSTEEN SECURE VIEW

This is a good tool for working with digital camera phones. The license dongle needs to be put in and then the software is run. The interface is easy to use and one can download the call logs, SMS messages, pictures, and other data important for an investigation. The SVProbe feature has a gallery feature that lets the investigator see all the pictures and choose some with GPS data to map on Google Maps. One can see where pictures were taken. There are other features to map all people in the contact book with the call log and quickly see what people spoke with the owner of the camera phone and how often. Secure View is used by many investigators; it comes with a cables kit, and they provide good support.

TOOLS FOR THE DIGITAL CAMERA PHONE INVESTIGATOR: GUIDANCE SOFTWARE ENCASE

This is a great tool but expensive in many people's opinion. The interface is great and the support for the product is excellent. You get what you pay for. There are options to create a forensic image of both allocated and unallocated space and then examine it. There are many advanced features for this product and tests and certifications that really allow one to become a certified expert. The software is great with an examination machine for examining nearly any digital device with the operating systems that the product supports. There is a lot written about Encase in books and magazines too. It should be easy to locate a community of tool users to ask questions to and help support others with their questions.

TOOLS FOR THE DIGITAL CAMERA PHONE INVESTIGATOR: ILook INVESTIGATOR SOFTWARE

This is a great tool that is free for law enforcement, military intelligence, and many government agencies with law enforcement missions. The software works well with digital media and allows people to image the data and recover files. It can be used with digital camera, hard drives, USB media, and other related items. The software has a strict set of terms and one must abide by them. For law enforcement, it seems to be a great deal. ILook Imager is one of the tools that examiners like because it can be put on a bootable CD or floppy diskette. This means that it can be placed on an examination machine and powered up. Depending upon the examination machine, its ports, the bios, any connected storage area networks (SANs), and so on, it may be possible to image the camera and/or external media cards and send a digital image of the seized material to an external USB device or networked device.

TOOLS FOR THE DIGITAL CAMERA PHONE INVESTIGATOR: PARABEN'S DEVICE SEIZURE

This is a great tool for some PDAs, some GPS devices, some cell phones, and some digital cameras. It allows for the seizing of the digital media and then has what seems like a hive structure where one can click on to get more details on files. This is a good go-to tool to try to get the camera phone or digital camera files. It would be logical to get the cables kit so that if one gets a camera phone, one might be able to quickly proceed with the investigation and not wait 1 week to order a cable from the camera or camera phone manufacturer.

SURVEY OF TOOLS: ABC AMBER IMAGE CONVERTER

It is the digital evidence examiner's duty to preserve evidence and keep it original. However, there are times when digital images are in a proprietary format that is not viewable. For those cases it is necessary to change the file type to another file type. With some of those freeware converters out there on the net, it is possible that some information may be lost during certain combinations of file format conversion but it should not impact the investigation because the same basic picture should be there. Some CCEs have discussed the use of the ABC Amber Image Converter because it is useful for translating one format to another more popular format that an investigator can view and print. This program is reasonably priced and was created by Yernar Shambayev. Some of the file extensions supported are BMP, WMF, EMF, ICO, JPG, GIF, PNG, TIFF, PCX, PCC, DCX, PBM, PGM, PPM, and TGA [17]. It should become obvious now why a digital evidence examiner needs to take a class on file systems. The class is also useful for knowing where the metadata exists in the file too.

THUMBNAILS AND THUMBNAIL VIEWER: DM THUMBS

Whenever picture files are stored in a directory on a Microsoft FAT file system device and viewed as thumbnails, a thumbs.db file exists. This file contains a group of small versions of the pictures in the directory. This picture size is small in Windows XP and previous versions, namely 96 × 96 pixels and could be considered a low-resolution miniature of the larger picture. When a picture is deleted, there is still a copy of the picture in thumbs.db. This is important in the prosecution of sex offenders who remove child pornography from their directories before being caught and arrested. DM Thumbs is a useful tool for viewing thumbnails of pictures that were deleted from a directory. In a digital camera, there is often a button with a picture of a garbage can that causes the file to be marked for deletion and is no longer listed with the other files. DM Thumbs was able to show the thumbnails of pictures in a directory on my computer (see Figure 2.1). DM Thumbs is an important investigation tool that mobile digital forensic examiners should consider purchasing for their incident response toolkits. One can find discussions about it online so that helps.

SURVEY OF TOOLS TO INVESTIGATE CAMERA: CASE STUDY OF ADVANCED IMPORT CAMERA (2009)

South Korean cameras are at least 1 year ahead of American cameras because the consumer electronics market is technologically more advanced than the U.S. market as was discussed in Chapter 1. I taught a class on cybercrime to visiting students from South Korea. The students brought cameras with higher-megapixel resolutions than were available in the United States. One student took approximately six hundred pictures with her camera during the week of classes. On the last day, the student was upset that her camera malfunctioned or perhaps someone intentionally or deliberately deleted her pictures. It appeared that all the pictures were lost. The student was upset since the pictures included both class trips and class material. I then instructed the student to go to the class graduation party

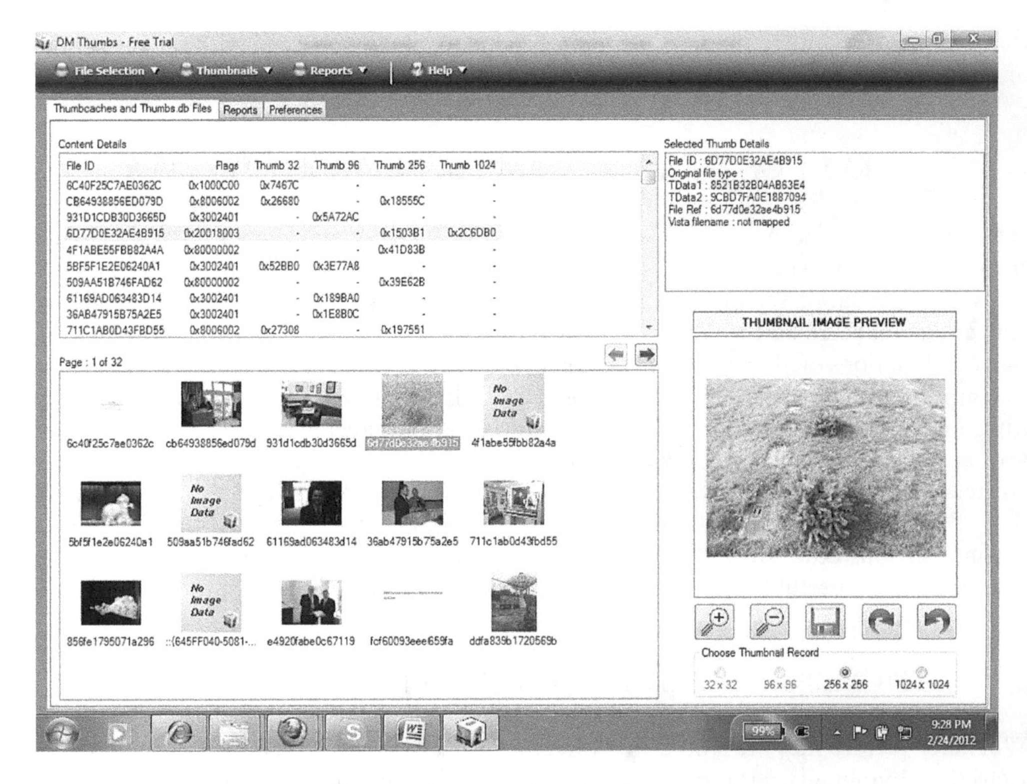

FIGURE 2.1 DM Thumb: Thumbs.db viewer program.

and I would bring my examination machine. There was no chain of custody form used since there was no policy infraction investigation. The student said that the camera did not have wireless connectivity and that there was no danger of anyone putting a virus on the camera. Since it was not a criminal case or policy infraction, the camera was not kept locked up in a secure location and a Faraday bag was not used.

I first ran antivirus and antispyware programs. The examination machine was not connected to the Internet and all connectivity was disabled for the sake of the student's privacy. A program called "Recover My Files" was then run. Depending on certain factors concerning the nature of the data loss, the program can often recover pictures from laptops, PDAs, desktop computers, or digital cameras that use the Microsoft FAT file system (see Figure 2.2). The camera was connected to the examination machine by a USB cable. Recover My Files was started and the camera was powered on. Then the option for fast file search was chosen.

The process was repeated three times and in the third time, 550 files of high-quality JPG format pictures were recovered. This was surprising because the camera was more advanced than those in the United States and the software recovered so many high-quality pictures. There is an option to select the pictures for recovering and then putting them on a CD. This was done. They were then opened with a viewer and the resolution was good and one could magnify the picture many times until pixilation occurred. The Recover My Files tool seems like a tool that should be considered for inclusion in all the computer forensic examiners toolboxes since it works well and seems easy to use. Usability and cost are items that should be considered in addition to the data recovery function as criteria for purchasing.

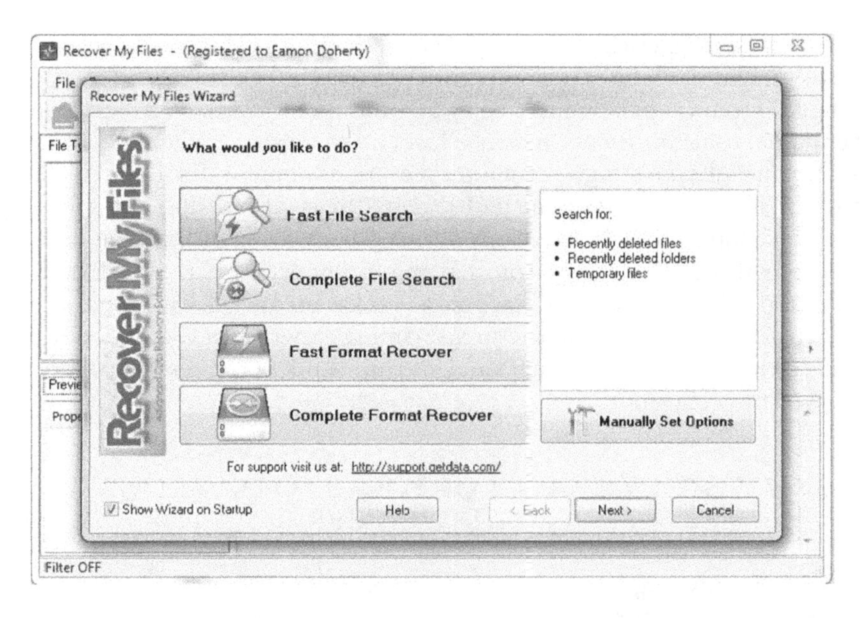

FIGURE 2.2 Recover My Files.

CASE STUDY: DATA CARVING WITH DATA LIFTER 2

There were still approximately 50 pictures that were recovered and not viewable. The thumbnails could not be viewed. It seemed prudent to use a data carving tool to recover these pictures. I have met law enforcement professionals who use Data Lifter version 2 and within this suite is a tool called "File Extractor Pro" that they often use. This tool was run and it recovered the other 50 pictures. The File Extractor Pro tool examines headers and footers and supports logical disk access for sector by sector search or just unallocated clusters [18]. These fifty pictures were returned to the student before leaving the university. The tool had a lot of useful documentation on the help screens. It is important to have an intuitive interface but the inclusion of good supporting material is important too.

CASE STUDY: DETERMINING THE LEVEL OF RESOLUTION LOSS OF THE RECOVERED PICTURES

Had the student had more time before leaving the university before returning to South Korea, she was interested in determining the average quality of the pictures that were recovered compared to pictures that were never lost. One could get a crude statistic by the following equation:

$$\frac{\text{Total bytes of 10 pictures that were taken on camera A and recovered}}{\text{Total bytes of 10 pictures that were taken on camera A and never lost}}$$

Suppose the student took 10 pictures and the total was 100 megabytes. Then the student selected 10 pictures that were recovered and added up to 80 megabytes. If we use the formula, we get 80 into 100 and get 80% quality. Therefore, 20% of the picture quality was lost. Perhaps some picture quality was lost, perhaps not. This would be a good experiment for the student of digital camera forensics to pursue.

The digital forensics person must be aware that if one opens up a picture and saves it as another name, some information can be lost and the resulting file size is smaller. As an experiment, you can take a high-resolution picture of about 630 KB and open it up in a viewer and call it pic1.jpg. Then open pic1.jpg and save it as pic2.jpg. Then, open pic2.jpg and save it as pic3.jpg. After 10 iterations of this process, you will see that approximately four kilobytes will be lost. The reason that this is important is that evidence should be preserved and if evidence was processed in a certain manner, losses could occur. There are many experiments such as this picture renaming experiment that one can give to students to do so that they learn concepts and start to think about testing concepts and not taking everything on faith.

CASE STUDY: SURVEY OF TOOLS TO INVESTIGATE A DIGITAL CAMERA

Suppose the police investigate someone that allegedly used the Meade Captureview binocular camera to take pictures of someone changing. The police would ask the person to surrender the binocular camera, put the item in a Faraday bag in case of connectivity concerns, and then fill out a chain of custody form. They would also keep the device locked up and in a temperature-controlled area on the way to the eforensics lab.

CASE STUDY: USE OF USB WRITE BLOCKER

Once in the lab, they would run the antivirus, the antispyware, and verify that the examination computer was not connected to anything else via Bluetooth, cable, infrared, or by another other type of connection. Then they would connect the USB cable to the camera. Some examiners might be afraid of creating temp files and altering time stamps on the camera. These examiners may elect to use a Tableau product known as "ForensicPC Ultimate Write Block Kit." This write blocker device can protect anything that connects to a USB port from being written to. This is a hardware write blocker.

Some people who do not want to spend the money to purchase a write blocker can configure the registry in Windows to make the USB port read-only. This is not difficult to do. One clicks on the button in the lower left corner and a small box appears. Then one types in REGEDIT. This means registry edit. Then one changes the double word to a value of 1. This location is off one of the hives as discussed in a different typeface below. This makes the USB port read-only. This is a type of software write blocker for Windows XP with Service Pack 2 or later or Windows 7 [19]:

```
HKEY_LOCAL_MACHINE\System\CurrentControlSet\Control\StorageDevice
Policies
```

CASE STUDY: ACCESS DATA FTK

Suppose the camera is connected to the examination machine now and we do not have to worry about altering evidence since the USB port is configured to be read-only. The Access Data's FTK or Forensic Toolkit program could be used to collect the picture files from the camera. If some pictures were deleted and no longer in the FAT, the pictures might be recoverable if the place where they reside was not written over. FTK does data carving and looks for headers and footers or various pictures. It also finds the chain of clusters to locate the parts of the file located in various parts of the disk because it is rare that files are contiguous on a disk. There is also an option in FTK to use a library function that contains the hash functions or digital signatures of a tremendous amount of child pornography. This saves the examiner a lot of time looking for such illegal material and if it is there, immediate violent crime charges can be put on the suspect's digital device. It was already established that the internal memory or the SD cards in a digital camera can be used as mass storage devices to hold documents, other people's pictures, PowerPoint presentations, or pictures taken with the camera.

Once the pictures are collected with Access Data's FTK, they can be burned to a CD. There are two versions of FTK. There is a free version that allows a limited number of files to be seized and examined. On older versions such as 1.7, there is a limit of five thousand files. There is also the full version that costs money and requires a license dongle to use. The full version is not limited and can allow the seizure and examination of a plethora of files from the device in question. When it comes to the justice system in the United States, two copies should be made. One is for the defense and one is for the prosecution. It is important to read about the legal system of wherever you live and wherever you do examinations. It is also worth reading many online articles about the e-Discovery process.

CASE STUDY: DETERMINING IF A PICTURE WAS CREATED WITH THE BINOCULAR CAMERA

A picture could be on a device because someone connected a USB cable between the computer and camera, and then dropped and dragged the picture file onto the camera. There are a few ways we can determine if the picture was probably not taken with the camera or tampered with. The creation and modifications dates are a good clue. When we right click on a picture that was seized from a camera, we should not see a modification date (see Figure 2.3). If we do, that is an indication that it was altered with a photo editing software. If the picture is opened up to view by double clicking it, the accessed date is altered but the modification date is not changed in Microsoft Windows.

If one clicks on the word details in Figure 2.3, there is information revealed about the camera that took the picture and the resolution. We can see the file size. There is also information about the file attributes. Some teachers will scratch their arm and mention the acronym rash to help students remember the following: read-only, archive, system, and hidden. Hidden means that the file is there in the directory, but it is not viewable in the directory.

If one goes through the tabs and scrolls down, there is also a digital signature or hash mark. If the file exists on the camera and on the person's computer, the hash marks should be the same. If the hash marks are different, it means the file was altered. The modification date and last accessed date should also give clues.

CASE STUDY: EACH CAMERA HAS IMPERFECTIONS AND LIMITATIONS THAT IMPACT THE PICTURE

People often know that in ballistics, a gun leaves a unique pattern on the bullet. Striations have often been seen on such popular television shows such as *NCIS* or *CSI Miami*. Some

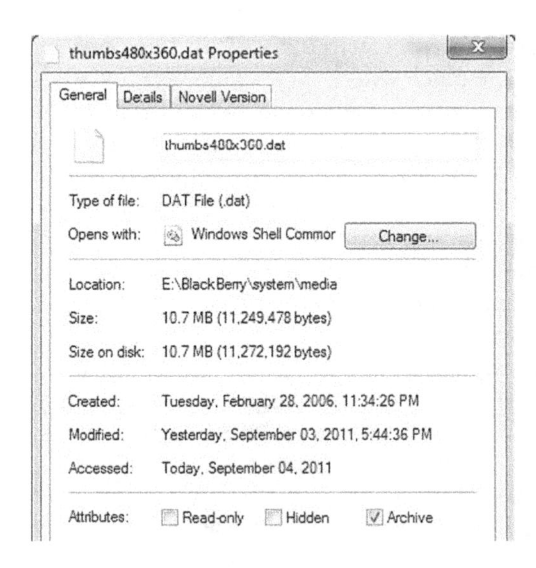

FIGURE 2.3 File properties of a picture.

digital cameras also have imperfections with certain CCD pixels that create a visible error in the picture [20]. If there were a group of the same make and model camera used to take pictures of the same subject, then imperfections can be found and this can help identify the particular camera from a group of others of the same type. The metadata embedded in the picture can often tell the make and model but it is the imperfections that often allow us to distinguish which camera.

The investigator should also examine basic factors such as file size and resolution. The Meade Captureview is a 2.0 megapixel camera. If the picture is found to be 3.0 megapixels, then the picture was placed on the camera or its SD card, or the picture was removed from the device, enhanced with a photo editor, and then put back. If the picture on the camera is in shades of brown, then it is probably an antique photo that was scanned on a scanning device and placed on the camera or its SD card.

THE INVESTIGATIVE MACHINE: DIGITAL CAMERA FORENSICS

The digital camera forensics investigative machine should have the sufficient processing power to run all the software necessary to recover pictures, break passwords, and create reports. That sounds simple but is actually a difficult task. The complexity is because of password breaking and the recovery of pictures. Most of the time passwords are very simple and can be broken with dictionary attacks or special dictionaries made from the slack space and unallocated space of the computer. Then there is that certain exception where a person might use upper and lower case letters, numbers, and a special character. If such a case occurs, then it is important to have a very powerful computer with a central processing unit with a very high speed in the gigahertz range. If there will be investigations regarding the recovery of numerous pictures from large-capacity media disks, then a high-performance workstation is also needed. The answer to what type of investigative machine is needed is directly related to what type of work will be needed to be done on it. Some corporate investigative units may start with a standard computer that everyone else in the corporation uses. There are some advantages to this. The first is that if there are fewer types of PC configurations around the corporation, then it is easier for the information technology department to support. If the investigative area receives a special high-performance computer with a large, high-resolution monitor, vain and petty employees who learn of this could create political problems in the organization. Some investigative units of the corporation may start with a standard computer and then upgrade the hardware if it becomes necessary. This sounds good but the budget for such an expense may not be there later or it may be difficult to justify. That is why many digital evidence investigators say to purchase the largest machine that one can afford because it will be outdated in a few years anyway largely due to the implementation of Moore's law. It would behoove the digital forensic examination unit to purchase the highest-end computer that they can afford and to get the highest-end graphics cards with multiple processors. This would allow the work to be shared and greatly reduce the time it takes to break passwords or carve data from unallocated space to recover pictures.

When it comes to peripherals, it is good to get every type of input/output device that could read any type of media that could possibly be encountered in the corporation. If there are people still using zip disks, 3.5 inch floppy disks, 5.25 inch floppy disks, and jazz disks, then it would be beneficial to have either internal or external device that can read that media and make it available for examination. If zip disk drives, jazz drives, and other external media devices use outdated interface connectors such as the DB 25 parallel port, then it becomes necessary to purchase a card with a parallel port that may be installed in the workstation. It is better, if possible, to purchase the external USB drive because of the decreased power demands concerning outlet space and electricity. The Digital Intelligence Fred System may be the answer with its write-blocked ports for SATA, IDE, Firewire 1394b/800, USB 2.0, and SCSI. The UltraBay II is supported by Tableau "tabphys" library and this so important. This means that if one has the Fred Workstation with UltraBay II, Tableau products, and Guidance Software's Encase, then a smooth seamless integration of forensic products is possible. Many people believe that it is worth the expense to have hardware, software, and examination tools that have a proven record of working together and have support in case something goes wrong. Who wants to be known as the person who saved money buying budget products that did not work together and the investigation failed?

If one is setting up a large-scale corporate lab, or law enforcement lab and will be archiving cases, then a 135 terabyte storage device might be the device to purchase. In 2012, one might wish to purchase a Dell DR4000 Storage Platform with De-duplication and Compression since the corporate examiners may wish to set up a SAN. SANs are used to both share cases and archive them. Dark fiber or leased optical fiber might be the solution for high-bandwidth sharing of cases. Fiber is necessary for sharing cases with numerous exemplars of multimedia, including high-resolution pictures. Some labs may choose to purchase, install, and maintain their own optical fiber, but it depends on the personnel available and any service contracts.

The incident response team may also seek to purchase or lease a mobile device such as the RoadMASSter-3 Portable Forensic Lab that is deployable in the field [21]. This equipment would allow the incident response team the immediate ability to examine digital cameras, USB flash drives, digital camcorders, cell phones, and hard drives on site. The device is sold with a ruggedized case much like the famed "Pelican Cases." The device also allows transfer of speeds to 3.5 gigabytes per minute. The device also contains a 15 inch color display and shielding from electromagnetic interference. The device contains five USB ports which is important for license dongles such as those needed with Guidance Software's Encase and Susteen Secure View.

One of the places where one can get free information about setting up a forensic computer lab is called Computer Forensics World, http://www.computerforensicsworld.com/index.php, and one can post questions for either students or computer forensic professionals to answer. It is important to get a variety of feedback because some people understand emerging trends and may suggest more work in an area such as breaking encryption. Then one may need the most powerful computer and graphics cards that one can get. If one had a large budget for a forensic workstation, a Forensic Tower IV Dual Xeon Quad Core—(FTK Ready) would be a good choice. It has three case fans and 1000 W power supply. The video card has 1 GB, Dual DVI Head. The two processors are E5620 Quad-Core 2.4 GHz and 12 MB cache. There is also room for 10 bays that hold 5.25 inch items. The unit also supports fast data transfer with its Intel gigabit controller 10/100/1000 Mbps transfer rates. The device is available from Forensic Computers Incorporated and also includes Realtek high-definition audio codec. It is also advisable to get the Tableau TACC1441 Hardware Accelerator which would help speed up the brute force and dictionary attacks. The TACC1441 also supports decryption of PGP SDA, WinZip 9, PGP SHA-1, and others. If one is going to spend a lot of money on such hardware, then an extended 3 year warranty is also good to get.

BLACK HOLE FARADAY BAG

There are times when an employee or visitor in a corporation or at a department store may be doing something questionable with a digital camera. The AR may be called and an incident response team dispatched. The person in question may be asked to surrender his or her digital camera for investigation. If the person surrenders it and it is turned on, then leave it on. However, evidence can easily be tampered with since so many devices have infrared, wireless, GPS satellite, or Bluetooth connectivity. The shielding has been specifically made to block the following frequencies: 900 MHz, 1.8 GHz, 2.1 GHz, and 2.4 GHz [22]. The best course of action is to leave the device on and put it in a Black Hole Faraday bag available from Data Duplication Ltd. in the United Kingdom. This bag has a filtered USB port on it. This allows the seized device which could include a digital camera, cell phone, PDA, or other mobile device to stay powered on before previewing happens. It is possible that there may be considerable time before a device is examined and that the batteries may die before previewing. That would be a loss because volatile memory would be unpowered and evidence lost. The Black Hole Faraday bag could be used in conjunction with a portable forensic workstation with a write-protected USB port so the device stays powered but there is no risk of connectivity with other devices through the bag.

Another bonus is that the Black Hole Faraday bag is transparent on one side. This means that the handheld device can be viewed and operated through the bag if necessary. The bag is 11.75 inches by 17 inches. Since the one side is transparent, it may be possible to put the device in a cradle and use a camera to manually photograph screens as one goes through the camera. Such a practice would only be done in an exigent situation where pictures had to be obtained quickly to save life or property and no proper forensic tools were available for use with the device. The opportunity to examine a device in a shielded environment is invaluable. This seems a much better option than just using a Faraday bag without a shielded USB port and having the bag bunched up with one wire exiting the bag to go to an examination machine or charger to keep it powered.

INCIDENT RESPONSE TOOLS AND SERVICES

There is a company that maintains a website with many security-related services and tools. The URL is http://qccis.com/services/incident-response/ and sometimes it is good to contact a company that can provide incident response service or consulting if an organization has a need. Below are a few of the tools available on their website or that can be sought from them.

CasesNotes LITE

The CaseNotes Lite program is for documenting notes of cases [23]. A digital examiner may have a VMware window open and is examining an image in that window. Then he or she may have another window with CaseNotes Lite running and put in notes in English, Russian, Greek, Italian, or Japanese [23]. The program is free and has been tested to work on Windows XP, Windows 7, and Windows Vista. This documentation tool works with any Windows time zone and format. It has AES 512 bit encryption with it so others cannot open it easily. There is also an MD5 hash tool with it to show that there was no tampering on the file.

FragView

This tool is available for download and is a viewer that allows many windows to be open currently so that the examiner may compare content. It appears to be a timesaver and is reported to be a good tool for comparing webmail fragments [24]. The ability to load various types of multimedia files and quickly navigate through Windows is sometimes invaluable.

GigaView

This tool is available for download and is good for reviewing various types of chat logs. The logs can be put in a comma-sorted variable format and then be exported for view in Microsoft Excel [25]. The ability to examine chat logs in a spreadsheet is valuable because the spreadsheet provides variable-sized cells for viewing.

VideoTriage

This tool is often used with seized video and allows the creation of thumbnails to be made [26]. The video can be played at five times regular speed and it is possible for the examiner to have thumbnails created at so many intervals of time. This allows the examiner to quickly get to the parts of evidence that are relevant without watching tremendous amounts of video. There is a statement about someone who had 860 long movies to watch. VideoTriage was estimated to save him 9 days [26]. As hard drives are getting larger and the number of movies available for download is growing, tools such as this are necessary. In an age of suspects owning multiple low-cost terabyte capacity hard drives, increased demands are placed upon investigators since what department has the manpower to sit and watch every movie? If movies are of an objectionable and violent nature, it may be difficult to sit through so much video without the aid of such a tool. Here is another instance where it is important to get the correct tool and have a person with the mental strength to do this type of work on a daily basis.

MY THOUGHTS ON THE NEED FOR TIME-SAVING VIDEO INVESTIGATION TOOLS

This is a topic that needs to be addressed in schools and in the job interviews. From my discussions with adult students going back to school, it seems that much of the public gets a job based on being at a certain place and at a certain time. The other consideration is if the applicant has the ability to perform the needed work. It would be better if personality type, mental health testing, and exposure to unpleasant things such as video, pictures, and people with deviant behavior would be considered. When I had this discussion with three detectives in a continuing education class on cell phone seizure, the detectives had a surprising answer. If people really knew beforehand what they were getting into, and knew of the things and people they were exposed to, there would be no people going into it.

THE INVESTIGATIVE COMPUTER AND PRECAUTIONS TO TAKE

The investigative computer should not be connected to the Internet. It should have digital forensic investigative, antivirus software, and be part of a SAN to share cases with other investigators sitting at workstations. The SAN is necessary for archiving cases and sharing expertise without having to run over to another examiner and crowd a workstation. If one works in a law office or large law firm, one may have to share the case with paralegals or ask another examiner about his or her opinion of possible steganography involving a digital picture. Rather than going from room to room and having people crowd around monitors, it is better to share active cases on a SAN. It is also a good way to access digital evidence from a previous case if there is an appeal. The SAN should also be connected only within the offices and not connected to the Internet. People outside the law firm should not be able to access the cases and possibly impact their confidentiality, integrity, or availability to others on the SAN. Each case may have its own folder or directory on the SAN.

PRECAUTIONS: BE PREPARED TO ANSWER MOST BASIC QUESTIONS FOR A DEPOSITION

The first precaution to take is to be able to answer the most basic question that a defense lawyer or fact-finding committee might ask. If one is doing digital camera forensics, the first question asked may be "What is forensics?" I once asked a student of digital forensics that question in a continuing education class. The student said that he saw all these television shows on forensics and knew what it was but could not express it. This demonstrates the need for a good definition. A website concerning health care careers said that "Forensic Science lies at the intersection of science and the law" [27]. Another important basic concept of forensics is known as "Locard's Principle of Exchange." John D. Wright tells us about change from contact between the victim and alleged assailant at the crime scene [28]. Other people have expanded that to mean any two things that come in contact will be changed. If a man takes a picture of a person with the digital camera, the picture of the person is in the camera and the file system on the camera has changed. There is a new file entry in the FAT. There is a new picture file in a directory. There may be a new thumbnail picture within a file known as thumbs.db on certain cameras with certain Windows type operating systems.

The second precaution is to show that you, the examiner, are qualified to do the examination. What if a fact-finding committee of lawyers ask, "What credentials do you have to examine a digital camera and how can you demonstrate that you are qualified to do this examination?" Many states are suggesting and may be soon requiring that digital camera examiners, cell phone examiners, and computer forensic examiners obtain a private investigator's license before they can examine any digital device. The process to get a private investigator's license differs greatly by what state you live in. Some states such as New Jersey require that you work under someone else's license for at least 5 years or have 5 years of investigative experience as a policemen before you can get your license. An application must be filled out and a background investigation should be completed. There have also been some discussions on the CCE list server that in the future more legislation may pass further regulating multistate investigations. Such investigations may mean that the investigating company or law office have employees with private investigator's licenses in all the states that an investigation is in. For example, if a person misused a digital camera in Texas, New Jersey, and New York, the agency needs to be employing private investigators who hold P.I. licenses in those states. It might be advisable to seek more than one P.I. license and include neighboring states. However, it is sometimes advisable to work for a lawyer who has passed the bar for that state because one can work under their license. Lawyers are also going to make sure that you and especially they do not get sued, so they may be good managers for both you and your investigation. Working for a lawyer will probably pay far less than working for a P.I. or on your own license. However, it should offer much more peace of mind and legal protection. It is often advisable to get a digital investigators insurance policy. A million dollars of liability may cost about five hundred dollars per year. The International Society of Forensic Computer Examiners has the contact information for such insurance people.

It is also advisable that the applicant take a class on private investigation such as many of those found online. The purpose for this is to learn how an investigation is conducted and to understand some of the laws that the investigator is bound by. Such courses also give great advice on how to dress for court, answer questions, and give a great overview of the legal system. The private investigator class also teaches a person the sources of finding information, interviewing people, determining the truth, and the basics of report writing.

PRECAUTIONS: GET CERTIFIED (ACE, CCE, CISSP, ETC.)

Texas State Technical College has a certificate program for becoming an examiner for cell phones, digital cameras, hard drives, USB drives, and other digital devices. Such classes give people a credential and teach both technical hands-on skills and theoretical training about the law. Any type of online classes, continuing education, seminars, and so on should be taken so that the person is as qualified as possible and can pass the *voir dire* process if they are to be an expert witness or even to be credible as an investigator.

Many law firms and private investigation firms will hire contractors or employees for performing mobile device forensics. This includes examining cell phones, smart phones,

PDAs, digital cameras, USB drives, flash drives, CD, diskettes, and the occasional pager/beeper. These firms often ask for prospective applicants to have a CCE, Certified Information Systems Security Professional (CISSP), or some type of certificate as from the Texas school previously discussed. The CISSP is a certification that from ISC^2. The acronym ACE is for Access Data's Examiner. This is important for showing that an examiner knows how to use FTK Imager and the Forensic Tool Kit.

If one is going to become a digital camera examiner, it is also good to take a class on digital cameras to learn about the various models, lighting, features, apertures, lenses, and photo editing. This helps the examiner understand if a certain type of picture could be real or if it is possible with a certain type of camera. Such a class may also help the examiner understand the types of effects that can be done with various photo editing software and how it might affect metadata. It also makes the examiner more credible in court if he or she must testify about the device and what was on it. A lawyer could always say something in court to raise doubts to the judge and jury such as, "We know you are certified to examine a digital camera, but do you know anything about using one?"

PRECAUTIONS: HANDLING EVIDENCE AND STATIC ELECTRICITY

The investigator should get a metal pole in the lab where he or she can touch it before doing an investigation. This allows for the discharge of static electricity. The Static Buster Model 201 wrist strap is also good for equalizing the static electricity on the person and not causing sparks on the equipment that can damage memory, circuits, or SD cards. It is also good to purchase an antistatic mat and place it on the floor in the work area for the same reason. In dry climates or in winter time, static electricity discharges are an issue and can impact the equipment. Why put oneself at risk for being negligent when being safe from static electricity can be done for less than one hundred dollars. Such preventative countermeasures also give the appearance of increased competence.

PRECAUTIONS: GET A GOOD IMAGE OF THE EXTERNAL MEDIA AND/OR INTERNAL MEDIA

If you ask any digital examiner what is the most important item in computer forensics or digital camera forensics, he or she will tell you that creating the digital image or clone of the hard drive/internal storage media is most important. This is the most critical stage because it is where the evidence is first collected and it is the first event that is documented on the chain of custody form. Throughout this chapter there are brief descriptions given of each product. One of the items that are mentioned is the ability to image the memory or storage device such as a hard drive or SD card. The image should be exportable to another digital device as a file or series of numbered sequential files.

PRECAUTIONS: FINDING EXPERTS FOR SPECIALIZED OLD DIGITAL CAMERAS

Suppose your investigation requires the examination of an antique digital camera such as the Apple Quick Take 100, what do you do? If you are a law enforcement person in a large city, perhaps another detective in the same city may have the expertise about examining the camera and you may call him or her. There may also be an Apple Computer Enthusiast group in that city and perhaps one of the members could be a technical consultant and may have the tools for removing pictures from the camera. Another possibility is that one of the Regional Computer Forensics Laboratories could be called. Then the detective could speak to a case manager and get the case assigned a priority. The New Jersey Regional Computer Forensics Lab in Hamilton, New Jersey is a good example of a facility that has specialists in examining PDAs, cell phones, Windows machines, and Apple products.

The CCEs belong to a group who send emails to the entire group of approximately 1000 CCEs on a daily basis. Sometimes a CCE will post a request for help in examining an antique obscure device or a new one. There is always someone with the expertise that is sought who can give advice on what to do, what tools to use, or some agency that will provide the service requested for a fee.

PRECAUTIONS: POSTING QUESTIONS ON BLOGS OR LIST SERVERS

There are many CCEs and expert witnesses who have sent emails on the CCE list server discussing a potential problem with posting questions about technical and legal topics. Some CCEs have said that defense lawyers have become CCEs and can now collect all the emails about questions posted on various topics Since the email is sent to everyone on the CCE mailing list, it can be collected legally by the defense lawyers who are CCEs. A few CCEs have sent emails expressing apprehension that their emails could be shown in court by an opposing defense lawyer who wishes to suggest to the jury that a certain examiner is not an expert in his or her field. Perhaps a list of questions on various eforensic topics might give the appearance of not knowing certain technical and legal concepts.

Because of this apprehension to post eforensic questions in a public forum, some examiners who wrote blogs online have resorted to using pen names or what might remind one of the Citizens Band (CB) radio handles of the 1970s. They also have disclaimers on their email stating that it is standard practice to question new changes in technology and the legal system and such questions do not indicate a lack of understanding of computer forensics. The debate suggests the intensity of the desire to win and the level of stress that must exist from the bantering between lawyers and digital forensic investigators.

The important point is that you should check with other investigators before you post questions and your real identity online. People can search your name and find every place where you have posted things. The Internet is also archived and one can use the "Wayback Machine" to find postings and websites no longer being hosted [29]. There is also what is known as the digital wind where items can be passed digitally from place to place and people have no control of all the places where something is posted.

PRECAUTIONS: FIND DIGITAL PICTURES WITH WRONG FILE EXTENSIONS

Many policemen will say that child pornography is considered a violent crime and that criminals who view, create, or distribute it will go to great efforts to hide it. One of the tricks people once did to hide pictures is to rename the file so that it appears to be a document instead of a picture. The file kids.jpg might be renamed kids.doc so it appears in the directory as a document. However, this does not work on Windows 7. The thumbnail of the picture is still visible with either the .doc or the .docx file extension. There are many tools such as FTK that look for these deliberate file extension mismatches. You should verify that your tool or tools do this because the prosecution or those conducting the investigation in the corporation could ask if your tools look for file mismatches.

PRECAUTIONS: NO BREAKS OF CUSTODIANSHIP OF EVIDENCE

The chain of custody form is very important for an investigation because it give all the details about a piece of evidence from the time it was seized until the time it appears it court. It has the case number, date time, details about the warrant, and the signatures of each examiner who handled the evidence, processed it, or received a copy of it. The chain of custody is for the integrity of the evidence so that it is credible for court. According to Mandia, Prosise, and Pepe, three writers of a book on incident response and computer forensics, evidence from a hard drive can be suppressed if the examiner did not maintain a proper chain of custody [30].

The chain of evidence form, also known as the chain of custody form, documents the entire process of how that evidence was kept secure and processed. The evidence should also have the chain of evidence custodians who keep the evidence locked up when it is not being processed or examined. The chain of evidence custodian should also be prepared to go to court if the organization's methods for maintaining the integrity of evidence is questioned [30].

PRECAUTIONS: DO NOT JUST RELY ON AUTOMATED TOOLS!

It is important to look at a picture and use some basic investigative skills to determine some facts. A student once brought in a scanned picture of his father, John Hogarth, a Navy Captain in Okinawa, Japan. The picture was labeled 1946. There was a second man in the picture named Fidel. The man was wearing a dog tag, wedding ring, and looked similar to Fidel Castro. Fidel also appeared to be between 18 and 30 years old.

Some students in the class perked up and became attentive. One man said that the man in the picture could not be Fidel Castro, President of Cuba, because he never heard of a Mrs Fidel Castro. Another student said, "Look at the beard, it is well trimmed, and all the pictures of President Castro online do not show a trimmed beard." The third student who was a former military man said that the dog tag was not American and thought maybe it was Fidel.

Members of the class also checked a series of biographies about Fidel Castro and found no biographies or articles that stated he was in Japan. They also saw that he was born in 1926 [31]. The man in the picture seemed a bit older than 20. The students also saw that he was in law school at age 20 and could not have been in Japan for a long period of time. Everyone knows that a law school in any country is so difficult and time away would mean failure. There are technical and nontechnical aspects of investigating a picture.

PRECAUTIONS: USING OTHER INVESTIGATIVE TOOLS SUCH AS BIOMETRICS TO SOLVE THE CASE

One person in the class suggested the possibility of adding all the pictures of President Fidel Castro to a Facebook page and then adding the picture in question. He then said that he would use a biometrics program such as Polar Rose to calculate the two hundred fifty facial measurements on the known pictures of Fidel Castro and compare it to the picture in question [32]. The students said a combination of technical methods, historical documents, and then questioning could be done to solve the identity of the man in the picture. The student said that someone could just write a letter to the Cuban Embassy, include a copy of the picture, and just ask President Fidel Castro if he was in Okinawa, Japan in 1946. Each clue is like a piece of a puzzle and we get closer to getting a full picture as the investigation continues.

PRECAUTIONS: CHECK HR FOR A SIGNED CAMERA POLICY

It is also important to recognize if a camera is issued by the company to the employee. Let us suppose that a person in a department store was accused by a coworker of taking a picture of customers with a binocular camera across the room in a partially open changing room. The incident response team was dispatched to the scene and the lead investigator spoke to the AR and asked the employee to surrender the property of the company, namely the binocular camera. The person who used the device could have used it as binoculars or a binocular camera to take close-up pictures or video. Binocular cameras have internal memory and can also take a SD card.

Perhaps a chain of custody form would be filled out in case the policy infraction becomes a criminal investigation. The AR is in-charge of corporate investigations and leads the incident response team [33]. Everyone associated with the investigation would be acting within the parameters of the incident response policy. The AR would most likely call human resources (HR) before the examination and make sure that the reported device, the binocular camera, had a signed policy on file so there would be no problem in taking it and examining it. If a piece of equipment is issued by the company, it is smart to have a policy for that piece of equipment. If the employee was not issued a binocular camera and no policy was signed, then it becomes a little more complicated to seize it because it is the employee's personal property.

PRECAUTIONS: KNOW WHEN TO INVOKE THE SILVER PLATTER DOCTRINE

If there is a misuse of a digital camera at work, a call to the general counsel might be in order. The general counsel would probably look in the employee handbook and there might not be a rule about bringing in cameras and taking pictures. However, if that misuse is something such as taking pictures in the corporate gym's changing room, it is no longer a policy investigation and becomes a criminal investigation. The Video Voyeurism Law of 2004 is a Federal law that prohibits the taking of pictures in any place where there is an expectation of privacy such as a changing room or tanning salon. At this point, the police would be called since a Federal law was violated and it became a criminal matter. At this point, everything would be turned over to the police or FBI and this would be known as the Silver Platter Doctrine [33]. Because it is now a criminal investigation, no further investigation could be done by the company because it is a criminal matter where due process applies and the Fourth Amendment applies. Further investigation can only be done by law enforcement and they must get a search warrant. If the incident response team continues investigating after the Silver Platter Doctrine is invoked, they would become agents of law enforcement which is illegal [33].

PRECAUTIONS: CHECK FOR COUNTERFEIT AMERICAN MONEY AT CRIME SCENE

John Wright tells us that computers make it possible for unskilled criminals to create high-quality color counterfeit bills [34]. The resolution of digital cameras is so good that people can take a picture of a hundred dollar bill and then use a high-resolution color laser printer to print out money. This would be difficult to get away with because the paper for American money is watermarked and uses a high-quality paper with security threads in it. However, some criminals have washed off American dollar bills and used the paper as a base for printing the old style hundred dollar bills. These bills look great to the untrained eye. In Madison, New Jersey, a person was arrested for possessing counterfeit bills. The reason the policeman knew they were counterfeit was because the bills all had the same serial number [35].

The Secret Service was started in 1864 to investigate counterfeit money in the American Civil War. Later, its mission was expanded to include protection of the POTUS, President of the United States. The U.S. Secret Service investigates many types of white collar crimes as well as the forging of coins, bonds, paper currency, and the counterfeit products for financing narcoterrorism.

PRECAUTIONS: WET BATTERIES AND CAMERA PHONES/DIGITAL CAMERAS

Sometimes a cell phone camera or digital camera will use a special battery that is rechargeable and costs up to thirty dollars. This type of battery can get wet and then cannot be recharged anymore. This can happen if condensation occurs in a bag or if the digital camera is moved when it is raining and the device is not in a waterproof container. If the battery gets wet, then there is a pink dot that will appear on the back of the battery. Investigators must look for such things when seeking to power up a device. It may be necessary to call up a battery specialist organization such as Radio Shack and ask for a replacement. If a battery is extremely rare, there are services in the People's Republic of China who can rebuild a battery by replacing the cells. However, such considerations take time and this may not be possible, depending on the case. A pink dot can be seen in a wet battery.

WHY STEGANOGRAPHY IS IMPORTANT TO DIGITAL CAMERA FORENSIC EXAMINERS

The digital camera contains resident memory and secondary memory in the form of XD cards or SD cards. The digital camera takes digital pictures which may seem innocuous.

However, both types of storage on the digital camera can be easily accessed with a modern desktop or laptop computer. Pictures can be altered on the digital camera with programs such as JP Hide and Seek and the picture does not have to ever leave the camera [36]. The original picture can be the container for embedded documents. It is the perfect place for the spy to hide documents without arousing suspicion.

PRECAUTIONS: USE THE BEST EVIDENCE RULE

I was once a cybercrime student in a class. The adjunct professor for the class was also a detective who had arrested a sex offender in Bergen County, New Jersey. The story was that when the case was about to go to trial, the alleged sex offender did not want anyone to see the child pornography images that he was arrested for. He then said that the digital images were a configuration of 1s and 0s and that he wanted the contents of each file, namely binary, to be printed and given to the judge and jury.

Luckily, "any printout that is shown to reflect the data accurately is an 'original'" [37]. This is from the Federal Rules of Evidence 1001. This is important that a printout of the picture can be used and not just the binary because it would greatly impair the justice system and make it difficult to bring people to justice.

PRECAUTIONS: WEEDING OUT THE WRONG TYPES AS INVESTIGATORS

Not every type of career is for everyone. A person who is extremely afraid of fire should not be a fireman. Others who are afraid of heights and flying should not become an airline pilot. One of the issues that is rarely if ever spoken of is about who should or should not go into digital camera forensics or computer forensics. I was once contacted by a man who had a successful academic computing career. The man was extremely intelligent, from a rich family that sheltered him greatly while growing up, and deeply religious. He watched shows such as *CSI Miami* and *NCIS* and wanted to do a career in digital forensics. The man wanted to go into eforensics because he found the television shows exciting and his career so boring. He was also low paid and there was no further path to go in his current job. It was basically doing the same things day after day in the computer lab with only the cost of living raises if he was lucky.

I knew that the man was deeply religious and was extremely bothered by any pornographic image. When I was a student, a part-time professor who was a full-time computer crime detective once told the class that out of a 100 hard drives that he examined, only 3 did not have pornography. Some of the cases that detectives deal with are with sex crimes or the creation, distribution, or viewing pornographic images. I also heard that many of the detectives are advised that mental health professionals/counselors are available to discuss the ill effects of exposure to images that they see. I told the man if he was serious about a career in this field, he should first speak to a detective and ask about the type of disturbing material that computer forensic/digital camera forensic examiners see. Then afterwards, he could make an informed decision about whether to continue with eforensics as a possible career path.

Janis Wolak and Kimberly Mitchell wrote a paper after being funded by the U.S. Department of Justice, OJJDP through the Internet Crimes Against Children (ICAC) Training and Technical Assistance Program. Their study included personnel from "511 agencies who had been exposed to child pornography during investigations of crimes involving child exploitation" [38]. An alarming "35% of ICAC Task Force Participants and 10% from affiliates had seen problems arising from work exposure to child pornography" [38]. Wolak and Mitchell also reported that 90% of the ICAC Task Force were somewhat or very concerned to exposure to child pornography [38]. One examiner said, "Intrusive thoughts during intimate times with my wife all but discontinued that part of our relationship for a long period of time. Going to a therapist on my own helped put it in perspective" [38].

After talking to the detective, he realized that he could not handle seeing the pornography or seeing real sex crime cases. This was real life and not television. The man said that he only considered the excitement of solving difficult cases, the technical challenges, the travel, and the changing of a career from boredom to excitement. He never considered the impact that sex crimes or the viewing of child pornography would have on his psyche. The man decided to get his doctorate and become a professor of computer science. The psychological suitability of this type of career is never discussed in academia. Perhaps academia can be distracted by students with good test scores and the need to fill seats to keep programs afloat.

There is some academic literature that gives hope to trying to match people for cyber investigator careers. Shinder and Tittel discuss the personal attributes of what makes a good cybercrime investigator [39]. They discuss all the positive attributes such as interest in computer science, love of lifelong learning, inquisitiveness, and many other traits. However, I would like to have seen a discussion of traits or habits that suggest one might be happier in another career.

PRECAUTIONS: KEEPING CURRENT ON DIGITAL CAMERA FORENSICS

The High Tech Crime Investigative Conference meets once a year and is a good place for learning about mobile device forensics. It is also an excellent place to meet vendors and talk about new products. Sometimes an investigator can become a beta user for new products and suggest changes that help the investigator and help the vendor make a better product which means more sales. At the 2010 HTCIA conference in Atlanta, Georgia, there were county prosecutors from various states and defense lawyers who gave unofficial opinions and great insights into legal issues in digital forensics.

The Computer Forensics Expo is a good place to network with professionals and there is always a group of people there who do mobile device forensics. These are often conference tracks that specialize in legal issues and others that concentrate on the technical issues.

REFERENCES

1. Lally, E. F. (February 1961). *Abstract—Mosaic Guidance for Interplanetary Travel.* Easton, PA: American Rocket Society.
2. Sasson, S. (October 16, 2007). We had no idea. Plugged in. URL accessed September 15, 2011. Retrieved from http://pluggedin.kodak.com/pluggedin/post/?id=687843.
3. *Patentstorm.* Electronic still camera. URL accessed September 15, 2011. Retrieved from http://www.patentstorm.us/patents/4131919/claims.html.

4. Bellis, M. (nd). History of the digital camera. About.com.inventors. URL accessed September 15, 2011. Retrieved from http://inventors.about.com/library/inventors/bldigitalcamera.htm.
5. Haag, S., Cummings, M., Phillips, A. (2007). *Management Information Systems for the Information Age*. 6th edition, p. 54, New York, NY: McGraw Hill Irwin Publishing, ISBN 978-0-07-305223-6.
6. *Meganslaw.* URL accessed September 22, 2011. Retrieved from https://www.meganslaw.com/.
7. *Spectorsoft.com.* Computer and internet monitoring software. URL accessed September 22, 2011. Retrieved from http://www.spectorsoft.com/products/SpectorPro_Windows/index.asp.
8. Cohen, K. (2007). Digital still camera forensics. *Small Scale Digital Device Forensics Journal*, 1(1), 1–8.
9. Yaglielowicz, S. (2008). The new face of amateur porn. Xbiz.com. URL accessed September 15, 2011. Retrieved from http://www.xbiz.com/articles/97733.
10. Shinder, D., Tittel, E. (2002). *Scene of the Cybercrime, Computer Forensics Handbook*, p. 19, Rockland, MA: Syngress Books, ISBN 1-931836-65-5.
11. The Standard of Japan Electronics Information Technology Association (JEITA) Standard CP-3461. Retrieved from http://www.jeita.or.jp/japanese/standard/book/CP-3461B_E/#page=1.
12. ZAR. Digital image recovery. URL accessed February 13, 2012. Retrieved from http://www.z-a-recovery.com/digital-image-recovery.htm.
13. *Tawbaware.* Thumber digital image processing software. URL accessed February 10, 2012. Retrieved from http://www.tawbaware.com/thumber.htm
14. Kessler, G. (2004). An overview of steganography for the computer forensics examiner. *Forensic Science Communications* 6(3), 1–27.
15. *Tawbaware.* Thumber: Frequently ask questions. URL accessed February 10, 2012. Retrieved from http://www.tawbaware.com/thumbfaq.htm.
16. *Tawbaware.* Camwork. URL accessed February 10, 2012. Retrieved from http://www.tawbaware.com/camwork.htm.
17. *ABC Amber Image Converter.* URL accessed September 25, 2011. Retrieved from http://abc-amber-image-converter.lastdownload.com/.
18. *Datalifter Computer Forensic Software.* Products. URL accessed September 19, 2011. Retrieved from http://www.datalifter.com/products.htm
19. Moore, T. (February 15, 2010). How to window Motersho. URL accessed September 9, 2011. Retrieved from http://motersho.com/blog/index.php/2010/02/15/howto-set-usb-drive-to-read-only-windows-xpvista7/.
20. Lanh, T. et al. (2007). A survey on digital camera image forensic methods. *Multimedia an Expo, 2007 IEEE International Conference in Beijing*, pp. 16–19, ISBN 1-4244-1016-9.
21. *Image MASSter.* RoadMASSter-3 Portable Forensic Lab. URL accessed January 19, 2012. Retrieved from http://www.dataduplication.co.uk/pdfs/RoadMASSter-3%20Ap07.pdf.
22. *eDec Digital Forensics.* Black Hole Faraday Bags, URL accessed January 19, 2012. Retrieved from http://www.dataduplication.co.uk/pdfs/DataBagBrochureS.pdf.
23. *QCCIS.* Casenotes lite. URL accessed February 9, 2012. Retrieved from http://qccis.com/resources/forensic-tools/casenotes-lite/.
24. *QCCIS.* Fragview. URL accessed February 9, 2012. Retrieved from http://qccis.com/resources/forensic-tools/fragview/.

25. *QCCIS*. Gigaview. URL accessed February 9, 2012. Retrieved from http://qccis.com/resources/forensic-tools/gigaview/.
26. *QCCIS*. Videotriage. URL accessed February 9, 2012. Retrieved from http://qccis.com/resources/forensic-tools/videotriage/.
27. Forensic science overview. Explore health careers.org. URL accessed September 21, 2011. Retrieved from http://explorehealthcareers.org/en/Field/22/Forensic_Science.
28. Wright, J. (2007). *Crime Investigation*, pp. 6–7, Parragon Books, U.K., ISBN 978-1-4054-93333-8.
29. Way Back Machine. Internet Archive. URL accessed September 18, 2011. Retrieved from www.archive.org.
30. Mandia, K., Prosise, C., Pepe, M. (2003). *Incident Response and Computer Forensics*, pp. 200–203, USA: McGraw Hill, ISBN 978-0-07-222696-6.
31. *Biography.com*. Fidel Castro biography. URL accessed September 18, 2011. Retrieved from http://www.biography.com/articles/Fidel-Castro-9241487.
32. Doherty, E. (2011). Teaching digital camera forensics in a virtual reality classroom. *The 2011 International Conference on Computer Graphics and Virtual Reality (CGVR'11)*, July 18–21, 2011, Las Vegas, NV, accepted as a Regular Research Paper (RRP), that is, publication in the *Proceedings and Oral Formal Presentation*. ISBN 1-60132-193-7.
33. Nelson, B., Phillips, A., Enginger, F., Steuart, C. (2004). *Guide to Computer Forensics and Investigations*. Boston, MA: Course Technology.
34. Wright, J. (2007). *Crime Investigation*, pp. 176–177, Parragon Books, U.K., ISBN 978-1-4054-93333-8.
35. Doherty, E. P., Stephenson, G., Fernandes, J. (2005). Computer security and telerobotics for everyone. Indiana: Author House. ISBN 1-4208-9682-2 (sc).
36. Cocchiaro, N. (2005–2008). Stegui. Steganography. URL accessed September 24, 2011. Retrieved from http://stegui.sourceforge.net/intro.html.
37. Volonino, L. et al. (2007). *Security, Computer Forensics Principles and Practices*, p. 17, 508, USA: Pearson/Prentice Hall, ISBN 0-13-154727-5.
38. Wolak, J., Mitchell, K. (November 2009). *Work Exposure to Child Pornography in ICAC Task Forces and Affiliates*, Durham, NH: University of New Hampshire, Crimes against Children Research Center.
39. Shinder, D., Tittel, E. (2002). *Scene of the Cybercrime, Computer Forensics Handbook*, pp. 133–140, Rockland, MA: Syngress Books, ISBN 1-931836-65-5.

CHAPTER 3

PDAs and Digital Forensics

PDA HISTORY

It is necessary to define a PDA before we can discuss its history or how it is used in the workplace or as a personal computing device. If we do not have a common definition or understanding of something, then it is not possible to discuss it. As you progress through the discipline of eforensics, you will find that it will serve you well to learn some common definitions of devices, tools, and legal concepts. It is also good to learn a brief history of each device, who invented it, what operating systems a device uses, and some tools to examine it.

The PDA can mean a personal data assistant to some people or a personal digital assistant to others. One of the great authorities of computer telephony and telecommunications, Richard Grigonis, author of the *Computer Telephony Encylopedia*, said, "The PDA is a natural evolution of the paper based day organizers of the 1970s and 1980" [1]. OldComputers.net reports that CEO John Sculley of Apple Computer "coined the term PDA" for the small handheld computer known as the "Apple Newton Message Pad" that was available in the 1990s [2]. It seems safe to conclude that the PDA is a personal digital assistant. This appears to be the accepted acronym.

There are sometimes informal discussions about mobile devices in schools, investigative agencies, and corporations. Some people express a sentiment that cell phones have replaced the PDA and perhaps the PDA can only be found in a museum next to pagers. "In 2011, does anyone still use a PDA?" This was a question posed to me in 2011 by a probation officer for the State of New Jersey. The question was asked because the officer was considering whether to take a continuing education class. I said that I used a HP PDA with Windows CE 5.0 and others that I knew also did. Some people use a PDA with an internal wireless card to access the Internet, do email, and read attachments. I said that many people who use PDAs do so because the screen is bigger than a cell phone screen and it is easier viewing for an aging population in the United States. According to Marist Poll (2009), 1 out of 6 Americans has a PDA [3]. Based on that statistic, we can then calculate the number of people owning and probably using a PDA. The United States Census says that there are approximately 312 million people living in the United States as on December 28, 2010 [4]. That means approximately 250 million people in the United States own a PDA. That sounds as if it is very relevant to be able to examine a PDA and that mobile forensic examiners should have a write blocker, a forensic imager, and an examination program for PDA devices. When I posed the question to the same source about cell phones, the answer was approximately 19 out of 20 Americans had a cell phone which made the statistic to be over 310 million people. Obviously, the need to be able to investigate cell phones is greater than PDAs but it still demonstrates that there is a needed capability.

In the summer of 2011, advertisers are still creating a variety of PDAs and claim that millions are still sold every year. In 2012, if one goes to the Amazon online bookstore and queries the term "PDA" within the books category, thousands of different books will appear. There are books specifically made to fit on the screen of a PDA that are available for purchase. The subjects range from bioterrorism, to health care, and to robotics. The fact that all these books are still for sale specifically for a PDA demonstrates that there is a large population of people who use this type of device and that it is not relegated to the museum as some might think. Some people who have these PDAs may have committed a crime, be a party to a crime, or be the victim of one. The PDA with its wireless and infrared connections may hold some evidence to that crime that needs to be examined. PDAs hold the same information that laptops did some years back. In our previous chapter, we discussed an operating systems phenomenon known as feature migration. This means that over time, the programs and activities that were only possible on the large computers are now possible on much smaller computers. We said that this extra capability of increased processing and storage is due to Moore's law.

Evidence on newer PDAs might be the same as older laptops and desktops for some people. Such data might be websites surfed, email, and pictures. Perhaps some of the activities also show that the person who used the PDA researched how to commit a crime or communicated with others on planning one. Because of increased storage capacity, the presently examined PDA might have approximately the same size storage of a laptop of 10 years earlier.

HISTORY OF PDA AND POCKET SIZE ORGANIZER

The history of the PDA is not standardized and one can find different accounts of the history depending on what source you choose. Jansen and Scarfone at National Institute of Standards and Technology (NIST) special publication 800-124 said, "PDA appeared in 1993" [5]. They say that it had a word processor, appointment book, spreadsheet calendar, and address book. It also included wireless communications for receiving faxes, email, and text messages. However, some people who are computing enthusiasts in the United Kingdom might tell you now that the Psion Organizer 1 was the first PDA, though the term was not coined back then. It had a flat file database, calculator, and clock. The Psion was used by some academics in the United Kingdom but it only had two kilobytes RAM and four kilobytes ROM. You could argue that the Psion was only an enhanced calculator. It may be difficult for the average consumer to distinguish between handheld calculators, organizers, and PDA devices. People might say that the Psion Organizer 1 had no operating system and therefore could not be considered a PDA. However, you could try to argue in favor of the Psion Organizer 1 being a PDA because of its flat file database.

One of the ways to settle the history of the PDA is to look up the United States Patent for it. The Personal Digital Assistant was patented on June 24, 2003 by Quanta Computer Incorporated in Taiwan [6]. The patent number is U.S. D476,328 S [6]. The inventor of the device was credited to Chang-Ta Miao and Gwo-Chyuan Chen of Taipei, Taiwan [6]. The sketches of the device appear to be the PDA that is often for sale at consumer electronics shows.

The pocket size organizer was patented on September 1, 1998 [7]. The inventors are Jeffrey Hawkins, Robert Yuji Haitani, Malcolm Smith, and Gisela Schmoll [7]. The patent was assigned to Palm Computing and the U.S. Patent number is D397,679 [8]. Dictionary.com refers to electronic organizer as "a pocket-size portable computer for storing appointments, addresses, memos, and to-do lists; also called personal digital assistant, PDA" [9].

PDA HISTORY: PDA PHONE

Jeremy Roche and Joseph Hanlon described a phenomenon called the PDA Phone in 2008. This is a device that looks like a PDA with a full QWERTY keyboard but also includes the ability to make and receive phone calls [10]. The device also has an onscreen keyboard that can be used with a stylus. The device looks like a cell phone but is called a PDA Phone. Here is another example of where device categories are difficult to distinguish. The Nokia E71, Palm Tree Pro, i-mate Ultimate 8502, Blackberry Bold 9000, and HTC Touch Pro are touted as the top five PDA Phones by Roche and Hanlon. These devices have enhanced versions of the traditional calendar/organizers found in a Psion but also contain cell phone applications such as the phone book, call log, and SMS messaging. These devices are also useful for email, web surfing, and other social media applications.

LEARNING ABOUT PDAs/MUSEUMS

The Infoage Museum in Wall, New Jersey, has a large display of computers of all sizes, including a PDA. The exhibits often change. Their website can be found at the URL www.infoage.org. It would be worth checking in with them before a visit. They also have what they affectionately call the computer destruction lab where people can take apart computers and learn about hardware.

The Computer History Museum is located in the town of Mountain View in Silicon Valley of California. It is also an excellent place to see the entire history of computing and the various branches of mobile device computing that have become popular. They also have some videos that take one inside the museum. One of the PDAs that they have which is very interesting is the Vadem Clio C-100. This device appears to have the characteristics of a large PDA, tablet, and laptop. The unit was manufactured in 1998.

Many museums are located online so that the maximum number of visitors can see the collection at their convenience. One site that is particularly well organized is http://oldcomputers.net/. This website also has contact information and has a repository of equipment in California. This is important because if an examiner needs a piece of equipment or some technical help with some equipment, there are interested parties who may be contacted.

There are occasions when antique media needs to be examined. Consider in 2009 when I (CCE) was contacted by the head of a small credit union asking about the possibility of examining an 8″ floppy diskette from 1979 for possible information about an insurance policy and benefactors (see Figure 3.1). The oldcomputers.net website purchases collections of computers and is expanding its holdings. This group also has an Apple Newton Message Pad which was dubbed a PDA by former CEO John Sculley. There is also another site called www.pdamuseum.com that has an extensive collection

FIGURE 3.1 Dr. Doherty holding an 8″ floppy diskette.

of Palm and Sony PDAs online. The website is well organized into two groups for Palm and Sony. Then, each model is listed with a hyperlink which makes traveling to quite easy. Another online PDA museum is located at the URL www.pdamuseum.info. They feature many PDAs from Armstrad, Apple, Atari, Casio, Compaq, Ericsson, Fossil, General Magic, Handspring, IBM, Motorola, Palm, Phillips, Psion, Rolodex, Sharp, Timex, and Xircon. Each link contains a picture, technical specifications, and something about the device.

This is useful to anyone who is conducting an investigation and must look at one of these. The Amstrad Pen Pad PDA 600 which is featured on the website shows that the device had a stylus for selecting letters on the device. The stylus seems to be a unique characteristic of most PDAs. The Royal DaVinci DV1 PDA pictured in Figure 3.2 is unique because it has the onscreen keyboard and stylus that most PDAs have but it also has an extra stylus and a folding external keyboard complete with function keys. The device was available with either 2 or 4 MB of internal memory. In 2011, PDAs have gigabytes of storage. This again is an example of the actuation of Moore's law.

LEARNING ABOUT PDAs: PDA HARDWARE

The PDA consists of some components. The first is the stylus, a small pointer. The stylus is often a piece of metal or plastic with a pointed tip for selecting letters on screen. The length of the stylus is often 3–4 inches long. Each PDA also has a battery that can fall into three categories. There is a flat cell that is usually metallic and about the size of a nickel. Some older PDA/organizer devices may use the double AA batteries. Lastly, many of the newer batteries are larger, thinner, flat, and are marked as lithium ion.

The screen is usually a liquid crystal display (LCD) that is of rectangular shape of approximately 2 inches by 3 inches. The screen is often touch sensitive and often works with an onscreen keyboard. Newer screens have higher resolution, more pixels, and a large range of colors. There is a group of wires in a ribbon that connect to the motherboard. The

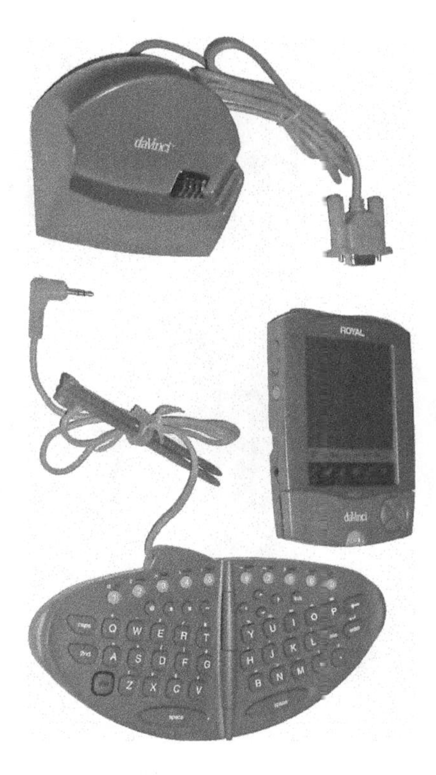

FIGURE 3.2 Royal DaVinci DV1 PDA.

motherboard has a set of metal paths known as a bus system. There is often a processor that is less well known those found in laptops or desktops. An example of a PDA processor is the Dell Axim x30:312 MHz Processor. PDAs usually have a group of spring-loaded buttons that act as switches for the human–computer interface.

PDA PROTOCOLS CONNECTIVITY AND OPERATING SYSTEMS

The PDA has more methods of connectivity than one might think. Some devices readily connect to other Microsoft Windows networks without the knowledge of the person who has the PDA so it is important to be aware of that and safeguard the PDA. If one is an investigator, involuntary connection with other systems suggests possible tampering and could ruin the results of an investigation. That is why investigators should move the PDA in a Faraday bag from the crime scene. The next section will discuss PDA connectivity in the forms of infrared, wireless, Bluetooth, and cables. Then, we will address the operating systems found on PDAs.

PDA CONNECTIVITY: INFRARED

This is a very interesting topic because there are many ways to connect a PDA to other computers or the Internet. Sometimes the connectivity surprises us. Here is an actual example to illustrate that. I was once using an old laptop with Windows XP in a lab and

FIGURE 3.3 HP iPaq.

practicing cell phone forensics for an upcoming continuing education class. I finished the assignment and was shutting the computer down. The security suite that I used asked me if I was sure that I wanted to shut down because others were logged on! This seemed strange. The wireless was turned off and no cables were connected to the laptop. Upon further inspection, the security suite said 16 others were logged on. Then I noticed there were 16 HP iPaq PDAs (see Figure 3.3) charging in the lab. The infrared from each PDA automatically connected to the laptop. This can happen if the operating system on the laptop is configured to automatically connect with incoming requests and security is liberally configured.

PDA CONNECTIVITY: WIRELESS

Many PDAs also have wireless connectivity. This is often 2.4 GHz on newer PDAs. The higher frequencies allow more bandwidth which means that newer PDAs with better processors and higher storage capacities can take advantage of the higher-resolution pictures and video on the web. Some of the older PDAs use 1.2 GHz as the frequency of connectivity. Shannon's theorem was introduced in a previous chapter. It stated that the higher the frequency, the more the bandwidth, and the more information that can be encoded on the signal. Lower-frequency waves travel further than higher-frequency waves. If a person was connecting to a public hotspot with the intent to do a crime, he or she could be much further away if a PDA with a lower-frequency signal was used. However, if a person used a directional antenna such as the cantenna or made a directional antenna with a coffee can, connectivity on either frequency might be done from a half mile away [11]. The limiting of connections is important to the integrity of the examination. That means that when seizing a PDA, it should be placed in a Faraday bag so there is no possibility of others connecting to it and tampering with the evidence. The examination machine should also be placed in a room that blocks RF signals from being eavesdropped in. The examination machine's connectivity should be blocked so others cannot interfere with the evidence or with the examination itself.

PDA CONNECTIVITY: BLUETOOTH

An interesting development happened with a blind student who was giving a presentation and someone who knew him sat quietly in the back of the room. The late person was amazed that the speaker addressed him by name and greeted him. The late person apologized but asked how he knew his identity. The cell phone that the late person had was broadcasting his name and the speaker had it set so that new devices discovered were read audibly. Sometimes people do not realize that their Bluetooth is enabled, broadcasting and unsecured.

It is possible for others with Bluetooth devices to connect, so we must be aware of that throughout the seizure and examination process. It is often said that Bluetooth devices must be approximately 10 meters or less to communicate with. However, if someone has a Bluetooth device where the forensic examination is taking place, it is a concern. One has to learn about all the methods of connectivity on the device that one is examining and also the machine that one is using. One must think of the CIA triangle of information security and try to keep the confidentiality of the examination, the integrity of the evidence, and only have it be available to those who are allowed to see it. Perhaps sharing the case with other investigators on the storage area network (SAN) would be the best option.

PDA CONNECTIVITY: CABLE

Many PDAs come with a docking station or cable. This allows users or examiners to easily transfer undeleted data, namely allocated files such as documents and pictures, to and from a computer and PDA. Some cables are proprietary and difficult to obtain. It is much better if one can get the original cable from the suspect at the time of seizure. Some PDAs such as the HP 48X have a docking station with a built-in cable. It is best to get these not only for conducting an examination but also for charging the battery. It is possible that a seized device such as a PDA may be a low priority and will not be examined until after the battery is dead. It may have to be charged while kept in a Faraday bag in a locked evidence locker.

The cables for the PDA may be as simple as something as a mini stereo jack that connect to a cable with an interface to a nine pin serial port as on the Sharp YO-P20H. Other PDAs such as the daVinci Royal DV1 use a docking station that has a built-in molded cable and nine pin serial port. Other PDAs use a mini USB port or some type of proprietary cable and connector. It cannot be stressed enough that it is so important to get a cable or docking station and external media that is used with it at the time of seizure. If proprietary cables are used, perhaps Paraben's Device Seizure cables kit or XRY Forensics' full investigation equipment including cables must be purchased. It may also be possible to Google search the seized device and purchase an after-market cable, an extra device with a cable, or the original cable that was preowned.

I would not suggest creating your own cable to specifications because it might be questioned in court and damage the findings of your investigation. Suppose the other attorney asked, "Are you certified to make cables?" You would have to say no. Then he or she might ask if there was a testing and validation plan with the cable. You would probably say no. Even if you did a testing and validation plan, the opposing attorney might create doubt that it was not rigorous enough and that the results of the investigation may be flawed. It is best to use certified cables from the original manufacturer or from a forensic cable kit.

PDA OPERATING SYSTEMS: MICROSOFT WINDOWS MOBILE 5.0 PREMIUM EDITION

This Microsoft Windows operating system feels like a lighter version of Windows XP and is based on the Windows CE 5.2 kernel. This type of operating system is found on some smartphones and comes installed on some PDAs such as the HP hx2490. This operating system has a set of functions that allow the user to scribble notes, use a word processor or a spreadsheet, view MS PowerPoint presentations, surf the web, operate a calendar, keep to-do lists, and check email. The applications installed within the PDA allow the user to do so many tasks.

There are also a set of video games that allow the user to recreate. It is important to mention recreation because there could be a need to examine gaming activity, especially if someone was supposed to be attentive at the time of a crash. In 2009, the Washington D.C. Metro crashed and there was some investigation to see if the driver had a PDA or cell phone and if inattention to the primary duty played a part in the crash [8]. In such an investigation, various log files of email receiving and sending would be checked. There would be a scan of the forensic image of the PDA and the examiner would check the temp files and regular files to see if the dates and times of file modification, deletion, or creation matched the time of the crash. High scores and evidence of game activity might be checked to see if it matched the time of the crash. Text messages, web surfing, and all activities would be checked for modification in case a friend tried to erase any evidence of distraction around the time of the crash.

The operating system allows the user to use a stylus and onscreen keyboard to write memos, conduct business by email, or go review MS PowerPoint presentations. The operating system also makes it simple to toggle on or off the Bluetooth, infrared, or wireless connectivity. The turning on or off of connectivity is done by touching the stylus to buttons representing Bluetooth, wireless, or infrared. The operating system had a display option that also let the user view the screen in either landscape or portrait.

The HP hx2490 operating system also has two places where the serial number for the device exists. It also exists on a faceplate behind the battery. If someone changes it in one place, they would probably not know to change it in all three places. The operating system also has a place that stores the owner's name and contact information.

PDA OPERATING SYSTEMS: PALM

The PDA operating systems could be said to fall into a few families. One of them is the Palm Operating System which includes Palm 1, Palm 2, Palm 3, Palm 4, Palm 5, and Palm 6. The 3Com manual shows that devices that used the Palm 2 operating system 2 had TCP/IP settings and could be used with a modem to call into an Internet Service Provider [12]. Palm Operating System 3 was good for the Palm III family of PDAs, including the Palm III, Palm IIIc, Palm IIIe, and Palm IIIxe. Some marketing literature of the time showed that these devices could easily be backed up.

Palm Operating System 4 worked with the Palm m500, Palm m505, and Palm m515. During this time frame, USB 1.0 connectivity was popular, and Palm 4 supported this. During the time of Palm 4, some PDAs allowed small-capacity SD cards to be used. If doing an investigation on a device that uses Palm 4, it may be a good idea to seek evidence on SD cards that could be kept near the computer or laptop that is synched with the PDA.

In the People's Republic of China, Palm OS 4.2 was available with certain Palm models and supported Simplified Chinese. It was very good for to-do lists and phone numbers.

The Palm 5 had a development suite that included capability for wireless. If the device has a Palm 5 or later operating, it is good to take precautions against connectivity. PDAs such as Palm V, Palm VII, Palm Tungsten T, and Palm VX used the Palm 5 operating system. The devices that used Palm 5 had faster processors and more storage than previous ones.

Palm 1000 and Palm 5000 devices used version 1.x of Palm. It might be a good idea to get a book about the Palm operating system if one might encounter these devices and learn about the file system structure and learn about how and where files might be mapped to storage. What algorithms might be used? This might be useful if you are going to testify as an expert witness.

According to the PDA museum at www.pdamuseum.com, the PDAs all included synchronization capability to put names, addresses, phone numbers, to-do lists, and calendars on a desktop or laptop computer. The Palm professional even let one store email on it. Since these synchronize with a computer, it would behoove the investigator to get a search warrant to include a desktop and/or laptop that was used with the device. The great thing about all Palm devices was that they allowed people to stop carrying heavy paper planners and phone books while still having that information available to them.

Sony also made a series of Clie PDAs such as the N, NR, NX, NZ, S, SJ, T, TH, and UX series. These units used the Palm operating systems. The interesting thing about many of these units is that they also had handwriting recognition capability. That means that people could scribble on the PDA screen with a stylus and it would recognize the writing which could be converted to standard text. However, some people also saved the scribbling. That means that one would have exemplars of a person's writing which might be useful for comparing against hand-written notes, signatures, and writing on checks.

There is also a science called graphology in which certified handwriting analysts can tell many personal things about a person from their handwriting. Mark Seifert has a text that describes this science in detail [13]. I once spoke to a lady who was a graphologist who told me such things as large handwriting is often said to show generosity while small close together handwriting suggests that a person is parsimonious. There are also other things such as going from the last letter of script backwards to crossing the letter "t." This may suggest that a person thinks a lot of themselves and that time is valuable.

LEARNING ABOUT NEW OPERATING SYSTEMS

An excellent method for learning about all the PDAs, cell phones, tablets, PDA Phones, and all the available means of connectivity and external storage is at the Mobile World Congress which takes place every year. In 2012, the Mobile World Congress took place in Barcelona, Spain in late February and early March. The 2011 exhibitors are listed at the following website: http://www.mobileworldcongress.com/2011-exhibitors. Mobile device examiners could go to this event and talk to the developers of various applications and operating systems and perhaps learn what tools or methodologies are available to seize data from a device. The examiner could also see what the latest bandwidth requirements are for the devices and learn about any new wireless frequencies these devices use. There may even be some new method of connectivity that is on the horizon that examiners should learn about.

Some years back, a police detective told me that he went to a suspect's house and had his examination equipment set up. He found something that was producing a strong signal but it did not appear to be a network. It turned out to be a wireless hard drive that was hidden in the attic. The suspect's mobile device connected to the wireless hard drive which contained all the illegal pornography. It was the first time he saw a wireless hard drive. If the detective went to a conference such as the Mobile World Congress, perhaps a vendor could have told him that someone could use a wireless storage device to store pictures or videos so as not to use up the mobile device's storage.

INVESTIGATIVE COMPUTER AND PRECAUTIONS TO TAKE

One of the important things that saves a lot of time in examinations is to put all the cables back in the correct place in the suitcase or bag. Each cable should have a twist tie and each cable should be rolled up. If a person teaches PDA forensics, 10 minutes can easily be lost untangling cables and finding the correct ones. Credibility can also be lost in the classroom if one appears unorganized. Sometimes it is also good to put cables in plastic bags and put a sticker on the bag to know what the contents are.

It is also good to check that the examination machine's software is licensed to the agency or person doing the examination. If it is not properly licensed, the results could be suppressed in court since it is considered "Fruit of the Poisonous Tree." If some of the software was illegally copied and used, the examiner might also be arrested and the agency would lose a skilled digital evidence examiner.

The examiner should also make sure that there is no connectivity by infrared, Ethernet cable, modem cable, wireless, Bluetooth, or phone line. The isolated machine should have the latest antivirus and antispyware run to remove any malware. The examiner should also try to look at the guidelines for Tempest and reduce the possibility of RF signal eavesdropping on the case. Perhaps, shielding can be placed on wires or in the room. The examiner should also make sure that his or her certifications are still valid just like the software. Lastly, if the case goes on the SAN, then the case should only be shared with authorized investigators. If additional expertise is needed, perhaps a call to the Forum for Incident Response Teams or the High Tech Crimes Investigative Association is needed to solicit an expert for the case.

INVESTIGATIVE COMPUTER AND POLICY PRECAUTIONS TO TAKE

The PDA examiner, the cell phone examiner, and the laptop examiner could all be collecting usernames and passwords, breaking encrypted files, reading other's email, and doing a multitude of other activities that could get most employees fired and sued. It may seem obvious but there needs to be a set of policies for the examiner as an employee at his or her office computer and a set of policies related to activities with regard to lawful investigations in the lab. An examiner may have to use hacking tools to hack a covert website that is used with a PDA and child exploitation case. However, the examiner cannot use such tools from his or her regular desktop computer. There needs to be a set of signed and enforced policies for an Internet connected computer where the examiner is a regular employee. There should also be a separate Internet line and policy for the research machine in the lab. The examination machine should be on a separate SAN and it too needs its own policies since employee examiners will be looking at child pornography, breaking encryption, and collecting others' usernames and passwords. If boundaries are not clear, an examiner could use his or her tools and work computer and break policies and lose his or her job.

In academia, the same could be true about policies with regard to hacking and collecting usernames and passwords from PDAs. There is also an additional problem in academia that many computers are managed by an information technology (IT) department that manages the network, installation of applications, and the Internet. Standard network images greatly reduce the work of the IT staff but the uniformity and lack of access to the registry and administrative permissions of the computers restricts teaching and learning in the lab. One important example is the access to the registry in order to make the USB ports as read-only. Students need to get practice accessing the registry and changing a bit on a double word so that they do not change the PDA when accessing it for examination. Student policies at most schools do not allow for the collection of usernames, passwords, and running software that defeats encryption on files. If one is going to teach computer, PDA, or cell phone forensics, then it is important to have policies exclusively for the lab and some type of exclusion clause for students so they can do their forensic activities at school in a fenced off environment without fear of expulsion or arrest.

It is also important in a school to tell students and staff about the boundaries of allowed forensic and investigative activities in an academic cybercrime lab. Many students and technology workers are never given indoctrination to the American legal system and think that there are few or no limits on people's privacy concerning electronic devices. The people who work or attend a school should also sign off on boundary policies because there are times when an ex-employee, staff member, or student will bring in inappropriate requests such as collecting the online header of a blog of a former spouse or love interest and then see where their present residence is from reverse lookups of IP addresses and email addresses. Someone may bring in a hard drive and ask if it can be recovered? Is it their personal drive? Are they allowed to view it? If recovered files reveal that there is child porn on it and the police are called, will that cause enemies at work or put the academic at risk for retaliation? All these type of concerns can surface at conferences where many type of academic, corporate, and law enforcement lab managers congregate. The managers can then share policies and discuss how they handled problems thus eliminating potential new problems for labs that adopt their policies and learn from other's experiences.

PRECAUTION: USE FORENSIC STERILE MEDIA

One of the most important things to do is to get an image of the device onto sterile media. Sterile media means that there is nothing on the media from a previous investigation, no malware, and no data remanence. The media should be wiped with a wiper that meets Department of Defense (DoD) 5220.22-M standards for data wiping. It is also good to run the latest antivirus and antispyware against the media to show that nothing is there that could harm the integrity of the investigation.

When choosing a forensic media, it is important to think about how long the data must last on that media because media has different life spans. A PDA that has evidence from a murder case may need to be reexamined many years from now, especially if it is one of these capital punishment cases that are appealed. It might be a good idea to keep such a case on the SAN which is actively refreshed and backed up. Jeff Rothenberg, an authority on media life span, said that CDs could last from 5 to 59 years [14]. If the evidence was burned to compact disks (CD), they have a limited life span depending on the quality of the media and at what speed the information was burned to the CD. High-speed burns on inexpensive CDs have a shorter life span than those that were burned at slower rates on high-quality CDs.

J. Rothenberg also brings up an excellent point about magnetic media. If we archive important files on magnetic tape systems, will we have the equipment to play the media back on later? J. Rothenberg says magnetic tape may have a life of 10–30 years but it needs a special player, and sometimes what is stored is in a certain format for the equipment [15]. Who will have such a player available and in working format? He makes a point which is rarely discussed; we should consider the ability to examine the archive in the future and not just look at the cost per byte of storage.

How do we choose the correct media? Digital examiners need to consider the cost, the ease to replay the media, and the life expectancy of the media. J. Rothenberg said, "Some gold plated/glass substrate digital optical disk technologies promise 100-year lifetimes" [15]. This may be the best option since optical disk players hold a large amount of bytes, are relatively inexpensive, are small, and the equipment for replay is readily available. This topic needs to be discussed more by all the branches of digital forensics. The topic needs to be ongoing because of Moore's law and the doubling of storage capacity approximately every 18 months.

It would probably be in the interest of the organization or agency that hired the PDA examiner to have a committee review the various types for PDA imaging media available on the market. The committee could review the present costs of media and consider the life expectancy of the media and the equipment needed to review the image of a former case. Perhaps the committee would have a member from the High Tech Crimes Investigative Association, the organization's general counsel, and one of the Regional Computer Forensic Laboratories. The fact that there was a committee reviewing the imaging process and media would most likely lower the cost of insurance and reduce the possibility of punitive damages from a lawsuit against the examiner and his or her organization.

The ZDNET's Australian branch had published an article on electronic media life spans in 2002 [16]. USB drives, DVDs, CDs, and magnetic tapes all have varying life spans depending on the materials used, any protective coatings, and how they were stored [16]. I once left some disks and a keyboard in a car in the summer time. The windows were shut tight and the vents were turned off. The disks were not readable and the keyboard was also twisted. One could then make an argument that they must be in an environmentally regulated storage place where humidity, static, and temperature is optimized for media longevity. It would seem that imaging each PDA that was seized to a hard drive would be the least desirable choice.

PRECAUTION: GET PDA EXAMINER CERTIFICATIONS

It would behoove the digital evidence examiner to get the certified computer examiner (CCE) certification from the ISFCE as discussed in earlier chapters. This is considered by many to be a vendor-neutral certification for FAT file systems and NTFS file system computers and devices. It would also be useful to get the certification of CISSP from ISC squared that was discussed previously in Chapter 2. Many federal and state government law enforcement organizations have examiners with CISSP and Encase certification. Encase is a world-recognized forensic software tool created by Guidance Software for seizing and examining data from computers, cameras, smartphones, and PDAs. It would also be useful to go to a Paraben Corporation's Device Seizure class and take a total of 4 days to obtain certificates for both the beginner and advanced classes. Sometimes there is also a special pricing so that students may get the software, training, and cables kit for one price This lump sum may be easier than trying to justify multiple purchases.

XRY Forensics also has a variety of forensic classes available in German and English for both Europe and North America. They offer a certificate of completion. Anything that shows an audit trail of a specific relevant body of knowledge is very useful. Many times symposiums have speakers who give demonstrations and talk about cell phone forensics. These are important to attend. Many have a certificate of attendance and show a continuing of education in the field of study. Sometimes there is a Marine Corps Antiterrorism Symposium at New Jersey Institute of Technology. Occasionally, this symposium has something related to mobile digital forensics or cyber investigator speakers. The previous was one of many examples of local low-cost education that one can get to keep active in digital forensics. Earlier in this chapter, we discussed the blurring of boundaries between the PDA and cell phone with the PDA Phone. There could also be strong arguments made that cell phone forensic training is also relevant to PDA examination.

PDAs are small computer systems. Many in 2011 are functioning at a capability that laptops did about 10 years earlier. Since there are very few specific PDA certifications, it would be good to get any mobile device forensics, computer forensics, or electronic evidence collection certifications that one has an opportunity to get. That being said, the Global Information Assurance Certification of GCFA (Global Certified Forensic Analyst) is a very good idea to obtain for people working within computer forensics who must do incident response. There are people in the incident response fields who do "Black Bag Operations" and go to companies after working hours to image a drive for a client. This type of individual can benefit by this certification because they first learn and then must demonstrate the concepts of preserving, collecting, analyzing, and reporting on evidence from both Windows and Linux computer systems. The test has 150 questions and one gets up to 4 hours to complete it. A passing score is 104 out of 150 questions [17].

One of the things that are worth noting is that this certification demonstrates that the successful candidate has a knowledge of the antiforensic techniques that attackers may utilize. One example of antiforensic technique is to use a tool such as Timestomp. Timestomp is a tool that can be abused by criminals so that access times and modification times are changed or deleted. The Defiler's Toolkit is another example of an antiforensic tool that can be used to change metadata and hide or destroy data [18]. The Coroner's Toolkit is an example of a tool that would be used in response to someone using antiforensic techniques. Many of the other certifications would benefit by making examiners learn about antiforensic tools and techniques.

PRECAUTION: WHAT ABOUT AMERICANS TAKING THEIR PDAs TO OTHER COUNTRIES?

Corporate executives should remember that the laws around the world are not the same as the United States. If a person is an American who is working in another country at an American company, they are bound by the laws of that nation. The only exception might be if the person is working at an American Embassy and that place is considered American soil. Let us now consider a twist on the previous scenario. Suppose the American with the PDA full of these bearskin baby pictures goes to England and is working at an American company in England. Perhaps the person had a year working in a corporate office and is returning home. He is at customs in the airport and the customs agent asks to see the PDA. The customs officer sees all these bearskin pictures.

Shinder and Tittel, the authors of *Scene of the Cybercrime*, say, "the United Kingdom has the Protection of Children Act of 1978, and Section 160 of the Criminal Justice Act of 1988, it is a criminal offense for a person to possess either a photograph or pseudo photograph of a child that is considered indecent" [19]. Are such a collection of pictures of children indecent? The American company in England needs to have a seminar for employees going to England. The seminar should define what constitute the boundaries of acceptable materials. The employee's family should be educated by the company ahead of time so they know what can be downloaded and transported on the computer from one country to another. It is too late when people are at the airport.

PRECAUTION: PDA EXAMINATION AND DISEASE

In 2005, I went to a digital forensic lab for instruction on PDA forensics. There were some PDAs in another part of the lab that were once examined and had no evidence on them, but we were told to stay away from those and they were going to be put back in the safe. The examiner told us that he wore disposable gloves when doing the examination on them. Students who were law enforcement technical types pushed the subject further and wanted to know why gloves must be worn. They asked if gloves should often be worn and in what circumstances. The examiner told us that some of the styluses had a smell of feces on them and could have been used as anal probes. Some of the police officers who investigated sex crimes also said that there exists "paraphilia" where people can be excited by inanimate objects.

The investigator told some of the people who were going to be PDA investigators that sometime in their career they may encounter devices that may contain pornography and be used in activities that many people could not imagine. When a person is an investigator, they can be exposed to a variety of people and activities that would shock the average person. If one is going to prepare investigators for the real world, they may wish to tell people that there are times when you might want to reduce the risk of contracting a disease and wear gloves. There may be times within your career when you may see images or learn of activities that you find shocking and may need to seek counseling through your employer.

PRECAUTION: ENCOUNTERING *PRIMA FACIE* EVIDENCE ON THE PDA

There have been many discussions among examiners about what constitutes child pornography and when to call the police. If one does not call when one encounters it, that is a crime. If one falsely reports it, one ruined someone's life and will face a life-ruining lawsuit. Suppose an examiner encountered a JPG scan of a Traci Lords men's modeling picture from the beginning of her career. She looked as if she was in her twenties and perhaps others thought so too. Later, it was discovered that she had a fake identification and was just under age eighteen [20]. Is that child pornography? This section will discuss the complexity of the problem of what is illegal pornography.

Sometimes it is difficult to define the borders of art, hobbies, and illegal pornography. The following will be a discussion of an example of how a mobile device examiner might question what constitutes child pornography and what is art. This example shows why legal definitions are important and why it is also important for digital forensic examiners to have access to a lawyer and a law library in their organization. By knowing some legal definitions and being able to discuss a situation with legal counsel, the digital forensic examiner will have more of an idea of what to do and when to call law enforcement.

Please consider this bit of history and then this situation. There was a time in the 1920s when it was fashionable to take a picture of your baby on a bearskin rug with no clothes, and having the baby's bottom face the camera [21]. This type of photo can often be found in an old family photo album. Suppose someone digitizes it and puts it on their PDA, it could be acceptable. How about if they have 30 of them from nonrelatives? How about if the person says that they collect old photographs and these were found in people's photo albums? Is it a hobby of antique photos? Is it child pornography because of the quantity of naked photos and bare bottoms from a collection of nonrelatives? Who makes the decision?

There is a 35 chapter document of almost 2000 pages called the *Attorney General's Commission on Pornography*. The document was written in 1986 under the President Ronald Reagan administration and created under the direction of U.S. Attorney General Edwin Meese [22]. If one reads this, it may give a direction on what constitutes adult pornography, child pornography, and what is art. The problem is that once a CCE finds child porn on a computer, the investigation is supposed to stop and everything taken to the police. If the person finds it and does not report it, the examiner could potentially go to jail. It seems logical that anyone who does digital forensics should have a lawyer on staff.

However, if a person makes a wrong accusation and rushes to act by calling the police, the suspect may be arrested, and then the suspect's life can be ruined. An arrest might cause a divorce and loss of credibility with neighbors. Some people are quick to believe the worst about someone, especially regarding sexual charges. Companies are sometimes brand and image conscious and could be quick to distance themselves from the alleged offender. Even when someone is exonerated, is there not some doubt? If a wrong call was made and a person was found innocent of having child porn, do you not think that a multimillion dollar defamation lawsuit against the examiner would follow? These are some of the issues that make investigation difficult and this type of digital evidence examiner as a difficult career. That is why it is important for people to take a class on the legal system, learn definitions, and to have access to a lawyer for work.

According to the Free Dictionary's definition and discussion of child pornography, "Thus, home movies, family pictures, and educational books depicting nude children in a realistic, non-erotic setting are protected by the Free Speech Clause of the First Amendment to the U.S. Constitution and do not constitute child pornography" [23]. The definition from Free Dictionary also said, "Child pornography is the visual representation of minors under the age of 18 engaged in sexual activity or the visual representation of minors engaging in lewd or erotic behavior designed to arouse the viewer's sexual interest" [23]. A lawyer would probably tell us that a dictionary definition is a good place to start but it would be best to look at the law, particularly the Child Pornography Prevention Act of 1996. A recent discussion on the CCE list server said that digital evidence examiners are not qualified to make rulings on what is child pornography. It is up to judges and prosecuting agencies to decide what is illegal and what is not. The CCE said it is best to work with a lawyer and to ask for his advice when encountering anything questionable.

SURVEY OF TOOLS TO INVESTIGATE A PDA: PARABEN DEVICE SEIZURE

Device Seizure is an excellent tool for the new examiner who quickly needs to examine some GPS units, PDAs, digital cameras, or cell phones. It is best to check the website and call to see if the device one is going to examine is supported. If your present version is not supported, then they may have a newer updated version with support of that device. Perhaps if you are a law enforcement officer and need it right away, technical support might be able to work on it within a reasonable time frame.

Basically, one downloads the software and puts the license key in the directory that they tell you to. Then you start the software and save the case to a file name. A wizard is used that asks questions about your name, address, agency, case number, and many other things pertinent to the case. If there is a JPG or GIF image of the agency logo, that can be added. This information will be added to a cover page and report area that accompanies the seized data. There is also a picture of a small mobile device that one clicks on to start the seizure. It is good to have the correct cable plugged in the PDA and the USB port of the examination machine at this point. The PDA may still be in a Faraday bag and only a cable is sticking out so nothing can connect with it by some wireless means. The device may now be powered on.

Then, Device Seizure will ask about getting a logical acquisition of physical acquisition. The physical acquisition is good to get because that gets all the allocated and unallocated space from the device. This can be saved and one may use a set of data carving tools such as Encase, Scalpel, Data Lifter 2, X-Ways Forensic Tool, and Access Data's Forensic Tool Kit to see what files can be obtained. Many people use a variety of tools to validate the results and make sure everything that could be captured, was captured. Exculpatory evidence proving no guilt should be sought as well as evidence showing guilt for prosecution.

SURVEY OF TOOLS TO INVESTIGATE A PDA: PARABEN PDA SEIZURE

PDA Seizure was a tool that was created by the Paraben Corporation to seize the data from PDAs. Paraben also had another product called Cell Seizure that was also discontinued. Both Cell Seizure and PDA Seizure have been incorporated into Paraben's Device Seizure. The program was approximately US$250.00 when it sold in 2005. There was a cables kit that sold for the same amount and also included a small toolkit. PDA Seizure offered very good support for the Palm Pilot and other PDAs that used an earlier version of the Palm operating system.

Sometimes, students in my class would make remarks about PDA seizing as an irrelevant skill. I would hold up a DVD of the movie called *Breach* and tell the class that the most important spy case in world history was solved because of evidence seized from a PDA. The class would gain interest as I would hold up a Palm III, the same type of PDA that FBI agent Eric O'Neill seized from the spy Robert Hanssen. Then the teacher and students would seize the data from PDAs in class and the students were happy to get the relevant information.

SURVEY OF TOOLS TO INVESTIGATE A PDA: AVANQUEST'S DATA RECOVERY PROFESSIONAL

One of the universal complaints from digital evidence practitioners, universities, forensic lab managers, and lawyers is about the price of tools needed to recover digital evidence. Avanquest has provided a low-cost tool for data recovery that can be used for many types of PDA, computers, memory cards, and digital cameras. The software can be used for partition recovery, lost folder recovery, and can recover approximately 300 types of file formats related to email, documents, movies, pictures, and sound files. The software is very easy to use and is menu driven. There are also help screens and a technical support number where one can ask questions. Support for a product is very important and if one can ask technical questions about recovery, the tool becomes more important. The file systems supported are FAT 16, FAT 32, VFAT, NTFS, and NTFS5.

Some PDAs use an ext3 or ext4 file system. The product did not mention these file systems, so one might use Encase for those. It is also good to have a data validation plan and use as many tools as possible to recover data. It is also good to see if one tool was able to recover more files than another. This type of validation is highly valued by the legal community. One of the features that should be very interesting is the disk cloning feature. I called the tech support personnel at 818-591-6245 on October 5, 2011 and verified that the cloning feature does copy the whole drive. The entire drive means that the tool can be used to perform a forensic image of both allocated and unallocated space. An examiner can never have enough forensic imaging tools with him or her because some software can go bad. Imaging is considered by many to be the most important part of eforensics and one often gets only one opportunity to image a device.

SURVEY OF TOOLS TO INVESTIGATE A PDA: RECOVER MY FILES

The tool called Recover My Files is made by the Get Data Corporation and works well with PDAs that use a FAT file system. Some PDAs will lose pictures, documents, or movie files due to a suspect damaging the FAT table. This tool can help recover those files from that type of action. This tool was discussed as a viable tool for digital cameras too. The company's website tells us that the product is useful with "Windows 98/ME/2000/2003/XP/Vista/Windows 7 and works with FAT 12, FAT 16, FAT 32, NTFS and NTFS5 file-systems" [24].

SURVEY OF TOOLS TO INVESTIGATE A PDA: GUIDANCE SOFTWARE'S ENCASE

Guidance Software Encase is an excellent tool for seizing data from many types of PDAs. The company also has excellent support if one is a new user and has to do an investigation. The software requires a license dongle to make it work. This is to reduce the possibility of software being pirated or breaking the license agreement by sharing it with unauthorized parties and thus put the investigation at risk. The software is considered expensive by some but it can be justified because of the level of support one gets.

Encase is also good because there are numerous classes and tests to pass that make sure that the examiner has both the theoretical knowledge and the practical hands-on experience to do an examination. Many of the classes are for a good part of a week at a training facility but this should be seen as an advantage and not a disadvantage. It gets a group of instructors and students to work together with the tool, the process, and the theory, with perhaps only a few outside interruptions from phone calls from the person's agency. The in-person training allows students to try to do examinations, make mistakes, ask questions, rectify deficiencies in knowledge, and then become proficient.

Many people express an opinion that the software interface is well designed and relatively easy to use. The menu-driven windows and wizards allow the examiner to point and click, fill in some information, and progress through the examination in a similar fashion as products such as Device Seizure. The tool allows one to collect all the allocated and unallocated data in an image. There are data carving tools to collect files and reconstitute pictures, documents, and other files. There is also a library that checks file signatures for hacker tools and child pornography. Special files can be saved in sequence and burned to CD. If Encase is used on a SAN or forensic server, then the image of the seized device can be stored in its own folder on the SAN.

The Guidance Software website tells us that in 2011, there are more than 6000 licenses of the product distributed among law enforcement, corporations, and governments. Since the product is well recognized and widely used, there is no problem with it being accepted by the legal community in court as an investigation tool. Encase passes the Frye test. It is important to use the methods taught in the class and understand the advanced features of the software so that one can appear credible in court and the results of the investigative findings are accepted.

SURVEY OF TOOLS TO INVESTIGATE A PDA: PALM PDD

Palm PDD has been integrated into Paraben's Device Seizure. It is licensed by Grand Idea Studio. This program allows a person the ability to capture all of the contents of storage of PDA for file analysis and data carving. Such a forensic imaging tool is valuable to investigators who must obtain email, lost phone numbers, and contacts, as well as calendar events. There is a website and point of contact information for the person who licenses the program out. A private investigator or law enforcement officer in need of imaging a certain Palm operating system device may wish to contact the website about certain models with any questions or concerns. The website for Palm PDD also makes available a white paper about the tool and the palm system internals which gives an in-depth look at file deletion, flash memory, and other hardware and software workings. This is important for anyone who must testify about forensic tools and how they work.

SURVEY OF TOOLS TO INVESTIGATE A PDA: XRY FORENSICS

XRY Forensics produces forensic software and cable kits for mobile device examiners that support almost 3000 mobile devices. The supported device list can be obtained by the company. It is good that they list the supported devices because often people can get confused with certain devices being PDA Phones or smartphones. As new mobile devices are released and supported, new cables will be sent to registered users. If the device is supported, the cables can quickly be found by the examiners and give them the ability to quickly collect the data.

There are a few different types of XRY products which is good because it gives digital examiners choices. This is important for people with small budgets. There is XRY Physical which allows the examiner to collect the forensic image of the mobile device. It is often said that gathering the image of the mobile device is the most important stage of PDA forensics. Then, one can use a variety of data carving tools from various forensic software vendors to recover files and pieces of data from the unallocated parts of the device. XRY Physical might be a good choice for a department with many data carving tools but no ability to examine mobile devices.

Let us remember that there are those PDA Phones that are PDAs, but very much like cell phones. XRY Logical is a great product for digital examiners who wish to collect all the allocated files found in the FAT file system and recover them without a SIM using the handset option. The SIM can be imaged separately using XRY Logical and choosing the SIM option. The program is launched and there is a wizard that lets you automatically recover the files even if you do not know if your device is a PDA, PDA Phone, smartphone, or camera phone. The software also gives you many options for connectivity. If you do not have the cable, perhaps you lost the one in the kit, then wireless, infrared, or Bluetooth might be an option. From unofficial opinions expressed at forensic conferences, tools that simplify the process and give people options usually receive praise from organization administrators.

The maker of XRY Logical also provides training opportunities for people who bought the product and are new to mobile device forensics. This is good because some people have a need to quickly become investigators but also need the knowledge and credentials so that they can be credible if the investigation goes to court. It is important to be able to pass the Frye test which means that your tools and mobile device forensic techniques are accepted by the legal community and your peers in mobile device examination. Therefore, any type of recognized training on recognized tools will help the digital examiner have his or her findings be accepted into evidence in court.

Version 6.0 of XRY has more file support for all the captured files and any attachments. In this version, one can easily click on email attachments and watch the video that the suspect watched or view the picture that he or she viewed. There are also more options for exporting the findings so it is easier for other parties to read during the discovery process. Version 6.0 also has more tabs to organize data into SMS messages, MMS messages, email, and chat sessions.

PDA FORENSICS AND INTELLIGENCE: CLEARWELL eDISCOVERY SOFTWARE

If a PDA was seized from a terrorist, then the device can be examined to see who his or her contacts are. Email addresses can be put in a reverse lookup site and names and addresses may be found. Messages may have message numbers and IP addresses showing an origin. The contact list in the address book, the email directory, and emails could be mapped with by hand but this would take a long time.

Investigators are using eDiscovery software such as Clearwell's eDiscovery software to quickly map all the parties communicating in the email and using organization tools to show who knew what. The software can also sift through millions of documents to find relevant ones. eDiscovery tools can organize the results of a digital investigation and then save a lot of time so that the investigator's time can be focused on analysis and not collating and organizing data. As devices have more capacity and communications get more numerous, eDiscovery tools become more of a necessity.

e2RETRIEVE

Many small organizers and PDAs use an ext2 file system. Occasionally, these systems are damaged and the file systems may also be damaged. There is a C programming language program known as e2Retreive that may be used to recover some raw data from the device.

e2fsck

Suppose someone had a PDA running Linux. Perhaps the power was lost and the device did not shut down properly. The file system became corrupted and there is an error. Perhaps important files cannot be accessed anymore. The utility e2fsck utility and its file system check feature might be able to repair the file structures. It will prompt the user what to do. The response "-y" means yes and will result in attempting to fix anything in spite of the severity [25]. The response "-p" means preen [25]. Preen results in automatic repairs without any type of system prompt [25]. The response "-n" means that the answer is no [25]. A response "-f" only forces the system to do a file check [25].

PILOT LINK

This is an open source tool that runs on a desktop and is used with a PDA. The program runs on MAC OS systems, Linux systems, and Windows. It has various utilities in it that are useful to the digital evidence examiner. Pilot-xfer is the utility that allows one to create a logical image and unfortunately it does not add hash values. This image can be scrutinized in a number of popular forensic tools such as Guidance Software's Encase or X-Ways forensics. Pi-getram retrieves the RAM contents from the device and Pi-getrom gets the contents of the read-only memory. Pi-getram is much like the utility called volatility that is located in Backtrack 5 since both collect the RAM contents.

PPWDump

There is a password recovery tool for Palm devices. The tool gets the encoded system password by going into the unsaved preferences database and starting at a certain byte and copying the string. Then an algorithm performs on the string and it is transformed to its original state, the typed-in password which is in ASCII, not Unicode. Once the password is recovered and the device is imaged. The image can be examined in the Palm emulator. There are many places online where this utility can be downloaded for free.

HARDWARE WRITE BLOCKERS FOR USB PORTS AND PDAs

There are a few different types of hardware write blockers on the market that one can purchase and use with the PDA and a laptop, so that when one makes a forensic image of the PDA, one does not create all types of temp files and change access time and dates on the file. The Tableau T8 Forensic USB Bridge is one such tool that connects between the laptop and the PDA being imaged. It was also tested by the National Institute of Standards and Technology in 2008. Wiebetech makes a nice hardware write blocker known as the Ultradock version 5.0 that can be used with USB ports to keep any changes from occurring on a PDA when acquiring a forensic image. Derek Newton, a writer for Information Security Insights, explains that the hardware write blocker is easy to explain in court and in a report to people who are not technical [26]. They also have blinking lights or prompts to show you that they worked and are simpler to operate than software write blockers [26]. The downside is that it is another piece of equipment that needs to be purchased, carried, and powered up.

THE DIGITAL FORENSICS FRAMEWORK

The digital forensics framework (DFF) was written in C++ as well as Python. Python is often used for scripting and applications like BitPim for cell phone forensics. DFF is open source and therefore the code can be scrutinized by everyone and all the tools that are developed and shared on the website become transparent since everyone can examine the code. The website also discusses how people can also contribute code and modules. DFF is a digital investigation tool but it can also be used as a platform for the development of new tools. The DFF can be used by corporate investigators, system administrators, and law enforcement detectives who do digital forensics. The tools are also used by criminal justice students, digital forensic researchers, and security people from around the planet. There is also an easy-to-use graphical user interface.

DFF is useful for systematically analyzing volumes, hidden items, file systems, and application data. The metadata can be extracted from examined items and copied into virtual read-only containers. If only copies of portions of data are exported to read-only containers, we should be assured of the integrity of the original data. The user interface supports more than one language. There is the GUI, but also a command line for certain tools that need it. There is a file browser similar to PTK and bookmarking is possible. There is a task manager to look at what concurrent processes are running.

One can look at images, text, and videos. It is also possible to get a time line analysis of items. This is very important to look at the events leading up to a crime, the crime time itself, and then what might be referred to figuratively as a postmortem look at things. There is also a hexadecimal viewer, so one can go into the file and view the metadata. The viewer can support both small and large files. The package is useful for carving files out of open space as well as the standard data recovery. There is also support for the reconstruction and analysis of the registry which can hold important information about users who were connected, devices connected, and how the user interface was configured. This framework can work with the usual FAT 16, FAT 32, and FAT 12 file systems. The real value seems to be that it can also work with ext2, ext3, and ext4 file systems which means that it can be used with some of the PDAs and organizers. There are not many tools for the ext2 and ext3 file systems of the PDA that can be used for data carving, gathering metadata, collecting aggregate data, and giving a timeline analysis. The fact that there is the source code and one can modify it means that it could be customized for the type of handheld devices that one might encounter in the future. The only drawback is that a law enforcement officer might have to learn C++ programming, how to use Python scripting, and learn to use an integrated development environment similar to .NET.

The good thing about DFF is that many researchers and law enforcement personnel are using it, so that there is also a community of users to ask questions. If foreign phones or new PDAs become available on the market, it may take a long time for the more-established forensic companies to address support for these items. Some models of mobile devices may never support them. If one becomes proficient with the framework, the IDE, and programming, it may be possible to support new devices long before others do in the United States. Since the DFF is also used by the law enforcement community, there is probably information about what cases it has been used with and then its acceptance for court should be easier. It should pass the Frye test. Since the tool is open source, any associated costs should be minimal. There is also ample documentation on the tool on a wiki. There is an IRC Chat on port 6667 and on the irc.Freenode.net server in a room called #digital forensic. There were approximately 25 people signed on when I tried it at night, indicating a good-sized support community. DFF is looking for volunteers to do some technical writing, linguists, module developers, and quality testers. It might be a good place for students to get some experience, meet some people, and then get into a forensic career.

WRITE BLOCKERS FOR PDAs WITH 9 PIN SERIAL PORTS

Many people consider the 9 pin serial port a legacy. When I was teaching computer science in 2003, I showed a 9 pin serial port cable and controller to a group of students from a developing nation. One student laughed and said that she had not seen one of those in her village in 5 years. However, there are still some people in the United States who use PDAs and organizers that utilize a 9 pin serial port. Digital forensic investigators who must make a forensic image of such devices have created a clever workaround to this problem. A USB cable is connected to a device such as the Tableau T8 and then a USB to 9 pin serial cable is used to access the PDA. Then a forensic imaging tool can be used to access the device without changing evidence. In 2004, Office Max in New Jersey was still selling PDA/organizers that utilized 9 pin serial ports.

The connection to the Sharp YO-P20H (see Figure 3.4) is fascinating because it uses a 9 pin serial port and stereo jack connection and is still for sale on the Internet in 2012. If one does not have a 9 pin serial port on one's examination machine, it may be possible to purchase a card with a 9 pin serial port to place in one of the EISA slots or just to use a USB port to 9 pin serial port cable from Radio Shack.

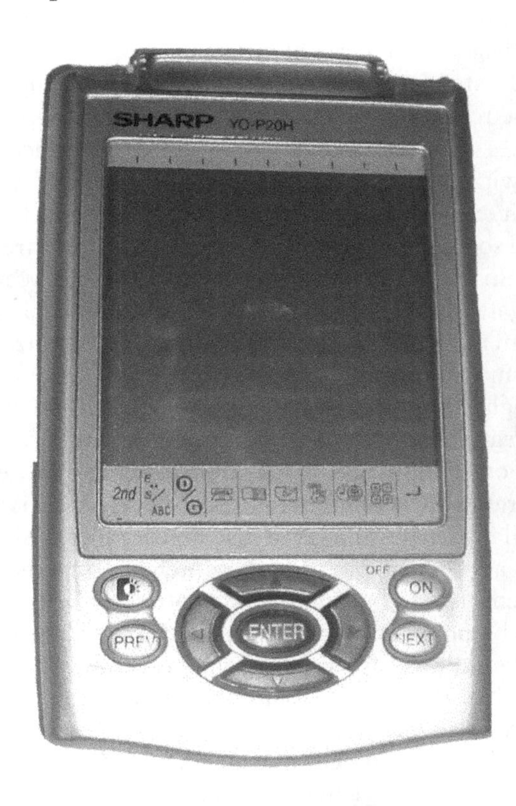

FIGURE 3.4 The Sharp YO-P20H.

USING VIRTUAL MACHINES FOR DEMONSTRATING THE ACCUSED'S PDA IN COURT

There is a program called Microsoft Virtual PC 2007. If installed on a Windows XP machine, it allows one the capability to open up a window on one's Microsoft Windows XP desktop and run another instance of an operating system. For example, suppose that a digital evidence investigator seized a PDA and created a forensic image of the PDA that was using the operating system known as Microsoft Windows CE 4.0. That forensic PDA image could be added to Microsoft Virtual PC 2007 and examined on the Windows XP machine. The virtual machine is a great way to look at the applications, pictures, documents, and multimedia just as the suspect would see it on his or her PDA. If a laptop with Windows XP is set up with a virtual machine and a forensic PDA image, then elements of performing an examination as well as some of the evidence can be shown in court to a jury. This can be done by running a data projector and having the laptop show everything on a wall.

FINAL THOUGHTS ON PDA INVESTIGATION

Investigators need to have the permissions to seize the mobile devices such as PDAs and examine them. They also need to fill out a chain of custody form and make sure that the evidence is never tampered with in person or remotely. It is important to have the correct investigation tools to make the forensic image from the mobile device and know how to use them. It is important to be able to pass the Frye test and use the tools and techniques accepted by both your peers and the legal community. The investigator may need to get eDiscovery tools as devices get larger and more documents and communications are too large for one person to collate in a reasonable amount of time. The investigator should look for both exculpatory evidence and evidence to prosecute someone.

SUGGESTED FURTHER READING

The NIST has a special publication called *Guidelines on PDA Forensics*. The publication number is 800-72 and is written by Wayne Jansen and Rick Ayers [27]. The document offers the reader a wide variety of topics on PDA forensics and gives sources of further reading. It is a good place for people to learn about PDA forensics to learn about the hardware, software, and uses of a PDA. It also gives the reader an insight into the various families of operating systems and the tools for examining the various types.

REFERENCES

1. Richard Grigonis quotation, author of *Computer Telephony Encyclopedia*, in a Telephone Conference on September 29, 2011.
2. *Oldcomputers*. Newton. URL accessed September 27, 2011. Retrieved from http://oldcomputers.net/apple-newton.html.

3. Marist Poll (July 6, 2009). Employment & age top factors in cell phone, PDA use. *MC Marketing charts*. URL accessed February 27, 2012. Retrieved from http://www.marketingcharts.com/interactive/employment-age-top-factors-in-cell-phone-pda-use-9678/.
4. US Census Bureau. US & World population clocks. URL accessed December 28, 2010. Retrieved from http://www.census.gov/main/www/popclock.html.
5. Jansen, W., Scarfone, K. (October 2008). Guidelines on cell phone and PDA security recommendations of the National Institute of Standards and Technology. NIST Special Publication 800-124. Retrieved from http://csrc.nist.gov/publications/nistpubs/800-124/SP800-124.pdf.
6. Miao, C. T., Chen, G. C. (June 24, 2003). United States Design Patent No. US D476,328 S. *Google*. Retrieved from www.google.com.pdf.
7. Hawkins, J., Haitani, R. J., Smith, M. S., Schmoll, G. (September 1, 1998). Pocket-size organizer. US Patent number D397,679. *Justia Patents Alpha*. Retrieved from http://patents.justia.com/1998/D0397679.html.
8. Goelz, P. (June 23, 2009). Metro crash: NTSB investigation. *Washington Post*. URL accessed October 7, 2011. Retrieved from http://www.washingtonpost.com/wp-dyn/content/discussion/2009/06/23/DI2009062301245.html.
9. Electronic organizer. (nd). *Dictionary.com's 21st Century Lexicon*. Retrieved September 27, 2011 from Dictionary.com website: http://dictionary.reference.com/browse/electronic organizer.
10. Roche, J., Hanlon, J. (2008). Top 5 PDA Phones. *CNET Australia*. Retrieved from http://www.cnet.com.au/top-5-pda-phones-240003981.htm.
11. Turnpoint. (August 2, 2011). Homebrew wifi antenna shootout. URL accessed September 28, 2011. Retrieved from http://www.turnpoint.net/wireless/has.html.
12. *Palm Professional Handbook* (1997). 3 COM, p. 38, Part # 423-0209-01B.
13. Seifer, M. (2008). *The Definitive Book of Handwriting Analysis: The Complete Guide to Interpreting Personalities, Detecting Forgeries, and Revealing Brain Activity through the Science of Graphology*. Career Press, USA, ISBN 978-1601630254.
14. Rothenberg, J. (June 1995). Magnetic Tape Storage and Handling, A Guide for Libraries and Archives, Dr. John W.C. Van Bogart National Media Laboratory.
15. Rothenberg, J. (January 1995). Ensuring the Longevity of Digital Documents edition of *Scientific American* 272(1), 42–47.
16. ZDNET.com.au (October 14, 2002). Tech guide: Storage media lifespans. *ZDNET*. URL accessed October 5, 2011. Retrieved from http://www.zdnet.com.au/tech-guide-storage-media-lifespans-120269043.htm.
17. *GIAC*. Certification: GCFA. Retrieved from http://www.giac.org/certification/certified-forensic-analyst-gcfa.
18. Wonko-ga (May 13, 2004). The defiler's toolkit. *Google answers view question*. Retrieved from http://answers.google.com/answers/threadview?id= 345604.
19. Shinder, D., Tittel, E. (2002). *Scene of the Cybercrime, Computer Forensics Handbook*, p. 20, Syngress Books, Rockland, MA, ISBN 1-931836-65-5.
20. Krajicek, D. (nd). Traci Lords. *TruTV*. URL accessed January 16, 2012. Retrieved from http://www.trutv.com/library/crime/criminal_mind/sexual_assault/traci_lords/5.html.
21. *Tvtropes*. Embarrassing old photo. URL accessed September 26, 2011. Retrieved from http://tvtropes.org/pmwiki/pmwiki.php/Main/EmbarrassingOldPhoto.
22. Department of Justice (1986). Attorney General's Commission on Pornography, Final Report, July 1986, U.S. Govt. Printing Office, Washington, D.C.

23. *The Free Dictionary by Farlex*. Child pornography. URL accessed September 16, 2011. Retrieved from http://legal-dictionary.thefreedictionary.com/Child%20 Pornography.
24. *GetData*. Recover my files. URL accessed October 1, 2011. Retrieved from http://www.recovermyfiles.com/.
25. *Beginlinux.com*. e2fsck. URL accessed February 2, 2012. Retrieved from http://beginlinux.com/desktop_training/comm/troubleshooting/353-e2fsck.
26. Derek Newton Information security insights. Write Blockers—Hardware vs Software. URL accessed December 27, 2011. Retrieved from http://dereknewton.com/2010/05/write-blockers-hardware-vs-software/.
27. Ayers, R., Jansen, W. (November 2004). Guidelines on PDA forensic tools. NIST special publication 800-72, pp. 4–60. Retrieved from http://csrc.nist.gov/publications/nistpubs/800-72/sp800-72.pdf.

CHAPTER 4

GPS Devices

INTRODUCTION

In this chapter, we will look at the invention of the GPS system by the United States and then discuss other nations such as China, Russia, and the European Union which followed later. This chapter will also describe some of the American and Russian GPS navigation devices that exist and then show examples of software that are used to examine such devices. There will also be a description of some of the education and certifications that GPS navigation device examiners have or would like to have. The chapter also contains a story that typifies some of the culture and humor one might encounter at a gathering of digital forensic examiners.

There are many occasions when academics and law enforcement personnel hear the name GPS forensics but there is another name for GPS forensics that may be overlooked by many. It is also called SatNav forensics. SatNav is an abbreviation for satellite navigation and the forensics component is the science of examining the evidence on a satellite navigation device [1]. The science can be applied to a GPS phone, a GPS fishfinder device, as well as to the small hardware devices used for navigation in boats, planes, and motor vehicles. The science can be applied to show that someone was not at a certain place or at a certain time for criminal investigations. However, a person with Alzheimer's disease can also be fitted with a GPS device and it can be examined later to see where they may have wandered. This may give search-and-rescue teams an idea where to look later if the person becomes lost and the GPS device is not functioning.

SatNav forensics or GPS forensics may show a person's daily routes with both time and date. This could be very important to show that someone did not violate a restraining order and get within a certain distance of someone's place of residence or workplace. Such a device could also be used to show that a person did not violate parole and only went from work to home and perhaps to a house of worship. However, the device would have to be required to be carried by someone at all times or be attached to the person as in the Martha Stewart parole situation [2].

GPS DEVICE HISTORY

Many people thought that the global positioning system was exclusively an American phenomenon that was a by-product of the 1960s space program. However, the former Soviet Union actually conceptualized it in summer 1976 and then launched a satellite for the system in 1982 [3]. One Russian website reports that 30 satellites will be in the Glonass system sometime within 2011 [4]. There is also a Chinese satellite system in operation called the Chinese Compass System and will have 35 satellites when it is finished [5]. It is also called the Beidou Navigation System [5]. There is also a Galileo System in the works for the European Union.

By 2012, there are a variety of operational GPS systems which means that there are consumer navigation products that may be used in vehicles and may need to be examined. In the United States and Canada, one can go to the store and see a variety of devices from companies such as Garmin, Tom Tom, Magellan, Mio, Insignia, and Kenwood. There are also Russian GPS devices on sale from the Russian Federation that use the Glonass system. The Automotive Portable GLONASS GPS Navigator Explay GN-510 ГЛОНАСС навигатор is one example.

The GPS is an acronym for global positioning system and is an American system that consists of a system of 24 geosynchronous satellites in the Earth's orbit. In any place on Earth, three or four satellites should be in view of a GPS receiver. GPS allows both people and machines to know their approximate or precise location at any time. The GPS system consists of two types of systems. There is the precise positioning system (PPS) which is used by the Department of Defense and the U.S. Military and is very accurate. Then, there is the standard positioning system (SPS) which some people call civilian-grade GPS and is not as accurate as the PPS system [6].

The big question for many people is why would other countries spend so much money to put satellites in orbit and create their own system when one already exists. It could be that only the U.S. Department of Defense has control over the PPS system and other countries do not want to only use the civilian-grade system that can be degraded to a high level of inaccuracy at anytime. China, Russia, and the European Union may see that it is a national security imperative to have their own GPS systems since smart weapons may use such signals too. GPS signaling on the civilian system (SPS) can be purposely degraded by the Department of Defense/military to 100 meters or about 300 feet [7]. That seems that degrading CA code on the Selective Availability (S/A) would harm civilian systems but keep military systems accurate.

GPS DEVICE HISTORY: MULTIENVIRONMENT GPS TRACKING DEVICES

In the early twenty-first century, it was possible to see various land-based GPS transponders sold by companies such as Ness Technologies that had magnetic mounts and allowed first responder agencies to track ambulances in real time. When vehicles are tracked, better dispatching of vehicles can be done in time of emergency. If one has emergency management exercises, vehicle tracking can be used as part of a performance gauge and to see if resources were used efficiently. By 2011, other companies had GPS tracking devices for any environment including air, land, or sea. These types of tracking devices can not only be used for resource allocation but also in criminal or corporate investigations. An example of one of these new devices that is now available to the consumer, corporate investigator, private investigator, and law enforcement officer is the LandAirSea TracKing® GPS Vehicle Tracking Device. This device has been sold from Radio Shack. This device can be used in the air, on the land, or on the water with a variety of vehicles. The device works in temperatures of –15°F to 185°F which is important since airplanes may encounter severe temperature changes in one trip. The device can be used to send out a beacon of its location every second. The receiver can then find out information on the device's direction, speed, and location. The device uses two AA batteries and can function to durations of 80 h on one set of batteries.

This type of multienvironment device is ideal for investigations where a suspect may use a boat, plane, or car. The device could also be misused by a stalker who tracks an ex-lover and then an investigator may need to do an investigation about the stalker's misuse of the GPS tracking device. The amount of uses, misuses, and types of investigations that may result are numerous.

If a digital examiner has to examine a GPS navigation device for the airplane, there can be some error because the device is used at a high altitude. Flight location has to be exact since people fly at night, in bad weather, and use instruments when landing at night. There are multitudes of Federal Aviation Administration (FAA) documents online or inline that discuss the Wide Area Augmentation System (WAAS). The WAAS system works with known ground points and GPS satellites through a Ground Uplink System (GUS) and provides an error correction system for the navigation system of the aircraft. The WAAS system can provide as much as 7 m of accuracy to the navigation data of an airplane [8].

GPS DEVICE HISTORY: GPS TRACKING AND WARRANTS

There exist GPS tracking devices that transmit a signal and allow someone with a corresponding receiver to obtain real-time location data for that tracking device that is difficult to see since and it can be just a small magnetic item that one sticks under the car. It is possible that the device may also have a small storage device that holds information on where a person drove and it could be examined later. The U.S. Supreme Court at syllabus U.S. v. Jones decided on January 23, 2012: "The D. C. Circuit reversed, concluding that admission of the evidence obtained by warrantless use of the GPS device violated the Fourth Amendment" [9]. It seems that this may have future implication for collecting data on a variety of mobile GPS devices. It goes to show how the law can change and private investigators and law enforcement people have to keep up with the law.

The problem with all these devices is that they were not even imagined by the founding fathers of the United States and therefore could not be addressed in the United States Constitution. The difficulty is that inductive reasoning must be applied to a variety of existing laws and commentary to come to a solution.

GPS DEVICE HISTORY: BRAILLE GPS DEVICES

By 2010, Michael Trei reported that there were devices that could be used by blind people for determining their GPS locations [10]. Suppose a car was carjacked and the police recovered it later. Perhaps one of the passengers was blind and would like to know where she or he was by having something that spoke where they were. The police should look for devices that may have been left in the car such as the "Braille Note GPS" [10]. The investigator should ask any special needs people about any relevant devices that keep track of location. Sometimes children have a cell phone with the family locator feature. If a phone was left on and in a stolen car, it may give the whereabouts to the car's location.

GPS DEVICE HISTORY: MILITARY-GRADE GPS JAMMERS

Once it was learned that the United States had GPS in the 1960s, the Russians tried to learn what it was and then built a rival system starting in 1976. They were not just satisfied with building a rival system but then started building GPS jammers. Zazona.com wrote that some jamming devices were sold by Russians to both the U.S. Army and the Iraqis in Baghdad [11]. These jammers could cause GPS navigation devices to be ineffective [11].

By 2011, it also became known that the People Democratic Republic of Korea had a GPS jammer cable of jamming GPS signals to distances of 100 km [12] as reported by *North Korea Tech*. *North Korea Tech* also reports that this jammer, if used near the border, could be capable of disrupting or wiping out signals in Seoul [12]. A possible disruption in signals for GPS devices could affect GPS forensic investigations in the future because signal data may not be available for the time of disruption. It becomes apparent that enemy use of military GPS jamming units prevents the military from knowing exactly where its planes, missiles, and drones are. Depending on where the GPS jamming was done, it could make investigations of embedded GPS systems difficult since the devices cannot accurately track where they were.

GPS DEVICE HISTORY: CIVILIAN-GRADE GPS JAMMERS

Anyone who grows fields of poppies or marijuana would not want drones or manned aircraft to be able to identify the location of a cash crop. Therefore, narcoterrorists—those who grow plants used to create illegal drugs to support terrorism—would have an interest in blocking GPS signals. The same groups would also wish to prevent border patrol aircraft or ground units to not be able to accurately describe the locations of drug caravans moving illegal drugs into the United States or across places such as Afghanistan. Individuals who may be under suspicion for drug trafficking or those people who have done nothing illegal but only want complete privacy from tracking devices would like to have consumer-grade GPS jamming devices.

By 2011, some type of GPS jammers for consumers became low cost and available online. There is a device offered for sale on eBay that is called a Mini Car GPS Jammer Blocker Anti Tracker Spy W/Switch. This device can be purchased from Hong Kong in the People's Republic of China for about $20, including shipping. These devices broadcast a signal within the 1500–1600 megahertz range for enough of a distance to disable tracking devices within the vehicle. A GPS jammer could be used by a cheating husband or wife who does not want any GPS device to record his or her trips. This is a countermeasure that investigators could encounter.

GPS DEVICE HISTORY: SIGNAL JAMMING FOR BOTH GPS AND GLONASS SYSTEMS

Some nations may wish to jam GPS signals from the United States, thus making American GPS devices unusable as far as knowing where they are. Then they may wish to use a competing system such as the Glonass system. However, Manuel Cereijo states that "Cuba recently acquired a capability for jamming U.S. GPS and GLONASS global positioning/NAVSAT signals, evidently based on a jamming system purchased from the Russian company Aviaconversia" [13]. The blocking of both sets of signals could even make devices such as the Automotive Portable GLONASS GPS Navigator Explay GN-510 ГЛОНАСС inoperative. If such jamming techniques were employed, then an investigation of where the device was during the times of jamming would not be possible.

GPS DEVICE HISTORY: GPS SPOOFING DEVICES AND TRUCK HIJACKING

My opinion is that a truck driver is more likely to be the victim of spoofed GPS signals than other types of people because thieves may wish to target their trucks. The spoofed signals may indicate traffic delays ahead and suggest a diverted path to a deserted road where an ambush may be waiting to kidnap the driver and steal the cargo. The GPS navigation device may be recovered later from the truck and the digital evidence examiner should be aware of spoofing so that he or she is not bewildered by times, dates, and routes that seem illogical.

In 2008, it was reported that there were some researchers at Cornell University and Virginia Tech who built a briefcase-sized device that sent out false signals to spoof GPS signals. Devices took the fake signals to be from the real GPS navigation satellite and put the spoofed coordinates in the GPS navigation device [14]. Such research complicates GPS forensic investigations because there is now the possibility that signals were spoofed and are not real. However, since such devices only exist in research laboratories and have limited range, it is probably not an issue. If there is a cargo theft investigation, it may be useful to know that such systems exist and data cannot always be trusted. It seems that in optimal conditions with the correct antennas and signal amplifiers, spoofed GPS could be broadcasted for long distances.

GPS DEVICE HISTORY: GPS FISHFINDER DEVICES

There are a variety of devices that use GPS that one can purchase today and have been available since the 1990s. One of them is known as the fishfinder. These devices use GPS and allow boaters the opportunity to mark locations where fish might be. An example of such a device is the Garmin GPSMAP 421 GPS Chart Fishfinder Combo with T/M. The device uses an SD card and routes and locations can be saved. Why might there be an investigation of such a device? The answer is perhaps because someone dumped something at sea such as a body, drugs, or toxic waste. An examination of such a device might show places for divers to look.

Another reason to examine such a device might be because of reports of taking treasure from a ship that might belong to an insurance company. Various places where the boat stopped may be checked against wreck sites to see if there was recovery of objects that are property of an insurance company.

GPS OPERATING SYSTEMS

There are many types of operating systems for GPS navigation devices. Some have more features than others. Some may be simple embedded systems while others are sophisticated operating systems such as Microsoft's CE 5.0. Let us start with a discussion of a unique Russian GPS device and then proceed to talk about the file systems and operating system on it. The Glonass GPS Navigator Explay GN-510 is a device that is available to Americans through online auction sites such as eBay. The battery life is reported to be about 3.5 h. Many GPS devices in the United States such as the Garmin Nuvi 1300 have a battery life of 4 h. The device comes with a battery, has a charger, and can be powered through a USB cable. The device has a CPU that is 400 megahertz and the CPU is labeled as ARM9. It uses a Chipset known as Mstar MSB2501. The external memory is quite large. It can take a micro SD card of up to 16 GB. It can also use an SD/MMC card.

The operating system is the interface between the human and the GPS device. It is a program that runs on the device and provides a layer of abstraction so the user can operate the hardware easily by pointing and clicking with a graphical user interface. The operating system frees the user from learning about complex hardware codes and dealing with the GPS device at the microprocessor level. The customized version of the CE 5.0 system allows the user to navigate a series of screens to collect or give information to the GPS device. One of the screens gives the user a map and tells which Glonass and GPS satellites are accessible at that location and then gives a longitude and latitude. It also gives the time, date, real-time speed, average speed, and altitude. The interaction can be done by touch screen for simple options such as clear all. The operating system for this device is broken up into six large areas such as video, music, ebooks and photo, system, navigation, and GPS information.

Since there are videos, photos, ebooks, and documents, just about any type of digital evidence could exist on this device. One has to understand what the device is capable of using or storing before one can fully investigate it. The operating system can use a variety of file formats that include ASF, AVI, WMV, MP3, WAV, JPEG, BMP, PNG, and TXT. This is not surprising since this device uses Windows CE 5.0 which is found on many American mobile devices. Some GPS devices sold in America such as the Tom Tom XL support approximately 56 languages and 35 different voices with both male and female voices. The Glonass GPS Navigator Explay GN-510 operating system is reported to operate in three different languages. That means that there must have been support packs added for language support. The languages on the screen may be configured to appear in English, simplified Chinese, or printed Russian Cyrillic letters. The device also uses a five inch LCD screen which is bigger than the American screens and may be that is why the battery lasts less time. The refresh rate for the screen is 1 s.

The Russian device also uses a Glonass module which utilizes 20 channels with a frequency range of 1.6025625 GHz to 1.6155 GHz. A person can upload or download information with USB 2.0 speeds on a Windows operating system and it is compatible with the Mac OS 10 operating system. The device also can be used with a stylus for those with large fingers and has an audio plug in case one needs external speakers due to hearing issues.

The interesting thing about the Windows CE 5.0 operating system is that it is considered very open and modification can be done to the kernel level which is different than other operating systems that confine people with application programmer interface (API) calls. This means that devices with Windows CE 5.0 could be very different since customization at the most fundamental level is possible. Any type of expert witness or examiner should be prepared to face this issue in court by a well-prepared defense lawyer. Windows CE 5.0 supports the X86 architecture as well as other types. Many people first encountered the X86 architecture in the late 1980s and early 1990s with the 8086, 80286, and 80386 Intel microprocessors in the IBM PC XT and AT computers. Windows CE 5.0 can also be found on digital cameras, PDAs, point-of-sale cash register systems, and in small Internet devices such as a hub. It can have a footprint as small as 350 KB so it can work on devices with limited space. Developers download an SDK or software development kit from Microsoft and create their own customized embedded systems for a device and then pay licensing fees for each device sold. Microsoft's website provides ample information on the Windows CE 5.0 operating system but it is important to realize that companies that create their own interface for GPS devices can greatly customize the embedded system.

SURVEY OF TOOLS TO INVESTIGATE GPS NAVIGATION DEVICES

There are some popular programs to investigate GPS navigation devices. There is Paraben's Device Seizure software which can also be purchased by itself or with a cable kit. Berla also has Blockthorn 2. Access Data's FTK Imager and Forensic Toolkit are also used by some examiners. One can also use Encase version six to seize the data from the device and look at the data. One could also use the latest version of Helix to clone the drive and then use a program such as TomTology to examine the data. The rest of this chapter will focus on the need to investigate various types of GPS devices, the tools needed to perform the work, and the various sources of education that would assist in this line of work.

NEED TO INVESTIGATE GPS TRACKING DEVICES

There have been devices available on eBay for some time that allow a person to track the location of another elderly person, dog, car, or valuable object. One device is called "Real-Time Spy Mini GSM GPRS GPS Tracker Tracking Device." This type of device needs a cell phone SIM and then one sends it a text message to activate it. The person with the receiver can connect to a PC and see on a map where the device is at any time and where it has been. This type of device can help to locate a person with Alzheimer's disease who may have wandered from home. However, it may be used by a possessive person who is stalking a lover and an investigation of a computer or GPS device may be needed if there is a complaint.

NEED TO INVESTIGATE GPS NAVIGATION DEVICES

It is important to be able to investigate navigation devices such as Magellan, Tom Tom, Garmin, Kenwood, and Mio because they may hold important clues to a crime. Suppose a person is suspected of killing someone and dumping someone in the desert near Las Vegas, Nevada. The suspect's GPS device would be a good place to start with to try to find possible dumping grounds for the body. A skeleton in the desert with a bullet in the skull was one of the scenarios that I participated in investigating at the CSI Experience in the MGM Hotel in Las Vegas, Nevada. The case against the angry ex-wife would have been open and shut if she had a GPS device and it could have been examined for a trip to the desert where the body was found.

EXAMPLE OF PARABEN VERSION 4 AND A GARMIN NUVI 1300

A relatively new handheld electronic device that is often seen in the public domain is called the GPS navigation receiver. This type of device is used in cars, trucks, airplanes, boats, and can also be used by pedestrians. These devices contain maps. The GPS device collects a series of signals from various orbiting satellites and then triangulates them. The result is that the location of the GPS receiver is shown on the map.

Let us consider a real example that I tested. On the Garmin Nuvi 1300 receiver, there is a small onscreen icon that indicates the location and speed of the driver. This GPS device contains a 2 GB drive that uses a FAT 32 file system and appears as a flash drive when it is connected to a Microsoft Windows-based computer. This is known as mass storage mode. A person can use a mini USB cable to connect the Nuvi 1300 to the laptop or desktop computer. That means that pictures, documents, music, videos, and all types of digital media can be easily moved to the GPS receiver from the computer by a process known as "dropping and dragging." A person can also use the GPS device as a digital container.

The Garmin Nuvi 1300 receiver can also hold information concerning a person's trips, favorite destinations, and selected points of interest. If a person pairs his or her cell phone to the device via Bluetooth, then the GPS receiver may hold evidence of phone calls that took place while riding in the vehicle. Many people pair the cell phone and GPS navigation device so that hands-free calling is possible through the GPS navigation device while driving on the highway. Because phones and navigation devices are paired, some call entries on the phone and the GPS navigation device should be the same.

If a terrorist misused such a device, then the amount of potential digital evidence that could be in the device is enormous. Let us now consider the kidnapping of General James Dozier by the Red Brigade in 1981 [15]. If such a situation happened today, a GPS reciever in the suspected kidnapper's vehicle might provide all the locations that the car visited. These numerous locations might hold the key to where the kidnapped person might be held.

Paraben is a company that makes a program called Device Seizure 4.0. This program is very useful for acquiring the data on various types of PDAs, cell phones, and GPS receivers. I was able to use Device Seizure 4.0 to obtain some information on previous trips and favorite places from the Garmin Nuvi 1300. However, the data was in aggregate form and would need to be copied to a spreadsheet and paired with a tool such as Google Maps to make it easy to understand. Since a terrorist group could purchase and possibly use a variety of handheld electronic devices, it would behoove all law enforcment agencies to have one person with such a program to be able to examine any such seized device.

ACCESS DATA FTK IMAGER AND FTK 1.8

FTK Imager is an important tool for capturing an image of a device. Since many GPS navigation devices are small-capacity FAT 16 or FAT 32 storage devices with embedded operating systems, they are relatively fast and simple to image. Tchatchoua Nkwenja Mathias wrote a paper about GPS forensics at the University of Wales in Newport, where he successfully used Access Data's FTK Imager to image a navigation device and collect the data with the FTK forensic tool kit version 1.8 [16]. The aggregate data can easily be put in a spreadsheet and then mapped with Google Maps. Tchatchoua Nkwenja Mathias also uses Google Maps to map selected data points and examine the view at street level with the use of Google Earth.

HELIX 3 AND TomTology AND THE TOM TOM

Helix 3 Pro is available from e-fense for $239.00 a year as on October 15, 2011 [17]. This product is great for imaging the evidence on a hard drive, USB flash drive, PDA, or GPS navigation device. The program starts up and there is a button for acquire. This button should be pressed and then the data will be collected as long as the cable is connected from the energized GPS device to the USB port of the examination machine. Then an image file or series of files is created. The Helix program generates the MD5 hash which is important for forensic methodology and in showing that the device's data was not tampered with. There is also a wizard that prompts the investigator for information and helps generate the chain of custody form. This is very important for establishing a valid audit trail for the evidence so that it will be accepted by the legal community and be usable in court for digital evidence.

Tchatchoua Nkwenja Mathias had success using TomTology with his forensic image to extract the data. TomTology is a program that is made specifically for examining the Tom Tom and costs about £250 as on October 15, 2011 for a perpetual license. The software is made in the United Kingdom and uses the Queen's English as opposed to the usual American English found on most forensic software programs. This program decodes the data from the device into usable information which includes Home Location, Favorites, Recent Destinations, Last Journey Start and End Point, Stored Phonebook, Called Phone Numbers, Received Phone Numbers, Sent SMS Messages, Received SMS Messages, and the Location where TomTom was turned off [18]. The reason that information has to be decoded is that the data looks unintelligible when viewed with a hex viewer. It needs to be put in a readable form. A similar experience is if a person opens up a JPEG picture in MS Word and sees all hex characters.

TomTology is very good because it has a wizard which helps the investigator prepare a report about the evidence. The CFG files in the Tom Tom can often be recovered and trips can be displayed in aggregate form. This data can be manually paired with Google Earth and Google Maps, allowing investigators to see former routes and the home location of the suspect. TomTology can often recover deleted logs of phone calls where the cell phone was paired with the GPS navigation device.

TomTology and Helix 3 Pro have options for paid support which is important for investigators who need reliable researched answers that they can use for reports that will go to court. It is very important to have credible answers that can be trusted for court so that there is no level of doubt about evidence. Support is also important for technical problems with software or if one has a device that is not supported and one wishes to ask them to update the tools so that the device will be supported.

BLACKTHORN 2 GPS FORENSIC SOFTWARE SYSTEM

The Blackthorn 2 GPS Forensic Software and cables can be obtained from the Berla Corporation. The software is very easy to operate. The first step is to contact the company, purchase the software, and then to download the software from the vendor's website. Then, one double clicks on the setup file, which is only 266 MB, to expand it in its own directory. One follows the wizard, agrees to one or more policies, and then it is installed in only a few minutes.

The Blackthorn 2 website states that their product supports Tom Tom, Garmin, and Magellan serial and USB GPS navigation devices. It is possible to acquire the device setting, address books, routes, waypoints, tracklogs, favorite locations, and other data if the device was paired with a cell phone. Cell phones can be paired with GPS navigation devices through a wireless medium known as Bluetooth. Paired devices allow users to receive and initiate phone calls while driving. A paired device has a phone book and call log that can be obtained. Blackthorn 2 also shows the MAC address and the phone number of the paired device, thus establishing proof of the connection. If there was a

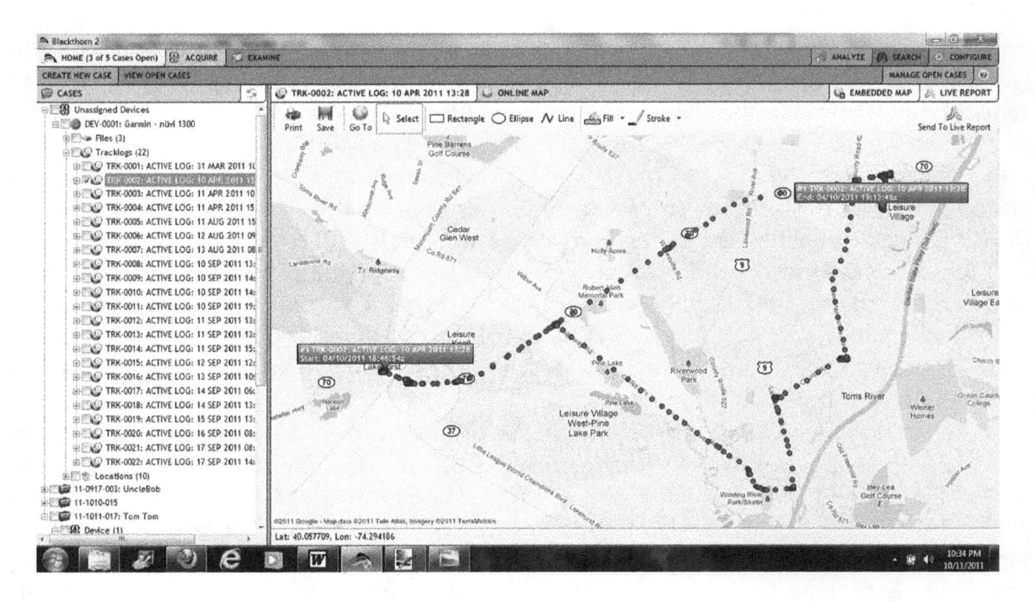

FIGURE 4.1 Blackthorn 2 allows investigators to seize evidence on trips taken.

paired phone, more evidence such as pictures, and stored documents could be obtainable. An example of a short trip within Leisure Village and to the neighboring town of Lakehurst can be seen in Figure 4.1. Each dot on the graph shows a place that the vehicle traveled at a certain time, date, and speed.

One of the features that is very useful to an investigator is the geolocation search. If a person was a suspect for a murder, for example, you could check if the victim's location was in the GPS device that was acquired in the investigation. All acquired data is given an MD5 and SHA1 hash mark to simplify the integrity process of the seized data. It is also very easy to acquire the data from the GPS device. One connects a cable from the USB port on the laptop to the GPS device. The GPS device must be powered on. The Blackthorn 2 program is started and the investigator uses the interface to select an option to acquire the data on the GPS device. When it is done, the information about the device's serial number, model number, and other unique identifying data is visible.

If there were trips taken with the device turned on, it is most likely that trips were obtained as can be seen in Figure 4.1. It is possible to look at trips at a very high level or at a street level. It is also good that individual points marked with both geolocation and vehicle speed can be viewed. The results of the findings can be exported in an XML format. This makes it convenient to share the report on other systems.

PORN DETECTION STICK

One of the things that digital examiners must keep in mind is that the GPS navigation device can act the USB mass storage device. Even though it is a GPS navigation device, it acts as a big flash drive when one connects it to a computer. That means that one can drop and drag child pornography, for example, on the device. There could be potentially thousands of files to sift thorough and Paraben's Porn Detection Stick can enable the digital examiner to quickly check the GPS device. The stick can be used if it is in one USB port on the examination machine and the GPS device is plugged in another USB port.

The device uses algorithms that scan the content for exposed skin and shapes, and give less than 1% false positives [19]. Since possessing child pornography is a violent crime, the use of such a device in some cases may be a quick and simple way to obtain *prima facie* evidence needed for an arrest. The device is often used by law enforcement and has a proven track record. The device is also very useful to probation officers who must check the electronic devices of paroled sex offender clients on a regular basis. Contact with minors or the possession of child pornography will violate their parole and put them back on the express track to jail or prison.

SURVEY OF GPS FORENSIC TOOLS: NETWORKING WITH GPS FORENSIC INVESTIGATORS

It is important to have a working relationship with the community of investigative professionals so that one can receive help when one needs it and quickly solve a case. Perhaps one comes across a Russian GPS device in a case and has no clue what software can examine it. Someone at a security luncheon with ASIS International may be able to point you to the correct vendor for such a tool. Groups such as the American College of Forensic Examiners, the High Tech Crime Investigative Association, ASIS International, and the International Society of Forensic Computer Examiners are good to join. Investigators, professors, security professionals, and some students will share knowledge and discuss security trends at a luncheon. People will discuss hiring, new equipment that helps with digital forensics, and many other topics that cannot be adequately discussed online where people are afraid of something being used against them.

Many of the technical investigators have their own culture, terminology, and humor that one should become familiar with. Many digital evidence examiners have a wide belt with an assortment of pouches, cables, and tools. This is called the "Batman Belt." It is not uncommon to see a group of investigators in a room who just met to aim the face of their watch quickly at all the other investigator's watch faces. There were some sounds that reminded one of a boing from a spring. This was how the group of investigators effortlessly exchanged business cards through the infrared beams on their USB watches.

GPS TOOLS AND EDUCATION

I have attended Homeland Security activities and discussed education with some government employees. Government officials would often like to have specialized classes on topics such as GPS forensics, for example, but cannot pay large amounts of money for classes and cannot fill big classrooms because few people need these skills. Many people at schools are interested in furthering the education of government employees if there is a profit in it. Schools are businesses with high expenses and already provide a lot of community outreach and often cannot do more. A teacher who would like to teach GPS forensics might have to spend close to $5000 to purchase the Blackthorn 2 software and the cables, take a 4-day class, fly to the class, and then stay at a hotel. There is the investment of time and money. Schools want to see a large return on the capital or it is not worth it.

The problem with GPS forensics is that it is a very specialized topic. It is difficult to fill a class for a period of time in one location because those who want the class live far apart and may work at different times. Some detectives work nights while others work day shifts. Some law enforcement people work what is known as rotating shifts. The year 2011 is also a time of fiscal belt tightening worldwide and unfortunately education and training is one area that is often first to be cut. This situation of both finance and logistics makes running such specialized classes unfeasible. What is needed is an online education program for GPS forensics. This would not be difficult to do. Each person could use their own GPS navigation device and their cable. Perhaps they could get a 30-day trial version of a GPS navigation device forensic tool such as Blackthorn 2. They could also register for a class online and join a virtual campus such as Blackboard and Webcampus. The class could also be conducted in Moodle.

The class could be done with PowerPoint slides and discussion boards. A set of slides could talk about the types of GPS systems in the world, the history of GPS, and discuss a range of GPS navigation devices on the market. Then, students could meet and greet each other in the discussion board and network. This is important so that they know other investigators of GPS navigation devices and can ask each other questions about technical issues or tools. Another set of slides could also show how to run the software, collect the data, examine it, and create a report. The students could also ask the teacher questions online and perhaps take a test.

Lastly, the students could use the free 30-day version of the software and seize the data of their own GPS navigation device. Then, they could make a report. This could be uploaded online. If the online professor felt that the report was satisfactory and that the test was passing, a certificate could be made available for downloading. This could be a credential that might be used for showing competence in the subject in case one has to go to court. Payment for the class could be done by credit card or PayPal. Registration could be done online in a protected website or by telephone. This type of online class is also good for the military or defense contractors who may be working in remote places such as Afghanistan, Somalia, or Iraq. The overhead could be kept low since it is all online.

If credentialing requires verification of the person who is taking the test and training, a video conference component could be added. Students could be asked to set up a certain quality webcam with a minimum resolution requirement and then provide an identification card. The card could be a driver's license, passport, or government issued identification card. The class could also be set up to include a certain body of knowledge that would be acceptable to a variety of countries around the world.

OUTFITTING A LAB WITH GPS FORENSIC TOOLS

If one wishes to set up a lab for teaching law enforcement at a university, then one might use an existing computer lab and purchase 15 copies of TomTology to examine the data. The cost would be 15 packages at £275 multiplied by 1.58 equals cost in U.S. dollars ($6517.50). Perhaps a local company may write a grant to help or a grant for such training may be available on the National Register from a group such as the National Register. The imaging tool could even be Helix 3 Pro but at about $250 each with 1 year of support. Multiply $250 by 15 to get the total ($3750). Used Tom Tom GPS navigation devices could be obtained for about $50 each with shipping. The total would be $750 for the GPS devices and then add $50 for the USB cables ($800). The registry could be changed on each machine to be read-only, so labs would not have to purchase write blockers such as Fast Block. The total cost to outfit an existing lab of 15 computers for GPS forensics would be $11,067.50. This cost might be difficult to justify unless a large number of classes can be run through to recover the cost or if a grant is obtained. Sometimes the National Institute of Justice gives grants for such specialized training.

A less expensive option could include be getting three copies of Blackthorn 2 and three GPS navigation devices and run teams of five people through at a time. The roles of the students would be as follows: one is the examiner; one fills out the chain of custody form, another takes notes, and two people critique what is happening and compare it to a sample incident response policy. Then students switch positions and run the class again. Each person takes a turn at each role thus learning what is required of each position on the incident response teams and thus learning by repetition.

GPS DEVICES, TRAINING LABS, AND PRIVACY ISSUES

If one is running an academic lab for a high school or a college, privacy concerns with digital devices are an issue. A student may think that he or she is using a GPS device privately because there is no usernames or passwords. Many journal articles in *Cyberpsychology* and *Behavior* say that people will act differently when they believe that they are anonymous when using various digital devices, including computers. Students may operate the device by putting in a few known addresses or places such as their home address, some relatives, or a friend's house. They may also input addresses of other places of interest such as a pregnancy clinic or community center that could indicate an interest in an alternative lifestyle. The student may just want to research the quickest route to go to one of these places after school to speak to a counselor. The student may not be ready to discuss any of these issues with peers at the present time. It may be advisable to wipe the devices after each class or load them with a standard image so that the previous activities were not saved. If one does not perform such due diligence, later classes could do an examination and deduce by home locations, date and times, and other indicators such as friends that a certain person was using the device. Certain locations could show an interest in a certain lifestyle or suggest a certain condition such as pregnancy or indicate a trip to a place to terminate a pregnancy.

Suppose a certain student gets the digital device and figures out that a certain person is pregnant, perhaps going to place that is known to terminate a pregnancy, or something else and then reveals it. It could lead to verbal disclosure and perhaps some type of bullying. This could lead to a lawsuit that could potentially include the teacher, the lab director, and the principal. It therefore should be a suggested practice to review the devices after usage and restore the device to a common image with practice data.

It is important for administrators and school teachers to remember that the digital devices in the classroom are still usable for connecting to GPS satellites and researching trips. The mobile device for use as a forensic test subject in the classroom still has an active life as it is used by students. There are not enough staff and faculty to supervise the use of each device while instruction takes place. Therefore, a wiping of data after each use or restoring it to a common image may be the best option for both privacy and the possibility of eliminating malware. Students may also be asked to sign a nondisclosure agreement (NDA) about anything found on the device. It may also be necessary to have a user agreement policy that states that each student will use a tool to remove any trails of activity that were performed on that device that day.

While we cannot anticipate every abuse of investigative software tools and mobile devices, it might be good to brainstorm with students and let them tell you what possibility of mischief and political incorrectness could be done. This would be a good proactive initiative that would please the risk manager at any school or large organization. Perhaps it would be good to bring the risk manager to a class and see what concerns that he or she has voices. The time to voice concerns is before an incident occurs.

LEARNING ABOUT LOW-BUDGET TOOLS FOR GPS FORENSICS

Sometimes students who are graduating college and looking to work for a private investigator or lawyer will ask where they can get low-cost or no-cost tools to practice doing GPS forensics before seeking a job. GPSForensics.org maintains a website that lists the freeware tools as well as the tools that cost money and come with technical support [20]. The GPSForensics.org website is a good clearinghouse of information with regard to digital forensics [20]. Since this specialty of digital forensics is growing so quickly and is ever changing, it is important to know of a place to go to seek this specialty information. The tools that cost money are Blackthorn 2, Paraben's Device Seizure, and TomTology. The free tools are Map Send Lite, Easy GPS, and GPS Utility [21]. This is known as freeware and not shareware which means it is really free. When using any of these tools, it may also be good to use a USB hardware blocker such as the one from Tableau or change the registry to make the port read-only so evidence cannot be changed.

MAP SEND LITE

It is great to have a program such as Map Send Lite that gives the examiner the freedom to see data from the Magellan navigation devices and then convert it to another format. The program does the regular GPS forensic tasks such as allowing one to see routes, waypoints, and alternative views of the data. The good part of the program is that it is Magellan specific and if your Magellan GPS device does not seem to work well with other tools, this one might be the best match [22].

EASY GPS

The Easy GPS program can be obtained at the URL http://www.easygps.com/ and it runs on the Microsoft platforms of Windows 7, XP, and Vista. On November 21, 2011, version 4.35 of Easy GPS was available. The program is reported to work with a wide variety of GPS receivers for the car or boat. Some of the GPS devices are chart plotters and others are fishfinders. The Garmin Nuvi, eTrex, Dakota, Colorado, Megellan, Oregon, SporTrak, and Rino are just some of the brands you can use with this program. The program is available for download on the website. The software makers say, "EasyGPS is the fast and easy way to upload and download waypoints, routes, and tracks between your Windows computer and your Garmin, Magellan, or Lowrance GPS. EasyGPS lists all of your waypoints on the left side of the screen, and shows a plot of your GPS data on the right. Use EasyGPS to back up and organize your GPS data, print maps, or load new waypoints onto your GPS for your next hike or geocaching adventure" [21]. EasyGPS also makes a paid version that includes topographic maps, areal photos, and works with Google Earth. This would seem to be the tool for someone who will be doing this seriously as an investigator after learning the basics with the free tool.

EasyGPS does not presently run on Linux or Windows 2000. It is important to read the system specifications before downloading the software to make sure one has an adequate processor, hard drive space, and the correct operating system to utilize this. It would be advisable to put all the freeware programs on an intranet so all the examiners could run the program and then share cases on a storage area network within the confines of the lab. Since it is freeware, there are no licensing issues.

GPS UTILITY

This program known as GPS Utility is free but requires more knowledge about maps and perhaps the basics of GIS. Their website says, "The program converts between different map datums and many coordinate formats (Lat/Long, UTM/UPS, country grids, etc.). Information can be filtered and waypoints sorted according to specified criteria. Route and track statistics are available and can be transferred into other programs for analysis (i.e., spreadsheet programs)" [21]. The feature that many investigators find useful is that one can take data from one device's format and change it to another. This is very important if you have a GIS expert in your organization who needs to put data in a particular format for their program and make maps showing routes and waypoints for you. One of the frequent problems encountered in computer forensics is to take documents from one obscure format such as PFS Write and change it to something more modern such as Microsoft Word. Quick View Plus is a software program that can be used to view obscure data formats.

GPS BABEL

Another tool that may be downloaded and used on a variety of systems is GPS Babel. This program allows one to take all the waypoint, routes, and tracks data extracted from a GPS navigation tool and change it to one of many formats. This may be necessary if you have mapping tools and your expert needs the data in a certain format. The program also works on a variety of operating systems that include Microsoft XP, Vista, Windows 7, POSIX, Mac OS/X, Linux, Solaris, FreeBSD, and OpenBSD [23]. This is important because some academic labs also like to use older computers to keep costs down and many people feel that Linux systems perform well than Windows on older hardware.

It is important that we have a common set of definitions for speaking about GPS forensics because the same word may have a different meaning somewhere else. *Waypoint* may mean different things in different disciplines but in GPS forensics, it means a place in time and space with latitude and longitude. This is true for boats, cars, and those walking. However, since airplanes now have GPS navigation devices, waypoints can also include altitude in addition to latitude and longitude. Waypoints may also be associated with a buoy on the water or a GPS location that is marked as a point of interest. This may be an important consideration in a smuggling case or as a dead drop site or prop point in an espionage case.

Another term that often confuses people is route. People often say that a route is a state highway that is an odd number if the highway runs north–south and even number if it runs east–west. While this is true, a *route* in GPS forensics is considered to be two or more waypoints on the same path. In geometry, we studied a line consisting of at least two points. If we studied wildlife, tracks show where an animal walked or ran. *Tracks* in GPS forensics means a recorded trip in laymen's terms. If one is researching a case with treasure hunting at sea, one may encounter GPS navigations which mark *vertical datums*. These are locations that have a latitude, a longitude, and a vertical depth. GPS Babel could be an important tool for changing a vertical datum to a format that your mapping tool understands.

QuakeMap

QuakeMap is a program that shows a world map and allows you to see where the latest earthquakes were. It can also be used with a PDA and allows hikers to put in their present GPS coordinates and then choose waypoints and create a route. This route can be exported to a PDA from the desktop or laptop. When a person goes hiking, he or she may use the program as a type of interactive map to help him or her get to the destination.

If there is an exigent situation such as in the case of a missing hiker, perhaps the police or family can get the computer of the hiker to a digital forensic examiner. The examiner could then print out the route and this could be used as a guide to find the missing hiker. Many airplanes have to file a flight plan before flying. This includes a takeoff time map of where he or she will fly to with exact paths, and then a landing time. QuakeMap can be used as a type of flight plan for hikers.

DSI USB WRITE BLOCKER

Document Solutions Incorporated has created a program called USB Write Blocker which allows digital forensic examiners to make the USB port read-only. It puts an icon on the lower icon tray that says it is active and that the port is in a read-only state. That is important so that if an investigator connects a GPS navigation device, none of the time stamps or metadata of files will be changed. The device works with Windows 7, XP, and Vista. This program is a much safer alternative to altering the registry. It is also cheaper than purchasing a Tableau USB write blocker.

VIRTUAL GPS 1.39: GPS SIMULATOR

Zyl Soft has a software product called Virtual GPS 1.39. It is a GPS simulator that can be run on the desktop or laptop computer and emulates a connected GPS receiver. Sometimes it is necessary for an investigator to conduct an experiment and recreate a situation to try to determine if a piece of equipment failed or how a spoofed signal transmitted by freight thieves would have been impacted by a GPS receiver. For such high-level research, a product such as this may be necessary. Virtual machine (VM) environments are becoming the mainstream for investigating malware. VMs are becoming more popular for investigation because there is almost no possibility of impacting the research computer and the VM provides a safe computer lab environment to test software.

DISCUSSION WITH A MOBILE DEVICE AND COMPUTER FORENSIC INVESTIGATOR

I spoke to a digital forensics investigator who was recently downsized after being in a corporation for approximately 9 years. The person did not want to be identified, so we will refer to him hereafter as Mr X, and some of his insights were helpful because of his range of investigations at corporate sites in Latin America, the United States, and the Caribbean. Mr. X thought that being a CCE, ACE, CISSP, and having Encase Certification were good for mobile device forensics as well as computer forensics.

He also said it is important to learn what cell phones, computers, GPS devices, and PDAs your corporation or agency gets and then collect the manuals and technical specifications on them and file them. Find out a number that you can call for technical support about any of these devices, the programs they use, and the operating systems they use. It is also important to find out where you can get extra batteries, parts, and cables for these devices so if there is ever a problem, you can power up the devices and seize the data.

Mr. X also told me that it is important to find out what court-recognized digital forensic tools can be used to investigate the devices that the company supports. At some point, there will need to be an investigation of them so it is better to find out all the available tools for those devices and budget for them. No tool does everything perfectly and it is also good to use multiple tools to validate the results looking for exculpatory data and prosecuting data. Mr. X also said to look into what training and certifications are needed to use these tools that support the digital devices used in the corporation or agency. He also said that a sign of downsizing for digital investigators may occur when the corporation does not want to pay to keep your certifications or tools current.

Mr. X also said to keep your certifications and resume up to date because many corporations put limited resources into investigation and their tools. Many corporations prefer to keep crime quiet and customer confidence up and will write off losses as business expenses rather than investigate policy violations and other events. Mr. X also said to take a file systems course and learn about the file systems of all the devices that your organization keeps actively used by employees.

Lastly, Mr. X said to have a copy of all the policies that are active in your organization and have the phone number of HR and the general counsel in case there are questions. Never go outside your authority and investigate a device that you do not have permission to seize the data from. The lessons that can be learned from speaking with digital evidence investigators at conferences, symposiums, and workshops are invaluable since they can give you the benefit of their years of experience.

EXAMINING A TOM TOM GO 700

Let us look at how a digital evidence examiner might examine a Tom Tom Go 700. He or she may first make sure that his or her signature was put on the chain of evidence form for the reason of examining the device. Then the examiner would accept all cables, documentation, power supplies, batteries, the Tom Tom Go 700, and any media cards. Hopefully, the device had pictures of the state in which it was found, including all cables plugged in. Those cables should be color coded and marked so the scene can be recreated in the lab If the device was on, the screen should have been photographed so that any evidence about the last activity done could be seen. Care should be taken when bringing the device to the lab because dampness, heat, cold, or large magnetic sources such as a strong radio transmitter can damage the evidence. Some police cars may have a hot or cold trunk in summer or winter and the police radio may be kept in the trunk. The radio is a source of radio frequency waves which might affect evidence.

The devices would be kept locked up and in a Faraday bag if necessary because the Tom Tom Go 700 is known to have Bluetooth capability. The examiner would also get the supporting documents and manuals for the device online and write down some notes such as the RAM being 64 MB and the storage device being 2.5 GB. The examiner would also note the serial number on the device and chain of evidence form. If the device was turned off, collecting the volatile data in RAM would no longer be a priority. It is commonly known that investigators seek to recover the most volatile data first and the on/off rule is that if it is found on, leave it on. Some GPS devices are known to have a lockout feature if they are powered down and access later may not be possible.

According to the Best Practices for Seizing Electronic Evidence by Price Waterhouse, the GPS device may contain stored data of text, images, and maps [24]. It may also contain Internet access information and information about received calls since many devices are paired with cell phones [24]. The device may also have current or recently deleted waypoints, routes, and timelines. The examiner would not want to just push buttons and manually go through the screens because it would change many access times on files and create temp files and literally cause a lot of evidence to change.

The examiner may elect to image the device first with a tool such as Encase. The examiner could also choose to make sure that the USB port had a write blocker connected to it and then used a tool such as OSFclone and OSForensics to create the image and examine the files. The examiner may also have chosen to use Blackthorn 2 from Berla to quickly and easily see the phone logs, routes, and waypoints in an easy to understand map form. The examiner may have elected to image the device with Paraben's Device Seizure which is reported to have a software write blocker and not change evidence. Device Seizure can collect aggregate data with waypoints and routes. The aggregate data can be examined with other tools later. Getting the image is crucial.

BEST PRACTICES FOR SEIZING ELECTRONIC EVIDENCE AND RADIO SIGNALS

The best practices literature referred to on the previous page also suggest looking for evidence from two way radio signals. We already mentioned that the GPS may have been paired with a cell phone and call log information may exist but there is also a TMS on many devices [24]. The TMS is an acronym for the traffic message channel. This channel is encoded frequency modulation (FM) signals which provide the device with local traffic information. This type of signaling also tells the GPS device about an accident, a closed highway, or some type of emergency and then other routing information is given.

If a strong false TMS message was transmitted by cargo thieves, the GPS device may give a truck driver wrong information such as a highway is closed ahead, and take this deserted desert road. That isolated area may be the location where freight thieves ambush the truck and then steal the contents. If the GPS device is found later, there may be information in RAM or unallocated space about the TMS signal that could link the thieves' radio equipment to the GPS device. Information about false signal programming could also be present in the home computers and terminal node controllers of the thieves.

If the thieves had Bluetooth devices active with the identification feature on, there may be a log of the thieves' devices linking with the Bluetooth. Other devices in the car such as EZPass toll transponders may also have data that shows the route.

GPS FORENSIC CREDENTIALS

If a person was going to do GPS forensics, the quickest way to get started in my opinion is to purchase Blackthorn 2 and take a 4-day class with it. In this manner you have a tool, practice with the tool on real systems, and a credential. Then, one can add knowledge and credentials as opportunity permits. A class on file systems would be useful since there are so many types of files that the GPS navigation devices can utilize for displaying or speaking aloud. A class on operating systems would be useful, so that one understands the features of the Windows CE 5.0 operating system and how files are added as well as deleted.

If time permits, one should purchase Paraben's Device Seizure which allows the digital examiner to examine a variety of mobile devices, including some GPS navigation devices. Paraben also has a Device Seizure class that one can take to quickly understand the file systems of mobile devices and learn about mobile device forensics. One of the important considerations is that there is testing and a credential. This is important for those who have to testify in court and whose work will be scrutinized. Often, Paraben offers a package that includes a discounted price for training, the software, and the cable kits.

It would also be a very good idea to get the CISSP from ISC squared. This credential is often sought after by digital examiners and network security personnel. Many law firms seek digital evidence examiners who hold a CISSP and Encase certification from Guidance Software. Encase certification is good because someone with a GPS navigation device could have deleted files or hid them on it since it acts like a mass storage device at times. It would also be a good idea to keep track of books that one reads on the subject of GPS device investigation and Encase. When one goes to court, the defense lawyer may ask about the 10 most popular books on the subject. An expert may be doubted if he or she is not familiar with the literature for such investigation and devices.

It would also be advisable to get a vendor-neutral certification such as the CCE, certified computer examiner so that one could be credible examining a wide variety of devices. The CCE is also very good because it stresses fundamentals such as chain of custody, file systems, and reporting. They also mention practical concepts such as using a Faraday bag to block connectivity. This is important because GPS navigation devices have Bluetooth and wireless. Certain models may also have infrared connectivity.

GEOGRAPHIC INFORMATION SYSTEMS TRAINING

If one is going to perform many investigations on GPS navigation devices, it might behoove one to take a class on GIS systems. Such a class would explain the types of maps that exist and give the examiner the confidence to explain various contours and legends on maps. It would also allow them to answer confidently and with integrity to a lawyer's line of questioning about a map with points that is presented as evidence to a jury. A lawyer could easily ask difficult questions on magnetic north versus true north and how that is represented on the map. Other questions about how GPS points may or may not be accurate on the map could also be addressed. If the examiner presents data and has to answer "I do not know" too many times, it could impact his or her credibility as well as the case. It is better to be overeducated and be confident rather than to cave in from a line of difficult questioning. ESRI, the makers of ArcGIS, offer a variety of classes on map making and reading.

ESRI also teaches the use of their software products for creating maps. This might be useful if the examiner has a data set of perhaps one hundred or more data points and needs to show it on a map for the jury. There is a whole set of complex vocabulary with words such as "Trans Mercator Line" that regard to maps that would be useful to know if he or she is to present in court. The ESRI GIS classes would also be useful for the examiner who must examine cell phones and explain where a person was at a particular time. There is a data set of the location of the active cell phone towers in the United States that can be downloaded and mapped in ArcGIS. Sometimes, it is necessary to show the location of a tower on a map and explain how a signal is triangulated and a person can be put within a general location on the map within a certain time frame. It is also good to know about the geometry of the triangulation and be able to discuss the accuracy and inaccuracy of the alleged location of the caller.

MOBILE FORENSICS CERTIFIED EXAMINER

The mobile forensic certified examiner (MFCE) is a good certification for any examiner who must collect, preserve, analyze, and report on evidence found on mobile devices. The certification consists of a series of projects and a final exam. The certification requires one to have proficiency in the forensic tools, methodology, and reporting of evidence on mobile devices. Mobile devices could include a wide array of digital devices such as cell phones, smartphones, PDAs, organizers, and GPS navigation devices. The wider certification of examining mobile devices may be easier to justify to the management in 2011 than just a GPS forensic certification. More information can be obtained about the MFCE from their website [25].

GRADUATE CERTIFICATE IN DIGITAL FORENSICS: RWU

The Roger Williams University offers a graduate certificate in digital forensics that can be taken as part of a master's degree or for its own sake. However, one of the courses in this program focuses on a range of mobile devices which could include GPS devices. This course has a title of "CJS 544 Forensics III Mobile Device Forensics." The main point is that a series of classes test and demonstrate the proficiency of the student in learning about preserving, collecting, analyzing, and reporting on evidence from an array of mobile devices. The classes teach the student in the use of digital forensic tools and techniques for mobile devices that are accepted by one's peers in the legal and investigative communities. This is important because the investigative techniques used on GPS devices must pass the Frye test in order for evidence to be accepted for use in the U.S. Courts.

MASTERS IN DIGITAL FORENSICS AT THE UNIVERSITY OF CENTRAL FLORIDA

The University of Central Florida has a masters in digital forensics program that has received positive feedback from many people who are CCE. The 30-credit degree program has course offerings from many departments. This is good because it draws upon the strengths of each department which is important for a discipline such as digital forensics which bridges computer science, engineering, legal studies, and criminal justice. "The

MSDF degree is a collaborative effort between various UCF academic departments—Electrical Engineering and Computer Science, Forensic Science of Chemistry, Criminal Justice and Legal Studies—and the National Center for Forensic Science" [26]. The university website also states, "The National Center for Forensic Science is a State of Florida Type II Center and a member of the National Institute of Justice Forensic Resource Network of the Department of Justice, serving the needs of state and local law enforcement and forensic scientists" [26]. This is very important for demonstrating a working relationship with local law enforcement who not only bring their expertise to the classroom but also further their theoretical knowledge from the university faculty.

This type of degree is important because it also emphasizes the legal aspects, the technical aspects, and the presentation of evidence aspect in the courtroom. A person has to have the technical skills to understand how the evidence evolved on the device in question, but an examiner also needs to demonstrate to the court how this evidence was not tampered with and supports or nullifies the charges against the defendant. The degree is about all aspects of digital forensics and not just GPS devices.

CERTIFIED ELECTRONIC EVIDENCE COLLECTION SPECIALIST CERTIFICATION

It is very difficult to get GPS forensic credentials because there are so few that seem to exist. However, the GPS navigation devices could be considered small embedded systems, a type of computing device, and thus any electronic collection certification could arguably be of significance. The International Association Computer Investigative Specialists (IACIS) offers a certification that is very important to anyone who must collect digital evidence from a mobile device, computer, network, or networked media device. It is critical to learn about the proper methodology to collect digital evidence. If one does not understand the correct methodology and reasons for doing this, then the process can become tainted and the whole investigation is compromised. Many books, security certification boards, and private investigators state that the most important component of computer forensics is getting the forensic image and hash of the evidence. Then the chain of custody form has to be updated and the evidence be secured from wireless or physical tampering. The evidence paperwork should be filled out properly and special precautions for protecting the evidence from magnetism, fire, or water must be taken.

This Certified Electronic Evidence Collection Specialist Certification (CEECS) is conferred upon those individuals who successfully finish the CEECS course and pass the examination. Students who complete this course also learn to tag evidence properly and then place it in an evidence bag. The bag is also marked with evidence tape and signed and dated. This process is often referred to as "Bag and Tag."

SANS: MOBILE DEVICE FORENSICS (FORENSICS 563)—CERTIFICATE OF ATTENDANCE

A person could request a certificate attendance after participating in the 5-day class knows as Forensics 563 or Mobile Device Forensics from the SANS Institute [27]. The first 4 days focus on cell phones, SIM cards, and the use as well as limitations for mobile device forensic tools [27]. The fifth day offers practical training in GPS forensics with creative scenarios using both the Garmin and Tom Tom devices [27]. Certificates of attendance are good, especially if one keeps a copy of the syllabus and the topics taught in the course. If there were quizzes or tests given in the course, then any good grades in tandem with a syllabus might be usable in stabling a competency in a certain area such as GPS forensics and cell phone forensics.

SUGGESTED FURTHER READING AND REFERENCE SOURCES FOR GPS

There is a lot of research and collaboration that goes on with regard to digital forensics. Investigators will often post a question on a CCE listserver and academics or practitioners may volunteer an answer or a reference to an answer in a journal or book. It then becomes very important for examiners to be able to read books and reference them in a hurry. This is important so that the examiner knows some fact about the operating system or its structure and can complete the investigation and write a report. Getting books on technical subjects has become increasingly difficult as many small bookstores have closed. Large bookstores such as Borders were once great sources to quickly get a variety of technical books, but they too have gone out of business. Barnes and Noble bookstores have greatly reduced their computer and mobile device book holdings over the years and adapted their inventory to a changing market.

Many libraries have reduced budgets and can no longer purchase or house the ever-changing technology books that are written on operating systems, networks, digital forensics, and mobile devices. Some digital device examiners have complained that books are often outdated as soon as they are written. Therefore, it would be considered a bad business decision for bookstores to purchase and resell many technology books. Therefore, many digital device examiners have resorted to purchasing a multitude of books on mobile devices, operating systems, and investigation tools that they may use to investigate cell phones with GPS or GPS navigation devices. The books must be shelved since they are often used. Another problem examiners face is that they cannot discard the old books because either their colleagues or they themselves often need to examine older digital devices.

As time goes by, the digital examiners work area of books and tools grows exponentially and not in a linear fashion due to the plethora of new devices that are released each year. Many digital device examiners have also received complaints from their bosses that the number of books, journal, and training videos needed for reference is unpredictable. The unpredictability has a disadvantage in that it makes budgeting for books and other media difficult and planning for storage equally difficult. Many examiners have tried to address the problem of storage, predictable costs, and instant access by joining online book services such as Safari Books Online which at the time of this writing have 17,000 books and media.

Some digital examiners have said that Safari Books Online is good because they do not have to wait a week for a book and can download it. Many online booksellers have overnight shipping on books while others have to locate it from an associated reseller who may take a week to get it to the buyer. The online books allow the examiner to download it immediately and use a search engine to find what they need quickly. The increased access to knowledge can often result in an investigation of a mobile device to be done sooner rather than later and with less cost of billing hours to the client.

It seems from anecdotes expressed online or in person that online book services with large holdings of technical books may solve the problem of ever-growing bookshelf demands, runaway book costs, and the need for instant knowledge to speed the throughput of casework. Online book services such as Safari Books Online are increasing as well as online holdings of libraries such as Fairleigh Dickinson University in New Jersey or DePaul University in Chicago.

REFERENCES

1. *Satnav Forensics*. URL accessed November 8, 2011. Retrieved from http://www.satnavforensics.com/index.php.
2. *Wisegeek*. What is house arrest. URL accessed November 8, 2011. Retrieved from http://www.wisegeek.com/what-is-house-arrest.htm.
3. *Astronautix*. Encyclopedia Astronautica, Glonass. URL accessed October 10, 2011. Retrieved from http://www.astronautix.com/craft/glonass.htm.
4. *Rianovosti* (December 29, 2008). Russia to set world record with 39 launches in 2009. URL accessed October 10, 2011. Retrieved from http://en.rian.ru/russia/20081229/119210306.html.
5. Barbosa, R. C. (October 31, 2010). Chinese Long March 3C launches with BeiDou-2. *NASAspaceflight.com*. URL accessed October 10, 2011. Retrieved from http://www.nasaspaceflight.com/2010/10/long-march-3c-launches-beidou-2/.
6. Newton, H. (2002). *Newton's Telecom Dictionary*, p. 331. CMP Books, New York, NY, ISBN 1-57820-104-7.
7. Ward99 posted by Timothy on March 19, 2003. US may reduce non-military GPS accuracy. *Slashdot*. URL accessed October 16, 2011. Retrieved from http://news.slashdot.org/story/03/03/19/0340247/US-May-Reduce-Non-Military-GPS-Accuracy.
8. Statement from FAA. (February 26, 2006). WAAS and Its Relation to Enabled Hand-Held GPS Receivers. URL accessed February 27, 2012. Retrieved from http://gpsinformation.net/exe/waas.html.
9. Supreme Court of the US Syllabus (October term, 2011). Syllabus United States v. Jones Certiorari to the United States Court of appeals for the District of Columbia Circuit No. 10-1259. Argued November 8, 2011—Decided January 23, 2012. URL accessed February 27, 2012. Retrieved from http://www.supremecourt.gov/opinions/11pdf/10-1259.pdf.
10. Trei, M. (September 20, 2010). Braille GPS helps the blind to navigate in unfamiliar places. *Dvice*. URL accessed January 15, 2012. Retrieved from http://dvice.com/archives/2010/09/braille-gps-hel.php.
11. *Zazona.com* (April 1, 2003). Army pays Russians, Russians sell to Iraqis. URL accessed February 27, 2012. Retrieved from http://www.zazona.com/NewsArchive/2003-04-01%20Army%20pays%20Russians%20Russians%20sell%20to%20Iraqis.htm.
12. *North Korea Tech*. Report: Stronger GPS jammer developed. URL accessed October 8, 2011. Retrieved from http://www.northkoreatech.org/2011/09/09/report-stronger-gps-jammer-developed/.
13. Cereijo, M. (nd). Cuba and Information Warfare (IW). *Amigospais-guaracabuya.org*. URL accessed January 15, 2012. Retrieved from http://www.amigospais-guaracabuya.org/oagmc207.php.
14. Schneier, B. (September 17, 2008). GPS spoofing. *Schneier.com (Web blog)*. URL accessed February 27, 2012. Retrieved from http://www.schneier.com/blog/archives/2008/09/gps_spoofing.html.

15. CBS Evening news (December 18, 1981). Headline: Dozier kidnapping. *Vanderbilt Television.* URL accessed on December 27, 2010. Retrieved from http://tvnews.vanderbilt.edu/program.pl?ID=274898.
16. Mathias, T. N. (April 30, 2009). The forensic examination of embedded device such global position system (GPS). BSc (Hons) Forensic computing thesis for the University of Wales, Newport. Retrieved from http://ril.newport.ac.uk/Mathias/finalYrProGPS-ForensicsExamination.pdf.
17. *Efense.* Helix3Pro. URL accessed October 15, 2011. Retrieved from http://www.e-fense.com/helix3pro.php.
18. *Forensic Navigation.* URL accessed October 15, 2011. Retrieved from http://www.forensicnavigation.com/#/products/4527490520.
19. *Paraben sticks.* URL accessed October 11, 2011. Retrieved from http://www.paraben-sticks.com/porn-detection-stick.html.
20. *GPS Forensics.org.* URL accessed November 23, 2011. Retrieved from http://www.gpsforensics.org/techniques_freeware.html.
21. *Easy GPS.* URL accessed November 23, 2011. Retrieved from http://www.easygps.com/.
22. *Magellan.* URL accessed February 27, 2012. Retrieved from http://www.magellangps.com/Maps?redirect_count=1&did_javascript_redirect= T.
23. *GPS Babel.* URL accessed November 24, 2011. Retrieved from http://www.gpsbabel.org/.
24. Best Practices for Seizing Electronic Evidence version 2.0 from Price Waterhouse Coopers, The document was created as a project of the International Association of Chiefs of Police Advisory Committee for Police Investigative Operations and Price Waterhouse Coopers LLP, and the US Secret Service.
25. MFCE. URL accessed November 9, 2011. Retrieved from http://www.mfce.us/contact.html.
26. *MSDF UCF.* URL accessed November 9, 2011. Retrieved from http://msdf.ucf.edu/general.html.
27. *SANS.* Security training mobile device forensics. URL accessed November 8, 2011. Retrieved from http://www.sans.org/security-training/mobile-device-forensics-4896-tid.

CHAPTER 5

Corporate Investigations on a Netbook

AUTHORIZED REQUESTOR LEADS INVESTIGATIONS

In a corporation, there can be many departments and people engaging in healthy competition. However, sometimes people may try to eliminate the competition for a raise, bonus, and perhaps put themselves in a better place in case of future layoffs. That is why there is an impartial person known as the authorized requestor (AR) who is in charge of investigations. The AR may be the chief intelligence officer or chief security officer [1], who dispatches the incident response team (IRT) and makes sure that the team follows the rules known as the incident response policy (IRP). The AR interacts with the human resources (HR) department to make sure that the employee, who is the suspect, had signed all relevant policies, and that everything was filed properly. The AR also interacts with HR and general counsel to make sure that the incident response team has all the guidance that they need to conduct the investigation properly.

The AR has to have a committee meeting with people from information technology, human resources, the general counsel, and key people from management. They need to create many policies and decide how to govern their organization's technology and make it work for them in an organization. Policies set the rules of behavior and sanctions need to be discussed for noncompliance. Policies help organizations manage the use of technology to be productive and not be a headache that people misuse to cause the organization endless legal problems. Investigations on netbooks or any other devices occur when policies are not followed and there is an allegation of misuse. Many policy violations also quickly become criminal investigations when it is found that the law is broken. This chapter is about corporate investigations but it is impossible not to talk about criminal investigations because what starts out as shopping on eBay on company time may lead to the discovery of someone misappropriating property or surfing child pornography. That is where the investigation stops; everything is packed up for the police, and turned over to them when they arrive.

I have spoken to many students over the last 10 years as a university professor and was surprised to find out that many students who work full time have workplaces with a lack of policies and a common language to discuss the policies. Therefore, an AR trying to create Internet usage policies, telephone policies, laptop/notebook policies, incident response policies, and mobile device policies may first have to start off a committee with the following dialogue to create a common vocabulary of technology and then get everyone to create the necessary policies.

The AR may say that it is first necessary to define a netbook before one can create policies for one or investigate the misuse of one. Many people have different ideas about what netbooks are and it is easy to confuse these definitions with other small computing devices such as the handtop, subnotebook, palmtop, and laptop. It is also important to define these devices because it will be necessary to develop policies in the workplace to govern their use on the networks in the organizational environment. It is also important to realize that many of these devices are inexpensive and often purchased by employees and then brought to work. Can employees bring their own netbook to work, download data and forms, and then do work at home? Who is responsible for that data outside the office on that notebook? What if their home network is not secure and they get rootkits, spyware, keyloggers, and Trojan Horses and then connect to the network at work? Did they cause an incident? Can they be liable? Policies should address all these questions. Then, employees need to read these policies, understand them, and then sign the policies.

Throughout this chapter, policies will be discussed, equipment as well as tools will be discussed, and various investigative situations will be discussed. This chapter will include practical knowledge, theoretical knowledge, and anecdotes that are relayed to make one think about issues that are rarely discussed because they may be uncomfortable. Then the reader may think of how he or she may address this as a practitioner, educator, or manager in a law enforcement organization or corporation.

AR ASKS, "WHAT IS A NETBOOK? HANDTOP? SUBNOTEBOOK? PALMTOP?"

Suppose you were asked in a deposition, "What is a netbook?" Could you answer it? Could you distinguish it from the other small computing devices in the workplace? If you were in court, would such a question cause you to lose credibility or could you convince the judge and jury that you know what you are talking about. There are many types of small portable computers in the marketplace that have different names and may confuse people. Some people have what is commonly known as a palmtop. It is better that you know the different between a netbook, handtop, subnotebook, and palmtop. If you wish to learn more about these small computers, Stephen Bigelow has a very nice book that not only describes many of the various small computers but also tells you how to repair them [2]. Another book about palmtops was written by Ed Keefe [3]. Nelson, Phillips, Enfinger, and Stueart also discuss the various types of computing hardware in their books and give good working definitions [1]. They are worth visiting and have many footnotes and further sources.

HR, GENERAL COUNSEL, CIO

The HR people, the general counsel, and chief information officer (CIO) often have to get together to bring in law enforcement when something goes amiss. Each person has to supply the other with the knowledge and support in an investigation so that nothing goes wrong such that the suspect turns around and sues the organization as well as individual people. The HR, general counsel, information technology department, the incident response team, some management, and the CIO would be advised to define a netbook and create a netbook policy if they are distributing them to employees. If the company is not giving out netbooks but people are bringing in netbooks, this group would be advised to create a policy about bringing in outside netbooks and connecting them to the network. Consider the doctor who brings his personal netbook to the hospital and takes away patient data to look at later. Perhaps it is necessary in case there is an emergency call in the middle of the night and a nurse needs to know about what to do. However, what if someone hacks in the netbook and steals the information, HIPAA violation! Who is responsible? Can you investigate the hospital info on a private netbook? There are also many things that connect to the netbook; the next section considers the accessories for the netbook and that the policies also need to include these items.

If investigators and policy makers are considering how netbooks can be misused, they need to consider all the accessories that are available for the netbook. Each of these accessories could potentially be brought by an employee to the workplace and connected with the netbook to a wireless or wired network. A USB to Ethernet device could be purchased along with an Ethernet powerline adapter, for example. Then the wiring in the walls could be used for a covert area network with others who use the same setup to connect to power outlets in the building. The AR, the IT department, the general counsel, personnel, and HR all have to consider these things and proactively create policies before some creative person misuses equipment at work.

What complicates matters with the netbook is that it has a USB port which means that it can connect to printers, external hard drives, USB drives, 3.5-inch external floppy drives, and USB external FAX modems. That means that one can use the netbook to make calls on a phone line in the workplace and connect to the Internet and chatrooms. That makes inappropriate chatting with minors through Internet connections via USB external modem connected to workplace phone lines. What about USB hubs? A person with a netbook can connect a USB hub and then have a variety of the previously mentioned devices along with a webcam. That can mean videoconferencing or taking pictures and sending information along the wiring within a building to someone outside. There is a potential for espionage. Risk managers and others need to consider all these things when allowing outside technology in the workplace.

INCIDENT RESPONSE TEAM

The IRT is defined by Vacca and Rudolph as "a group of people with responsibilities for dealing with any security incident in an organization" [4]. Vacca and Rudolph say that there need to be four types of members on the IRT [5]. These roles are public relations officer, human resources representative, and management [5]. The public relations officer is necessary because incidents can often get incorrectly reported and then become an embarrassment or cause a lack of consumer confidence to an organization. Negative perception can affect revenue or perhaps impact those who decide on funding for the organization. Public relations officers usually coordinate a response with management and the legal representative. The press needs to report a story if they cover an event and it is best if the IRT can perform damage control and get the accurate story dispersed.

Many organizations such as ASIS International have conferences and monthly meetings. It is often a lunch time topic to discuss how information technology incident response teams are teaming up with physical security. This trend is important because there are so many reports of workplace violence. If an IRT is investigating an accusation such as a person looking at child pornography on their computer at work, the employee could lash out at the team and try to stop them. Adding physical security personnel is not only a good idea but might also lower one's insurance premiums.

The legal representative is important because events must be handled so carefully so that all policies and procedures are handled properly and the suspect cannot sue the IRT or the organization. The AR, legal representative, general counsel, and management must work together as a team communicating effectively and making sure that everything is handled in a fair manner and according to policy. If a person was found to be doing something criminal such as viewing child pornography, *prima facie* evidence, then the silver platter doctrine is invoked. This means that the investigation stops. Law enforcement is called. Whatever materials, paperwork, or electronic media that was collected is given to law enforcement. It then becomes a criminal investigation. In the United States, the Fourth Amendment now applies, and any further information must be collected with a search warrant since due process applies.

Management needs to be on the IRT so that they can authorize that the team has adequate resources for the event [5]. A server or computer may be needed to be taken offline. Computer files on the netbook may need to be viewed. Management needs to make those decisions so that an IRT member does not exceed his or her authority and then face sanctions. The management and HR can also work together to make sure the manager has the most accurate advisement concerning various policies on dealing with an employee who caused an incident. The management may also assign or recruit specialists to assist the IRTs.

The specialist may be a person who can collect a forensic image of a netbook. This person may need to bring hardware such as a LogicCube to forensically image the drive. They may also need to have a bootable CD or DVD with a program that lets them collect the most volatile evidence of RAM, and then the lesser volatile data such as the hard drive. This person may need a variety of tools and backup tools so they can collect the evidence in a timely manner on the first attempt. The specialist should also have protection from the physical security person in case the employee tries to impede the investigation. The specialist should also have special training not to take advice from the suspect because there may be a kill switch to erase data.

INCIDENT RESPONSE TEAMS WITH SPECIALIZATIONS

There are incidents in this chapter when the IRT is a group of law enforcement officers who may include someone trained in BDRA, also known as basic data and recovery. They may have also completed some basic computer forensics. They may be accompanied by a lawyer and computer crime detective from their local county prosecutor's office. They may also employ a civilian technical expert as part of the IRT if the investigation may include some antique piece of computing equipment such as an Amiga 2000. When a criminal investigation occurs, a search warrant is obtained, and a person may be arrested and a netbook and other digital devices may be taken.

There may be other incidents in this chapter when one discusses IRTs in corporate environments who are made up of specialists, management, human resources personnel, public relations personnel, legal representatives, and a security officer who may be an off duty policeman hired after hours. He or she is often called a peace officer and has more power than regular security personnel. It should become evident that there are situations where civilians may end up in part of a police IRT and vice versa. The corporate investigation of policy violations in the workplace are a common occurrence.

There are also private investigations that occur when a wife hires a private investigator to investigate a commonly owned family netbook or desktop computer to see if the husband is unfaithful. One can see that there are three types of investigations that may occur. Throughout this chapter, we will look at a wide range of IRTs and a wide range of examples that may be encountered. This is done to enlighten the reader to the complex large world of digital forensics that exists. The complexity of investigations and the skills needed increase exponentially if one considers a netbook that must be seized from a location where chemical biological nuclear explosive (CBRNE) weapons were used and then examined.

INCIDENT RESPONSE TEAMS (LAW ENFORCEMENT) AND DANGEROUS ENVIRONMENTS

I teach a variety of digital evidence classes that have students who are unemployed, law enforcement officers, firemen, private investigators, and visitors from other countries. Sometimes students will tell a story about something that happened to them or a colleague to illustrate a point. These students will often say that the motivation for telling these stories is to share common experiences that are often not discussed. Most of these students appear sincere but there is no evidence to back these up. These stories may or may not be true but are told to illustrate some of the dangerous scenarios that could exist and that IRTs should be ready for anything. I also go to a variety of conferences related to Homeland Security, high-tech crime, and computer forensics. Some stories are also told to me by credible sources to illustrate some of the dangerous environments that academics may never have heard about or could even imagine.

A man who was once a policeman and volunteer fireman related a story about a house that he and his law enforcement group had to visit. The story takes place in New Jersey. They had a search warrant regarding a person at a location who allegedly was selling a drug known as "crack." The house had a drainage ditch around it which made access to the house possible from the front door. When the officer and his partner approached the house, what appeared to be a drawbridge was lowered and some Doberman pinchers were released on the approaching policemen. The policeman shot one dog and the others stopped and ran back. Once inside the house, there was a false floor that when walked on, let one fall to the basement. Exposed razor blades were said to be embedded in the walls. There was also said to be a pit bull in the basement with its vocal chords removed so that the dog could not bark when law enforcement entered the house. The suspect was arrested. Drugs were seized, and the computer was seized. The environment was so dangerous that onsite analysis of the computer was not possible. Though such a story seems unbelievable, Erin Cardone from the Victoria News reported on an incident in Canada where the Victoria Police Department visited that was booby trapped, had cameras, and had a bomb. The house was also rewired to give unauthorized visitors electric shocks. The Royal Canadian Mounted Police's bomb unit was called in to remove the bomb [6].

Another student who was a former law enforcement officer said that he heard stories about criminal organizations that have large magnets embedded in the doorways. If IRTs were dispatched to that site and computers were seized, it was said that the evidence would be erased as the material was carried through the door. Donald G. McNeil Jr., a writer for the *New York Times*, wrote an article about magnetic resonance imaging

(MRI) magnets being so strong, that a police officer's gun "flew out of his holster" and across the room thus landing on the magnet [7].

IRTs need to review the intelligence reports that they have and then depending on the type of suspect and organization, plan accordingly. Seizing a computer from a crime family or drug cartel is much more dangerous than from a pedophile. Many policemen who have arrested pedophiles who download child pornography say that these people offer little to no resistance when caught and are often ashamed of their behavior. Many pedophiles often plea bargain using a guilty plea when they are caught to get as little time in prison as possible. IRTs in municipal law enforcement agencies often work with their county prosecutor's computer crime unit to get the technical specialists and the necessary equipment and personnel to safely accomplish the mission.

Many law enforcement officers who have been on IRTs to arrest online sex offenders have told me some amazing stories. They have spoken about arresting people who have what is commonly known as shrines of the victims who were stalked. Shrines may consist of many things such as telephoto pictures of the victim doing activities of daily living inside or outside the home. It might include garbage belongings or papers of the victim and online biographies. Some of the sites that computer crime IRTs visit may not only be dangerous as previously discussed but may also have exhibits of bizarre and disturbing human behavior.

If one examines the curriculum of schools and universities that teach mobile device forensics or computer forensics, little is discussed to mentally prepare the student for the future danger he or she may possibly face in law enforcement. Physical danger such as a booby trapped house is one thing but how does one prepare for the mental danger of being exposed frequently to those situations where sex offenders and stalkers have shrines of their victims or child pornography. If a school had a program that included all these uncomfortable topics, how could it be explained in a comfortable setting to audiences of young men and women? Perhaps it would scare prospective students away, cause a loss of revenue, and cause curricula to be abandoned. There are much informal viewpoints expressed on television programs that there are not enough trained cyber investigators now. A curriculum that exposes all the disturbing experiences that law enforcement digital evidence investigators face could further shrink the pool of students and result in a more severe reduction of trained investigators to arrest criminals and examine their computers. Perhaps many computer investigators will be hired by private investigation forms or corporate security departments and will never face anything more than child pornography or a violent employee. Perhaps it is not necessary to discuss the experience of computer IRTs in law enforcement.

If one wanted to have students gain experience with going in a "crack den" or visiting a computer stalker's workplace and shrine, it could be done virtually in a CAVE environment. The CAVE is a virtual reality environment that includes large screens on three sides, numerous speakers, and a head mounted display or special eye ware for the participant. Many CAVEs have been demonstrated at various cybertherapy conferences such as the one held in Gatineau, Quebec, Canada in 2006. Virtual reality is used as a component in many types of training, education, and sex offender rehabilitation. Virtually Better is an organization that can provide a number of virtual environments for training or as a component of a mental health treatment [8].

Some academics in virtual reality have been informally discussing a concept known as stress inoculation. Perhaps having students use virtual reality environments to train in stressful environments before they graduate would be one proposed solution to exposing them to such environments that they may face. Each incident response team may control a lifelike avatar and a series of tools in the virtual world. Each team of avatars may work as a team to go into a crack den and seize the computer, related media, document it, and fill out a chain of custody form. There may be fake floors, booby traps, and other devices to prepare students for the worst possible situations that exist.

INCIDENT RESPONSE TEAMS NEED A PHOTOGRAPHER

Law enforcement IRTs sometimes visit locations with a search warrant. The suspects at some of these locations have complicated arrays of computers doing parallel processing to break encryption and also have redundant arrays of independent disks. There may be tremendous amounts of networked devices, cables, and digital media nearby. It may be too time consuming to seize everything, figure out how it is connected, restart it, and then reexamine it unless there is proper documentation. Many times, the county prosecutor computer crime units will bring a photographer who will put various colored and numbered labels and stickers on computers, cables, and equipment. Then the photographer will take pictures. Then everything will be bubble wrapped, taped, tagged, and transported. This allows everything to be reassembled and examined later.

The photographs may be printed out later and everything can be easily reassembled in the computer forensics lab since a marking scheme and pictures were used. Most people have some experience with simple color coded tagging and assembling computers if they have bought a desktop computer, for example. There is usually a green connecter that goes in a receptacle marked with green on the back of the computer. This is standard for installation of the mouse. A similar purple schema is used for the keyboard. A picture is used for setting up the speakers, printer, monitor, webcam, and any other peripherals that may have accompanied the purchase.

INCIDENT RESPONSE TEAMS (RESILIENCE TRAINING)

It was already stated that many people who work for any type of law enforcement agency and become a member of an IRT for digital forensics might have to face physical danger at times, and psychological discomfort from disturbing environments with sex offenders. Perhaps these issues are discussed to some degree in the police academy or at briefings when one joins a particular unit such as the computer crimes unit or white collar crime unit.

It may be advisable for mental health professionals to revise the curricula of digital forensics to also include a mental resilience program. The United States Army started such training with a program called "BattleMind" and then improved that into a program of "resilience training" [9]. The resilience program provides various psychological tools and behavior modification techniques that may help some soldiers to transition to various parts of his or her mission or career. A similar program may be applicable to the various career tracks of digital evidence examiners. There may need to be tracks for law enforcement and civilian/corporate investigators. Both groups may need training to successfully endure rapid fire questions from lawyers during cross-examination in court or lengthy depositions.

Dr. Albert Rizzo is a mental health and virtual reality researcher who has performed and published significant research within the field of education, training, mental health, and virtual reality. He has a paper on virtual reality, caves, and stress resilience for the military [10]. However, the lessons of such research and techniques could probably be adapted and applied by mental health professionals and virtual reality experts to those digital evidence examiners who visited stressful environments when seizing netbooks or personal computers. The research could also be applied to other stressful situations such as preparing for deposition, testifying in court, and then the aftermath. Perhaps a company such as Virtually Better could seek a government grant and then have its programmers create a virtual law office where a lawyer avatar takes a deposition from an avatar who is a digital evidence examiner. The same virtual world could also include a courtroom where digital evidence examiners are given a variety of questions about their expertise and then endure a cross-examination. The avatar could be switched to a male or female lawyer with a variety of body types from small to large and physically intimidating. Perhaps there could be a slide bar to increase or decrease their volume of speech as well as their intensity of questioning and rebuttal. Perhaps the heart rate, breathing rate, and other indicators of anxiety could be measured before and after training sessions to assess the efficacy of such training.

INCIDENT RESPONSE TEAM SKILLS

It is often necessary for the IRT to image a netbook, laptop, or hard drive. These incident response team members may include a legal person/documenter, someone from physical security, and a netbook/laptop forensic examiner. The forensic examiner may find that a network connection and CD/DVD ROM may be broken intentionally or accidentally by the suspect. The examiner may have to open the netbook case and remove the hard drive. Many mobile computing devices contain slots for external cards but where might these cards be hidden? There may be an external SD card or micro SD card that is not connected to anything and hidden inside the chamber where the hard disk resides. There may be criminal evidence stashed inside the case itself too. Sometimes the examiner may also need to open the netbook case, disconnect the hard drive, and then remove it from the case. This is done so that the examiner may put the hard drive on an adapter and outside a device called the LogicCube. A forensically wiped new drive of the same or larger storage capacity may be put inside the LogicCube. Then the suspect's drive is copied byte for byte until the whole drive is copied or imaged.

IRTs could be made up of newly graduated students. Employers may expect their newly graduated IRT members to have both practical skills and theoretical knowledge. Therefore, let us address the topic of the proficiency of hand tools. One of the topics that seem to rarely be addressed in schools that teach computer forensics is the selection and use of tools to open netbook, laptop, or computer cases. One may also need to learn to disconnect components such as ribbons, hard drives, and then later reconnect and reassemble everything. This may not seem like an issue to some but it is often encountered on some level at universities where many foreign students study. While teaching at the university, I often encountered students from India and the United Arab Emirates who often had personal servants as well as household servants. Sometimes the students said that they never used a tool before because servants did the tasks for them in their home countries before arriving in the United States. Students have told me that college preparatory courses were valued in the home country but not knowledge and experience with hand tools.

Many of the foreign students from developing nations who come to the United States have to be rich to afford an American education because the exchange of currencies such as the Rupee or Rupiah to dollars is so unfavorable. This means that the teacher of computer forensics should budget more time to not only teach the lesson but perhaps to use tools such as manual or electric screwdrivers. Many American students often take college preparatory classes while still in high school. This study track often deals with intellectual topics and does not include classes such as wood shop or car repair. This means that many American students may also not know how to use tools. When one also considers the number of busy single parents who are supporting children, they may not have time to teach their children how to use hand tools.

Many of the computer cases and perhaps some laptops and netbooks have complex latching systems that take time to figure out how to open. Many netbook systems use screws that cannot be turned with standard Phillips head or flat head screwdrivers. The screw may have a head that needs a special screw driver with a complex star pattern on it. Sometimes special screwdrivers must be ordered and could delay the investigation by times up to one week. On some occasions, the screws are too difficult to turn with manual tools since they were installed with some type of electric or pneumatic tools. It is therefore important to have access to a small workshop where both manual and electric or pneumatic tools can be borrowed. YouTube videos or in-person instruction may be necessary to teach students to open cases, remove hard drives, and then later to reassemble everything. Chapter 1 of Adrian Kingsley–Hughes' book, *The PC Doctor's Fix It Yourself Guide*, addresses how to open cases, use hand tools, remove components, and be safe [11].

There may be also a need to teach students basic tool safety so that they do not burn themselves with soldering irons or stab their hand when they first put the screws back in the system. Some books such as Scott Muellers, *Upgrading and Repairing PCs* have part of a chapter devoted to the use of hand tools and safety [12]. Safety glasses and gloves are necessary for both eye and hand protection. I often have some students insist that there were some extra parts left over after reassembling a computer in the LogicCube hard drive imaging class. Students should take pictures of the inside of the computer as soon as they open a case so that they may see exactly where everything is located.

Some organizations such as the United States Army and some large law firms realize the need for computer forensic professionals to be able to work with computer hardware and tools and now either require or strongly encourage certifications such as the A+ Certification.

CompTIA offers the A+ Certification and consists of 100 questions. The candidate gets 90 min to complete it. The Department of Defense recognizes the certification and the Department of Homeland Security includes it in their computer forensics program. It is vendor neutral and really forces people to get into the nuts and bolts of the machines. The certification was originally sought by computer technicians but is increasingly sought after by computer forensics professionals. I took a class of cybercrime students to the Regional Computer Forensic Labs in Hamilton, New Jersey. At the time of the visit, all the digital evidence examiners that the class interacted with had the A+ Certification.

It is also worth taking an in-person A+ Certification class for some of the little tricks one learns to deal with low-tech situations such as losing a screw in the computer case. Some nontechnical people often suggest techniques such as using a stick with a magnet to feel around in the computer until one locates the lost screw. However, the A+ Certification teachers will tell you that it is better to locate the screw and use a pen-like tool with a push down top that releases claws that help one pick up the screw because it does not use a strong magnet. Magnetic wands are called DeGaussing wands and can be used to wipe data. Why expose your system to unnecessary risk?

In college, I saw someone get a shock from a charged capacitor. Many people do not know that certain electronic equipment has capacitors that may hold a charge for a significant period of time and degrade the charged value at a logarithmic rate. This means that a computer forensic professional faces some risk when he or she is handling everything inside the computer. It is best that the student take an A+ Certification class in person and learn about techniques for grounding equipment as well as reducing static.

The A+ Certification is getting worldwide acceptance as the need for computer forensic investigators grows worldwide. The test is now available in Arabic, Turkish, Spanish, traditional and simplified Chinese, English, Korean, and Japanese. In the old days, American computer forensic professionals flew to other countries to do work on high-level cases but to keep costs down and deal with the ever-rising rates of cybercrime, many nations around the world are developing their own computer forensics programs.

INCIDENT RESPONSE TEAM'S DIGITAL FORENSICS LAB

Some laboratories have a separate set of policies and storage area network for the computer and mobile device forensic labs. Some lab facilities that do not give the autonomy to a digital forensics lab may require that the forensic software be installed by an information technology department. Some academic and corporate digital forensics laboratory facilities require that the lab director or examiner initiate a request for examination software with a call to the IT department. Then a ticketed request to install the examination software is formalized. The reason for this perceived inflexibility is to control what is installed on the examination computer so that improperly licensed software is not installed. This can be a good control so that results from digital device examinations from unlicensed/improperly licensed software tools are not dismissed if they go to court by the legal principle of "fruit of the poisonous tree."

This process of initially calling in a request for software also allows the IT department the ability to know exactly what is installed on a computer so that they can fix it if there is a problem. Occasionally, some programs may share DLL libraries or utilize resources with concurrent programs that result in unexpected side effects. The IT staff can monitor what programs are installed and notify the examiner of potential issues. This type of side effect may be better understood with an analogy to a medical situation. A classmate in the 1980s had to take six pills for a heart problem and another six pills to counteract some of those side effects from the unexpected interaction of those pills.

In the United States Government, there are three branches of government known as the Executive Branch, the Judicial Branch, and the Legislative Branch. There are also a system of checks and balances to prevent abuses. There should be a system of checks and balances in a lab where activity is monitored and a rationale for various activities be put in a verbal or written report. In academia, digital forensic teachers report to an academic dean and may be on a technology committee with the IT department. Many law enforcement labs such as the New Jersey Regional Computer Forensics Labs in Hamilton, New Jersey have a lab director who brings a sense of governance for the activities and resources of the lab. Oversight and good leadership can often reduce mistakes. Depending on available resources, a digital forensics lab that examines netbooks may have Access

Data's Forensics Tool Kit (FTK), OSForensics, RecoverMyFiles, AvanQuest's Perfect Image, Guidance Software's Encase, Helix3 Pro by e-fense, and Backtrack 4.

INCIDENT RESPONSE TEAM'S TOOLKIT

The toolkit may consist of hardware, tools, and software. The hardware could be hand tools to open a netbook or desktop and then an interface for the hard drive to a computer for cloning. A LogicCube is a device that is used to wipe drives before receiving a forensic image. LogicCubes are also used to do keyword searches on unallocated and allocated space in a hard drive without making changes to it. It may also be used to clone a suspect's hard drive to a forensically wiped and cleaned hard drive. The LogicCube is a piece of hardware that is portable, lightweight, and simplifies the forensic imaging process. Another low-cost method of copying a drive is with a program such as OSFClone that is located on a bootable CD. The suspect's computer may be booted up to OSFClone and then an image or clone of the hard drive may be put onto an external device such as a USB 1 terabyte external hard drive.

Once the cloned drive is obtained, it will have to be examined with a tool or suite of tools. Some incident response teams and investigators may chose to hire an outside agency that has the tools and expertise to examine the forensic image. They may choose this option because they do not have the trained personnel and tools to perform the investigation. Keeping a forensic investigator up to date with certifications, training, and tools can be prohibitive for many organizations. Some places may look to transfer the investigation for a fixed cost. Other organizations may have the trained investigator and use a tool such as OSForensics. The next section will look at the issues concerning OSForensics.

OSForensics: A COMPUTER/COMPUTER MEDIA FORENSIC ANALYSIS TOOL

One of the frustrations expressed by students, school administrators, private investigators, and corporate investigators has to do with the price and limited lifetime of digital forensic tools. Many tools are only good for 1–3 years and then must be repurchased. This is understandable since there is a great deal of expense and time that goes into making these tools. The maker of the tools may also be subpoenaed to court to testify and not compensated. Tool makers struggle with trying to make a product and to make a living. The tool users struggle with shrinking budgets and smaller staffs, thus reducing any unnecessary expenses. Passmark Software has tried to deal with this situation by releasing a free version and a paid version of a computer forensic tool known as OSForensics. The paid tool gives more capability and there are support options available. The free version allows one to do critical analysis work on a system.

In the late 1980s until about the year 2000, software was often ordered and a set of disks or CDs arrived in the mail a week later, or within 3 days with expedited postage. In 2011, the norm is to go to a site, make a payment by phone, credit card, or PayPal, and then download the software. This reduction of time is critical for investigations that must be performed as soon as possible and reports must be given to a variety of parties as soon as possible. OSForensics has a tab on their website that quickly takes you to the download. The executable file for the application can be downloaded within minutes on a system with broadband. In the mid- to late 1990s, when dial up and DSL access was the norm, downloading software took hours.

OSForensics AS A TOOL TO ENSURE INTEGRITY OF EXAMINATION MACHINE

One option for computer forensic investigation machines is to wipe the drive after each investigation and then reload the image of the hard drive that includes the operating system and examination tools. This ensures that there is no malware or digital residue from previous examinations. Then one can run antivirus software and antispyware to make sure the computer is clean. It would also be good to wipe the media and check it for malware before using. OSForensics can take that level of security one step further by checking the operating system that was reloaded with hash file libraries that can be run in conjunction with the forensic application. OSForensics has various hash libraries for each operating system. One just has to download the correct hash set and make it active to use it. This allows one to quickly compare the digital signatures of the suspect operating system with the digital signatures of verified clean operating system files. A difference in digital signatures within a file means that the file is either corrupted or as malware or some hacking tool embedded within it.

WEB-BASED TRAINING THAT IS "ON DEMAND" FOR OSForensics

One of the trends for digital forensic training is to go for an on-demand web-based training. This means going to a password-protected portal or open access area to watch video, download MS PowerPoint slides, and read pdf-formatted documents. The reason for this trend is discussed at length in two academic papers published by me in the WorldComp 2011 conference [13,14]. However, to sum it up quickly, it is because investigators may live in remote locations and cannot easily attend a class, they have limited budgets for training expenses, and the need for the use of the tool is immediate and may not wait for a class to occur. OSForensics has eight free web-based training that is available on YouTube, a popular video website. The topics that seem most logical to start with are "Create & Install Hash Sets" video and the "Using Hash Sets and File Name Search." This will allow the examiner the ability to ensure the integrity as discussed earlier.

PASSWORD RECOVERY IN OSForensics

The password recovery tool is an important part of the OSForensics suite of tools. It allows one to check a browser for various URLs, usernames, and passwords. This is important because a person in a corporation may have been using a website for logging into and uploading intellectual property. This tool can allow the examiner to find the URL, username, password, and something regarding the activity. Let us suppose that a file was uploaded that was encrypted. There is a tool within OSForensics to select a type of encryption and use a brute force attack or dictionary attack on the file to recover the password and open the file. Custom dictionaries that may consist of data from unallocated space on the hard drive may also be used.

OSForensics CASE MANAGEMENT

The OSForensics software also has a tool for case management. This starts from selecting a tab and then using a wizard to answer questions about the investigator's name, agency, case number, and a location of where the results can be stored. As the investigator recovers files and activity, it may be added to the case for later printing or inclusion on a CD. Case management is an important part of computer investigation because one can quickly get overwhelmed by the amount of data that exists and forgetting a piece of evidence can severely impact a case.

BOOTABLE VERSIONS OF OSForensics

If the examiner encounters a suspect's computer, he or she does not turn it on. If a Microsoft Windows machine is energized, a series of temp files will be created and the hibernation file and swap file will be changed. According to an experiment performed in the lab at Fairleigh Dickinson University's Cybercrime Training Lab, some operating systems may have as many as 100 writes to the hard drive upon booting up. I took a LogicCube hard drive write blocker (paid for with a Department of Justice Grant), and connected it between a motherboard and a hard drive. The operating system attempted to write 98 times to the hard drive upon bootup but was blocked by the write blocker.

OSForensics can be put in a CD or DVD with various system files so that a bootable version of the forensic software tool can be made. Then, one can take this bootable CD and place it in the CD/DVD drive of the suspect computer and boot it up without changing evidence [15]. OSForensics can also be placed on a flash drive. The BIOS on a computer can be changed so that it looks first to the USB flash drive or CD/DVD player for an operating system before booting up. OSForensics also works on a variety of systems and the preliminary results of an investigation may possibly be exported to another storage device, depending upon the hardware of the suspect's computer and the configuration of the bootup CD/DVD or flash drive. The importance of a tool that can be used to examine a machine without changing evidence at bootup cannot be overemphasized. If the tool can be used to create a forensic image of the hard drive, this is also extremely important.

OSFClone

There is a free computer forensic tool called OSFClone that can be put on a bootable CD or DVD or flash drive. This tool can be used on a computer once the BIOS is changed to initially bootup from the CD or DVD or flash drive. Once the computer is changed so that it boots up to OSFClone, it can clone the drive and put the results in a dc3dd file format or the new Advanced Forensic Format (AFF). In the 1990s and early 2000s, dd files were used. However, an improved version of the original dd file format is the dc3dd file format. If one takes a class on operating systems and file systems, some of the reasons such as hashing files on the fly may be discussed. Brian Carrier writes a good forensic file system book that may further your knowledge on the drive cloning process and file system in general [16].

cs2025 FORENSIC SOFTWARE AND RECOVERY SOFTWARE FOR UNIX, LINUX, WINDOWS, AND MAC OS X

This forensic toolkit comes on a CD and is being sold on eBay on February 7, 2012 from Jerusalem, Israel. The eBay item number is 270344143431. The cost is $15 and the shipping is $3. The item is new and the vendor says 15 of these are available. This toolkit could be useful in an investigation concerning a netbook, laptop, or desktop computer. The website said that 78 have been sold already. Some of the tools included are Zipcracker, FileExtractor, Defraser, Open Computer Forensic Architecture (OCFA), Live View, and e2undel.

One of the tools that in the toolkit called ZipCracker is used on zip archives in order to help recover files. The tool is for Unix and Linux systems and uses the Gnome User Interface which many say is convenient to use. The tool is said to be distributed under the GNU General Public License. Another item in the toolbox is called FileExtractor. The tool is also said to be distributed under the GNU General Public License. The program is said to run on all 32 bit Windows Systems and Linux/Unix systems for the purpose of file recovery. Therefore, Windows 7 32 bit architecture would be good to run this on. The tool is reported to be good for floppy disks, flash drives, digital cameras, hard drives, and partitions. The program was created with Python which is also used in BitPim.

There are many investigations where someone in a corporation or home is reported to be looking at child pornography movies. They may have deleted the movie and some of the evidence by the time the police arrive. The only place that some of the video may still exist is in unallocated diskspace on the hard drive. The corporate investigators and police may wish to use NFI Defraser to carve out the partial video from unallocated diskspace. The program was made with C#, pronounced C sharp, .net and runs in some of the newer Windows environments such as Windows 7, Windows Vista, and Windows XP. Another item called the OCFA is included in the CD for the same type of investigation and was created by the Dutch National Police Agency. The OCFA gives the investigator a means of safely previewing the suspect's computer and performing searches. The investigator does need some Linux and Unix training before using this tool. It is not as simple as many of the point-and-click Windows tools.

DcFLdd is an imaging tool that is allegedly better than the regular dd program. Once the image of the drive is made, it could be previewed in LiveView. LiveView is another interesting tool that allows one to take a forensic image and boot it up in a VMware virtual window to run it. This might work well in Windows 32 bit systems. This is much like being right on the suspect's computer and having the same view of everything that he or she has. It is great because all types of files can be inspected and various applications can be run without making any permanent changes to the image. DcFLdd and LiveView are two good tools to capture a hard drive's data and preview it. These two tools should be in every investigator's toolbox.

The e2undel tool might be good for recovering data and deleted files from PDA handheld devices and organizers. The reason is that many of those handheld devices utilize ext2 Linux file systems and e2undel provides an interface that can be run on a Linux computer. The makers of the tool say that no specialized knowledge is needed about ext2 file systems to benefit from the tool. This tool which is also included on the disk is handed out under the GNU General Public License.

NETWORK MINER

This tool was made for Windows and could be useful for checking on a netbook connected to a network in an investigation on industrial espionage. This network forensics analysis tool can be used for packet sniffing and could also extract files from network traffic. It is specifically made for a Windows 32 bit system and the tool was created with C#.net. It can be given out through the GNU General Public License.

MAC-ROBBER

This could be a good tool for the Mac operating systems because it can be used in conjunction with another forensic toolkit known as Sleuth Kit to create a timeline of activity with files.

PLAC

This is a Linux tool that is made for Linux and Unix systems. It could be useful for data recovery and the forensic analysis of files.

RDD FORENSIC COPY PROGRAM

This program is for creating forensic images of hard drives. It was created at the Netherlands Forensic Institute for their own uses. The program is considered robust by many because it keeps trying to read even when it hits bad sectors. It could be very useful for imaging a partially damaged drive or digital storage device.

AIR

The Automated Image and Restore (AIR) program is also included here. It is a good tool that works on a network and allows one to image a tape drive or hard drive. AIR is also good because it allows for detailed logging which can be used with notes to show how an investigation progressed. AIR also allows for the splitting of images into files so that it becomes manageable.

TULP2G

This is a tool that is good for Windows XP and Windows 2000. Its purpose is for collecting data and decoding it. It is a classic tool and much is written about it online.

ODESSA

This is a type of incident response kit that can be distributed because of the GNU General Public License. Some people have said they like its versatility since it can be used with

Mac OS X, and also with the Linux and Unix operating systems, and Windows systems from Windows 95 to XP. The acronym stands for open digital evidence search and seizure architecture.

SLEUTH KIT

The famous computer forensic program called the Sleuth Kit is included on this CD. This could be a useful tool with PDAs and organizers since it can be used on ext2 and ext3 systems. It could also be useful for digital cameras and USB flash drives and hard drives since it can be used on FAT file systems. It also has tools that allow it to seize and examine data on the HFS+ and NTFS file systems. It seems to be a very versatile tool that should be in every incident response kit. Overall, this Israeli CD comes with many great tools at a great price and should be considered for inclusion in every incident response kit.

MARESWARE LINUX FORENSICS

There are a set of tools that exist and run on a Linux system for Intel processors. The tools could be useful to corporate investigators or law enforcement investigators who must search for evidence on a computer. Hash is the name of one tool that creates SHA or MD5 hashes on every file of a drive. Hashcmp is a tool that compares the hashes for two files or two set of files. This could be useful if a suspect in a corporate investigation accuses the investigator of tampering with the evidence. The original drive could be hashed again with hash. Then, the first set and the second set of Md5 hash numbers could be compared with Hashcmp. If they are the same, then the allegation is false. If there is a discrepancy, then further investigation is needed. Perhaps a second investigator from an outside agency is needed. A second investigator who has nothing to do with the case is needed. It may also be time to turn over the case to a law enforcement agency if there appears to be some tampering of evidence.

Bates_no is a program to associate numbers with filenames in e-documents. It could be used as a way to help manage all the records in a case. Strsch is a program that searches for character strings or sets of words in a file, or a group of files. The ability to locate specific character strings could help identify relevant files to an investigation quickly. The files with potential evidence could be copied to another disk and examined by a document reader. Catalog is a tool for cataloging every file that resides on a Linux system. This could be useful for organizing all the evidence on a computer that is related to an investigation. It is a common occurrence in computer investigations that people will encounter files that are in a format they cannot read. Perhaps their reader cannot view that format for text. There is a tool called U_toA that can convert the format of *ix text to a DOS-formatted type of text. Then, one can use another program called Md5 to calculate the hashes of the file. Another program that is useful is named boot. It is a small program that is used to reboot the computer. Diskimag is a program that creates a forensic image of a diskette. It can be used with Linux and Mac. Filsplit is a program for splitting a file. Another program in this suite is called Diskcat. It could be useful because it identifies the headers of files and catalogs every file that resides on the disk. Crckit calculates a 32 bit checksum. This is good for the error checking of transmitted files on a modem. It would seem that MD5 would be a better tool since it would be more precise.

Some operating systems have a way of keeping track of the time that is good for computers but not convenient for humans to use. The time is displayed in some type of format that is calculated from some type of epoch time. The first creation date of when a motherboard was first designed is one example of an epoch time. Another example of an epoch time might be calculated as the present time relative to a time when the first version of the operating system was released to the market. Some people say that certain versions of the Windows operating system have a number relative to the date and time which need to be converted to something that is more understandable and traditional, such as month, day, year format. Dater is a program that converts complex time and date formats and puts them in a more understandable usable format.

Any set of tools may have one or more utilities that are useful. It is good to collect them, evaluate them, and save the ones that are useful. A toolkit may be good for certain platforms and not for others. It would be advisable to have more than one case with various sets of hardware and software tools marked for various operating systems.

DRIVE PROPHET

There is a tool called Drive Prophet. It has a professional edition for sale on the Guardian Digital Forensics website for $395. The information about Drive Prophet is available online from a www.driveprophet.com and it is free for United States Law Enforcement Organizations (USLEO). The law enforcement officer has to get it from the National Repository of Digital Forensic Inventory (NRDFI). The tools made available on the NRDFI website are for official use only and not available for use by commercial vendors [17]. The NRDFI says that they have over a thousand documents to share with other law enforcement officers in the United States, Australia, Canada, England, and New Zealand [17]. If one wishes to join, it starts with an email to join@nrdfi.net and from there they will make available some forms that need to be filled and checked before access is given. It is also good to know that "the Defense Cyber Crime Center (DC3) and Oklahoma State University's Center for Telecommunications and Network Security (CTANS) are joint partners in sponsoring the development and operation of the National Repository for Digital Forensic Intelligence (NRDFI)" [17]. The NRDFI is a secure portal for law enforcement people to share ideas, tools, and documents.

Drive Prophet was tested on the 64 bit platforms of both Microsoft Windows 7 and Windows Vista and ran successfully [18]. The website reports it to be able to collect USB device history. Sometimes USB history is gathered from USBSTOR on a computer. The tool should provide useful information about the operating system type, its version, and the OS owner information. One of its good features is that it can create an HTML report of the evidence that can be burned to a CD. Many computer forensic tool vendors now use this format because it is convenient and is said to save space. The tool can also give reports about pictures and graphics as well as user information. The tool is also reported to provide Internet history and reports on Thumbs.db and Vista.Thumbs. Thumbs has information about pictures that were in a directory or still are. Drive Prophet can be run on a USB stick and does not need to be installed.

The tool is said to be useful to incident response teams and corporate investigators. It could be useful to IT departments that do internal reports or investigation. It is definitely useful for law enforcement and intelligence agencies of all types. The fact that it is made available free to law enforcement officers in the United States, Canada, England, New Zealand, and Australia through the National Repository of Digital Forensic Intelligence implies many good things. If many law enforcement people in many countries are using it, then that should provide a community of users who can help each other. That should also promote new synergies which will become useful as more cybercrime passes through international borders and law enforcement officers (LEOs) need to know other LEOs in other countries.

It seems that this tool should be included in any incident response toolkit that could be used to examine desktop, laptop, netbook, and notebook computers. The fact that it is free to law enforcement should be good for the small towns that cannot afford their own tools and since the interface is menu driven, training should not take too long. Perhaps a county prosecutor's computer crime unit could help the small town officer as a mentor in order to teach how to use the tool effectively and create reports. It is also good that the website with Drive Prophet has a video that one can watch and see how to use the tool.

ELCOMSOFT: ADVANCED PDF PASSWORD RECOVERY

The following paragraphs about Elcomsoft's products were written with information provided by the Elcomsoft software's website and intertwined with my opinions as an academic. There are times when digital forensic experts need to open password-protected pdf files, but do not have the password. The password can also be for copying, printing, and editing a file. It is important for digital forensic examiners to be able to print documents to show evidence on paper to juries, defense lawyers, and prosecutors. Therefore, it is important to be able to break passwords in order to print documents. It is also important to be able to copy pdf documents files without restriction so that evidence may be shared in the ediscovery process. Therefore, it is important to be able to break the password so that the file can be copied without restriction. The Elcomsoft Advanced PDF Password Recovery software works with early versions of Adobe Acrobat up to and including Acrobat version 9.0. Many people also produce PDF files with tools such as Microsoft Word and PDF Converter. PDF Converter allows people to add password protection and 128 bit encryption on all types of Microsoft Excel spreadsheets and Word documents and convert them to PDF documents. This Elcomsoft Advanced PDF Password Recovery tool also comes in Enterprise and Professional editions. It can remove owner and user passwords. The software in the tool is also written so that it makes use of advanced features in processors such as the multicore CPUs. The enterprise edition and professional editions can use GPU acceleration on 256 bit AES password-encrypted files by taking advantage of the NVIDIA graphics cards. An example of such a card is the NVIDIA Quadro 4000 by PNY Graphics. The recovery tool uses dictionary attacks before doing a brute force attack. It also uses variations on words. The brute force attack tries every possible combination of characters, letters, and numbers in that password string. It is also possible to use various templates or masks for running the tool. Perhaps we know that the password is all numeric, then we would make our mask to only try numeric combinations.

The software can recover passwords to remove restrictions on PDF files. The tool works with both 40 bit and 128 bit RC4 encryption and both 128 bit and 256 bit AES encryption. Many computer investigators are familiar with rainbow tables for breaking passwords but may not have heard of thunder tables which are what this software uses to recover 40 bit passwords in minutes. The software can also run in batch mode to work on many files concurrently. The thunder tables use special computations to reduce time.

The software vendor for this tool recommends not using it on a Mac running Windows 2000, XP, or Vista in Virtual PC or VMware.

The vendor gives a year of support without cost. That is good because updates are needed with all the new advances in encryption and that could affect PDF documents. The website also has options to change product information into Polish, Russian, Italian, Japanese, Chinese, Portugese, German, Spanish, and French. Version 5.0 works with PDF files created in Adobe Acrobat 9 (with 256 bit AES encryption), with multicore and multiprocessor support and hardware acceleration using NVIDIA cards. The professional version is $99 and the enterprise version is $399. The standard version is only $49. The software has a nice graphical user interface and seems intuitive to use. There is also a button for help and ample documentation is available to the user. There are also many short videos that demonstrate the product on YouTube. Here is the URL of one example: http://www.youtube.com/watch?v=duI7KTZUk5Q.

ADVANCED SAGE PASSWORD RECOVERY

There are times when digital forensic examiners are looking at notebook computers and come across what appears to be financial misdeeds in the corporation. The person may be putting money in a shell company or buying real goods that the company uses and shipping them to another location where others pick them up and sell them. Such transactions could be stored on spreadsheets such as Visicalc, Microsoft Excel, or Sage PeachTree Accounting software. Some truck drivers say that they have a set of books for the company and a set of books to show auditors. They may not match and be password protected. In a corporate or criminal investigation, it may be necessary to open password-protected files. The password or passwords may not be available, so software from Advanced Sage Password Recovery from Elcomsoft in Moscow, Russia would seem to be the correct tool to purchase. They were established in 1990 and have decades of experience. They are members of the Russian Cryptography Association and the Computer Security Institute, and are a Microsoft Gold Certified Partner.

The Advanced Sage Recovery tool uses a graphical user interface that is intuitive and instantly displays administrative passwords and user passwords for Sage PeachTree Accounting versions 2002–2011. The software is also said to support all editions of Premium Pro, Complete, and Quantum. There is also good documentation for the product and support online in the forms of blogs and newsletters. The online product information informs the reader that Sage PeachTree Accounting passwords are stored in a file called PERMISS.DAT. If one uses the recovery tool, all user and administrative passwords will appear in plain text. The tool can also be used to recover passwords on remotely connected PCs. If the tool is used with an ACT! Database, a user account can have its privilege escalated to administrator which removes restrictions. The tool is also said to be good for unlocking password-protected ACT! files with long and international passwords.

ELCOMSOFT INTERNET PASSWORD BREAKER

There are times when the digital forensic investigator encounters a netbook in the corporate investigation and must check a special username and password to get to a noncompany website that may contain key evidence about missing funds and misappropriated supplies.

Elcomsoft's Internet Password Breaker is the tool for recovering such usernames and passwords. The standard edition is $49 and available online from the company's website. The platform of the examination machine running this tool can be Windows 7 on both the 32 bit and 64 bit versions, Windows 2000, Windows Server 2003 and 2008, and Windows XP. Unfortunately it is not supported by Windows 98 SE which is the platform that many certified computer examiners use. The hard disk also needs about 1 MB of free space. Most corporate-owned computers have employees sign away all expectations of privacy so that email, websites, and files can all be easily examined by the AR and the IRT. Elcomsoft's Internet Password Breaker also quickly displays Internet passwords and obtains both login and password information that protects a variety of web resources and mailboxes for many email clients. The latest version of the forensic tools should help in the fast recovery of passwords, which were stored on a form with the Internet Explorer's AutoComplete information as well as plain text passwords protecting mailboxes, POP3, IMAP, SMTP, and NNTP news accounts and user identities in a variety of Microsoft email clients. All versions of Microsoft Outlook, Outlook Express, Windows Mail, and Windows Live Mail are supported, including Microsoft Passport passwords in Windows Live Mail. Elcomsoft's Internet Password Breaker retrieves passwords protecting PST files for all versions of Microsoft Outlook up to Office 2010 [19]. Once the passwords are broken, my opinion is that it would be advisable to use an examination tool such as Vound Software's Intella Team Manager and Intella Team Reviewer to distribute the workload of email investigation among investigators and graph relationships of people involved. The software is also good for bookmarking data and creating notes or annotations to the bookmarked evidence.

ELCOMSOFT WIRELESS SECURITY AUDITOR

This is good for recovering the WPA/WPA2-PSK passwords in plain text. Suppose a corporate netbook has a USB-to-Ethernet connector which is connected to a wireless router and a person pulls up in the parking lot with a directional antenna and laptop to collect trade secrets. If someone sees this and reports it, then an investigation is started. The AR would dispatch an IRT with a network card in promiscuous mode. The IRT would then run Elcomsoft's Wireless Security Auditor to break the WPA/WPA2-PSK passwords and enter the router to look at the security logs. Once in the router, there should be a log that shows concurrent connectivity between an outside computer, the router, and the netbook. The wireless security tool costs $399 for the standard edition but it would also be a good investment to purchase the professional edition because the program could be used as security audit tool and a network forensic tool.

The IRT needs to have 6 MB free on their hard disk and should have an ATI graphics card or NVIDIA card to help speed up the processing for encryption breaking. The operating system on the examination/incident response team computer needs Windows XP or Windows Vista 32 or 64 bit, or Windows Server 2003 or 2008. The team should also be running the professional edition and using the AirPCap adapter for sniffing the wireless networks. If the team is using the professional edition with GPU acceleration, and a computer with an ATI Radeon HD5970 graphics card, then checking over 10,000 passwords per second is possible [20]. A Core i7920 graphics card only allows about 4000 passwords per second to be checked. It is best to check the website and documentation to see what kind of hardware is needed and the possible results that one can hope to achieve [20]. The tool uses dictionary attacks with mutations first, before trying a brute force attack. Many software tools that do dictionary attacks will add short prefixes such as 123 and similar suffixes to words.

If the IRT finds any evidence of industrial espionage, then the FBI needs to be called immediately. The Secret Service Electronic Crimes Task Force (ECTF) could also be called. If local law enforcement was called they would most likely not have resources for such a crime and would call the county prosecutor's office and ask for the computer crimes unit or white collar crime unit to respond. If one of the suspects was a terrorist and the pilfering of trade secrets was used to finance terrorist operations, then the JTTF (Joint Terrorism Task Force) might also need to get involved in the investigation. Some students have asked if the AR can directly call the Central Intelligence Agency (CIA) on a matter like this and the answer is: "the CIA, as a foreign intelligence agency, does not engage in U.S. domestic law enforcement" [21]. However, there is an email form on the CIA website that lets them know of information that people feel that they should have.

ADVANCED ARCHIVE PASSWORD RECOVERY

Elcomsoft also makes Advanced Archive Password Recovery. The standard edition is $49 and the professional edition is $99. This software is very useful just as Access Data's Password Tool Recovery Kit (PTRK) is for recovering WinZip, PKZip, and WinRAR passwords. Elcomsoft has one product differentiator which is "guaranteed unlocking of archives created with WinZip 8.0 and earlier in under one hour is possible by exploiting an implementation flaw" [22]. This tool also supports a diverse group of encryption algorithms as well as compression algorithms and unlocks archives by utilizing legacy shrinking, imploding, reducing, and other methods.

Dictionary attacks are used first and then brute force attacks. GPU acceleration can be used with certain graphics cards in conjunction with the professional edition. The optimized code works best when the examination machine uses a new multicore processor. If something is known about the passwords that the suspect uses, then user-defined masks could be used. Another good feature is that the tool can be run in the background and with unused processing cycles. The processes can be interrupted and resumed anytime. This is not possible with many tools. The tool can be used with ACE, WinACE 1.x, Winzip legacy versions to version 8.0, RAR, WinRAR, ARJ archives, and PKzip.

This software can work exceptionally fast if one has a certain condition occurring on the suspect's computer. The condition occurs if there is an unprotected file called Bill.doc and it is accessible to the investigator in some directory. Then the same file Bill.doc is in a zipped archive file. It is then possible to get that unlocked zipped archive unlocked or recover the password by a method known as a "plain text attack" and get Bill.doc and all the other files in that zipped file.

ADVANCED OFFICE PASSWORD BREAKER

Elcomsoft has a product called Advanced Password Breaker. The program is useful for the digital investigator that finds password-protected documents on a netbook or laptop and must recover the passwords. The program is reported by the software vendor to support the early versions of Microsoft Word 97 and Excel 97. It also supports Word 2000 and Excel 2000. It also supports the Microsoft Office XP and 2003 documents that are saved in the 1997 and 2000 formats. The password recovery tool uses thunder tables and unlocks 100% of the password-protected Word 1997 and 2000 documents in less than 1 min. If rainbow tables are used, the software vendor reports that 97% of the password-protected Microsoft Excel 1997 and 2000 documents can get their passwords recovered in a few minutes [23]. Much of the work can be split among CPU cores if one is using the newer multicore processors. The tool's success is that it attacks the encryption which is 40 bits instead of the long password and much of the work is split for simultaneous parallel processing on other workstations.

There is a low-level optimizing code which seems to be something like assembler routines which also saves time. In any case, the investigator needs to be able to quickly recover evidence from password-protected documents so that the investigation can move on. The standard edition of the software is $99, the professional edition is $199, and the enterprise edition is $399. There is ample documentation for the software too and blogs online where people can chat.

INSTANT FACEBOOK PASSWORD RECOVERY

This program could be good in child exploitation cases and missing person investigations where there is a missing person and it is exigent circumstances. Perhaps the young man is known to check in with his or her family and neighbors daily but nothing is heard for a few days. Perhaps there is a netbook computer and the son lives in the basement and comes and goes as he likes but nobody has seen him. He may be known to date online and make girlfriends from Facebook. Perhaps the mother hires a private investigator who has Elcomsoft's Instant Facebook Recovery program on a USB Flash Drive. The program is free and instantly shows Facebook passwords that are cached. It can also show the password and login information for a variety of browsers. The tool also works with Internet Explorer 9, 8, and 7's enhanced security models. It also works with Firefox 4,

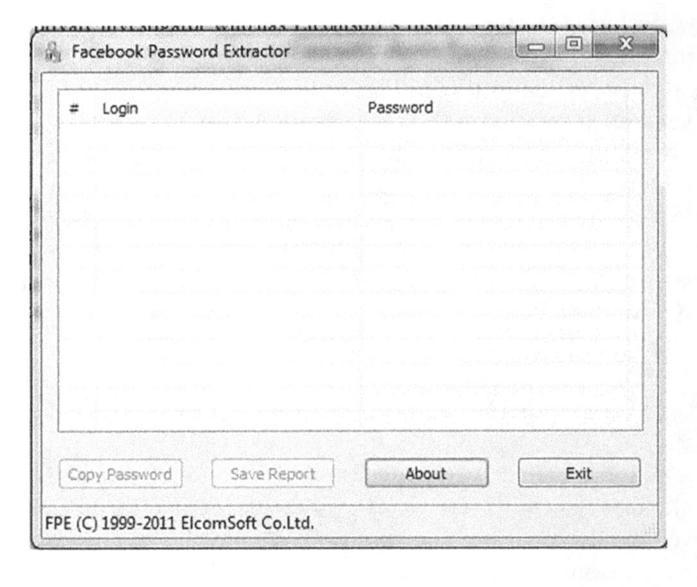

FIGURE 5.1 ElComSoft Facebook password recovery tool.

Google Chrome, Apple Safari up to version 5, and all versions of the Opera browser up to version 11.

The system requirements for this are very modest. It can run on Windows 98, Windows 2000, Windows XP, Windows Server 2003, Windows Vista 32 bit architecture, Windows Server 2008, and Windows 7. There is a simple report with a login name and password in a regular Microsoft Windows window as can be seen in Figure 5.1. The hardware needed is also modest. A simple Pentium processor should suffice. The download for the software is less than 5 MB.

ZEITLINE

Cerias makes a tool called Zeitline. This is a tool that in my opinion could be used in nearly any mobile device investigation because it is useful for creating a timeline of events. In a digital camera investigation, it could be useful to see when a group of pictures were taken and then deleted, for example. The timeline might be useful to show something such as the time frame in a set of pictures concerning a contract killing, and then deletion. It might be important evidence in a case. Zeitline is also useful in netbook investigations because it could be used to show events such as a firewall breached, an attack on some data, log file deletion, and escape. Timelines are extremely important in any investigation. Law enforcement and corporate investigators often wish to see the 24 h leading up to a crime and then the time afterwards. Zeitline can be used to save one time from manually creating a timeline of activities from a variety of logs.

ProDiscover IR SMART AGENT VERSION 6.7

This tool is useful for remotely imaging large groups of netbooks or laptops connected to a network as part of an investigation. Suppose there is a large-scale breach of data and many machines need to be investigated. The suspect machines could be ones that used modems and dialed in, or wirelessly connected to a company network. This tool allows for the remote sequential batch imaging of computer hard drives from a central location. A group of netbooks, for example, could be imaged over the network starting very early in the morning before the network bandwidth is almost unavailable. Then the same netbook imaging process might resume later after hours when the network bandwidth allows for the remote imaging of laptops both quickly and effectively. More can be read about this tool and its place in investigation at the following URL: http://toorcon.techpathways.com/uploads/SmartAgent.pdf.

ZeroView

An investigator may encounter a computer that he or she believes to be encrypted, but is not sure. There is a freeware tool called Zeroview from Technology Pathways. It is included on a bootable CD such as the free version of Helix 3. Once the system is booted to the CD, Zeroview will show the investigator the initial parts of the disk that will reveal clues to questions concerning whether the disk is encrypted or not. If it is encrypted, another certain set of tools and more expertise may be needed.

X1 PROFESSIONAL CLIENT

This program might be a very good application for the examiner's investigation machine for performing queries or text searches on a large number of diversely formatted files on a forensic image. Consider it for use with a large image with a terabyte of material to search. The GUI appears to be easy to use and the search engine could help increase productivity by allowing the examiner to quickly locate and read relevant emails, documents, and spreadsheets.

WINDOWS GREP

Here is a program with an easy-to-use graphical user interface (GUI) to help people look through large amounts of data for various strings of text and through zip files. There is a command line option for searching for character strings too. It is also possible to search binary files and exe files for bits of code which may hold important clues in malware investigations. There is a soundex feature for words that may be spelled in a variety of ways. Soundex saves one from manually searching every phonetic possibility for a word. There is an unregistered version and a version that costs money. Many people will tell you that it is better to get the registered program whenever possible because there is usually documentation and support that will save time. Many private investigators will tell you that in 2012, one of the most difficult parts of investigation is getting paid. There is financial pressure to keep costs down and do things fast. A small investment in a tool can often allow the digital forensic investigator to make better use of the tool and thus save time. The registered program for Windows Grep is $30 and one can use PayPal to send the money needed for the registered version. The website for this shareware program is http://www.wingrep.com/.

WinFingerprint

This is a security tool that was created with the Win 32 Microsoft Function Class.NET and it is useful to determine the operating system of a suspect's computer. The program is reported to be able to "enumerate users, groups, shares, SIDs, transports, sessions, services, service pack and hotfix level, date and time, disks, and open tcp and udp ports" [24]. This seems as though it would be a good tool for running on an image of a suspect's hard drive in a virtual window on an examination machine. The tool seems as if it could give a lot of information about the system as if one took a documenting snapshot of the system's configuration file. This might be a good tool for a hacking case or malware investigation. If one knows what hot fixes were installed as well as service packs, then this could possibly play a part in an investigation of liability. Network administrators are expected to patch vulnerabilities and do due diligence to implement countermeasures.

SAFARI BOOKS

I have recently heard certified computer examiners refer to their library as a tool because it provides knowledge, expertise, and references that can be quoted in a forensic report.

Many investigators say that they cannot wait for paper books to arrive at home after ordering them because they can take as much as 2 weeks for media mail. The investigators also say that the book's life is short and that their home library is growing too fast due to the frequent advances and changes in technology. This is a problem in the northeast United States where the cost of living and storage are both quite costly. If one lives in an apartment, then the weight of many books in a house becomes a serious issue. One solution to problems with weight, space, and short-lived books seems to be registering with an online book service such as Safari Books. A monthly fee gets an investigator access for downloading a certain number of books per month. Another benefit to online collections is the immediate access to large holdings of books. Online subscription access might be the way of the future.

Virtualization allows me the ability to run digital forensic images of many computer hard drives in various Windows on my examination machine and has allowed me to save space by discarding the old computers. However, with virtualization I have the benefit of many different systems. Now with online subscription access, I may be able to have the same benefit of having a large library at home on one's iPad or netbook. Then I may use X1 Professional Client or WindowsGrep to find what I need right away. The online on-demand subscription access coupled with virtualization may allow the certified computer examiner the ability to take all his or her tools, knowledge, and systems to every crime scene. One may have one laptop that is an examination machine with many tools and virtual systems installed. It may also have a downloaded library of computer forensics and operating system books. This may also allow a green solution, since less printed paper books are needed.

SILENT RUNNER MOBILE

Access Data's Silent Runner Mobile could be considered by many busy incident response teams to be a necessary investigation tool for collecting evidence from a netbook on the network. In a corporation where there is no expectation of privacy, the tool can be used to monitor network traffic. Perhaps someone who is sitting at a netbook at work is running their own business that does the same thing as the real company, but it is off the books for cash. Silent Runner Mobile would be a tool that would allow investigators to monitor and capture packets. The captured network packets can also be played back to hear phone calls and see documents and emails that were sent back and forth. The website says that it allows playback by demand. Silent Runner Mobile has a capability to let people utilize graphics to see the type of traffic that travels on their network. It reminds one of Wireshark. There is also capability to see chatroom activity. This has become important because some adult employees in the workplace may try to chat with minors outside the workplace and meet them for sexual activities. The tool can be used to monitor the network and collect evidence at the same time. Webmail and social media can also be analyzed. Many organizations have Facebook accounts but people may also operate their own personal Facebook accounts from work. This could be a problem if the activities are non-work-related and contain biased remarks or photos embarrassing to the professional image that the organization tries to uphold. Silent Runner Mobile could also be a good tool for Facebook and social media investigations.

The Silent Runner Mobile has the advantage that it is made by Access Data and has a lot of support. Since Access Data makes FTK, PRTK, eDiscovery tools, and Silent Runner Mobile, Access Data could probably be hired or find someone to hire to integrate all the evidence for preparation for court. It may be good to use one vendor for a complete solution for all the tools and support. This way everything is supported and nobody can blame another toolmaker for products not being able to integrate each other's data. One could probably go to various network security and computer forensic conferences to learn how other agencies have utilized the tool. It would also be useful to sit and listen to the wisdom of their lessons.

The tool could also be used in hostile environment investigations. In the 1990s, before I was a professor, I heard hearsay from fellow students about their day time jobs and workplaces where people had lewd screen savers on their computers and downloaded dirty pictures. If a tool such as Silent Runner Mobile was available back then, it would have been easy to catch employees who look at objectionable material on websites and download it for viewing or screen savers. However, as more investigative tools become available, people try to circumvent them by going to other websites that allowing tunneling to websites and obfuscate network traffic. People may also try to hide their identity and IP address at work. The legitimate need for obfuscation is if someone at work wants to be a whistle-blower about some bad practice in an organization. Consider this, "using proxy software (IP changer software) you can surf through proxy chains with any number of proxy servers in the chain to change IP address or hide IP address on the fly and provide IP address security and tunnel Internet activity through proxy servers" [25].

Silent Runner is also good not only for policy violation investigations but also to demonstrate cyber security standards so that regulatory compliance can be achieved. On the other hand, it is important to make sure that all the legal mechanisms are in place to legally use such a tool because listening to other's phone calls where there is an expectation of privacy sounds like a wiretap violation and someone could do some jail time. The program can be put on one central computer on the network and components can be distributed separately.

UnHide

This is an open source tool that could be used on laptops, netbooks, and notebooks. It has a purpose of locating hidden processes, rootkits, and TCP/UDP ports used by various types of malwares. The link for the site is http://www.unhide-forensics.info/. The link for the software download has moved once and might possibly move again, so it is better to Google search it to find the latest link. There is a Linux version and a Windows version of UnHide. This tool could be very helpful in solving an industrial espionage case where information could be covertly taken by a rootkit.

FORENSIC SOFT: WINDOWS FORENSIC BOOT DISK

There is a company called Forensic Soft that has created the Windows Forensic Boot Disk. This product could be very useful because it comes on either a USB flash drive or CD and allows the digital evidence examiner to bootup the suspect's computer to Windows PE. Windows PE is a stripped-down version of Windows. This environment has all the ports and storage devices configured as read-only, so nothing on any devices is changed and everything is forensically sound. Once the system is booted up to this new environment, one can go into explorer to preview all the netbook's documents, email, pictures, and digital evidence without changing anything. When the system boots up with this utility, it uses a RAM disk. The system loads up with many Windows drivers. This allows storage devices to be easily added for sending a forensic image to it. It provides NTFS support so that large storage devices can be added and one is not limited by the usual FAT 32 address capacity limits. A virtual write blocker is also added at bootup time with Windows PE, so that nothing is changed. A license dongle is needed or it goes into a limited trial mode. Trial mode allows the software to work with drives of 1 GB or less and the system only stays active for 15 min.

At bootup time, there is a great variety of resolution options. High resolution is good for seeing an entire application such as FTK on one screen. If one selects a lower resolution such as 600 × 800, this is good for having things magnified, but one may have to move scroll bars to see everything. After bootup, a wizard comes up that shows the time and date on the system and asks the examiner for the current time and date. This is good for showing the differential between the time and date on the system and the time and date where the examiner is doing an examination on the machine. There are prompts for the examiner's name, agency, and case number. The information about the system hardware, RAM, and connected devices will all be collected along with the system information and stored in a file. Before shutting down later, this log will be put on an external device such as a thumb drive.

Once the system is booted up to "Windows Forensic Boot Disk," a person can navigate menus to reach a drive directory structure. Then one could connect a large USB flash drive and make it so that it could be read and written. Then, one can use the menu system to navigate to a choice for preparing forensic media. It is here that one can wipe the large USB flash drive byte by byte. Then one can format the drive to be FAT 16, FAT 32, or NTFS. There is a quick format option too. The preparation is very easy to do in this type of environment. When this is done, one can put in another USB drive with Safe Tools. Safe Tools has FTK Imager on it and allows one to image the hard drive on the suspect's drive and put it on the large USB flash drive. This could take some time depending on the size of the suspect's hard drive and if the USB port is USB 1, 2, or 3.

The digital evidence examiner may not need to image the entire hard drive. Perhaps just using explore to get to the suspect's My Documents and then dropping and dragging it onto the large USB flash drive may be sufficient. Saving the log file on this drive may be fine too. Safe Tools includes limited versions of EnCase 6, X-Ways Forensics, and WinHex. If one has the license dongles for these programs, then the full version is possible. The other programs that do not require license dongles that are included with Safe Tools are FTK Imager, WinRAR, Irfranview, VLC Video Viewer, and Open Office 1.5. The Safe Consultant is $399 and the Safe Enterprise edition is $1199. They seem to have a nice phone and email support. The ease of use of the product would seem to make it a good addition to the incident response kit for the incident response team.

Another thing that is very good about this Windows PE environment is that there is a button to add more Windows drivers. Suppose a new type of storage device comes out; one can just load the driver for it by putting it on a CD or USB flash drive and loading it. Then one can just connect the new device to a USB port. Other tools can also be on USB drives and used. If one is a fan of Paraben, one could use that tool once the system is booted up. This program really provides a safe platform where one can add all kinds of tools and storage devices and safely preview and copy evidence. One could also use software encryption breaking tools or tools to find Facebook accounts.

Another thing that is very interesting is that the Windows Forensic Boot Disk and a seized image of a suspect's Windows XP desktop could be run in a virtual window. This allows for various types of testing of tools and experiments. This program could be part of a research program to teach laptop and notebook forensics to various groups of students, corporate investigators, and law enforcement personnel. The fact that so many different types of tools could be used in this safe environment without changing the original evidence makes it a good teaching tool. Forensic Soft also has some videos on its websites that can be used as a teaching tool or just to simplify the process of using their forensic software. The company also has a link about how to follow them on Twitter which shows that they are modern and highly interactive. This should also increase the ability to get technical questions answered if a problem arises.

One could also connect USB adapters that allow a connection with an SCSI drive, SATA drive, IDE drive, or EIDE drive. Then an examiner can use the Windows Forensic Boot Disk environment to image those drives to other types of connected drives such as the 1 terabyte USB flash drive. One is only limited by one's imagination and the set of adapters at hand to find creative solutions to forensic problems.

WINDOWS FORENSIC TOOL CHEST

The Windows Forensic Tool Chest is sold by Full Moon Software. This software does logging and puts the results in the HTML report. The software executes some forensic programs to collect information about a machine. The machine could be a netbook running Windows XP, Windows 2003, Windows 2000, or Windows NT. Many of the classically trained computer forensic personnel who used DOS prompts and command line tools should like Windows Forensic Toolchest because it does have commands and switches that can be manually typed. Outputs can be sent to a file. This tool chest also supports MD5 hashes. One command is "wft [-md5 filename] and that outputs MD5 checksum of FILE to stdout" [26]. The author's email is on the site and welcomes comments. Such openness to feedback can only be positive.

PARABEN'S SHUTTLE

It is nice to be able to recommend an investigation tool that is free and made by a reliable vendor for digital forensic tools. This software called Shuttle is downloadable from the Paraben website and can be used by law enforcement or corporate investigators to obtain the active memory or RAM of a computer that is on the network. One can see active processes running on that computer and then do a printscreen. Services such as ftp or telnet can be seen if they are active. Sometimes people misuse services such as telnet to tunnel to a computer in another country and then once in that country, open another telnet session to an additional country. This type of tunneling is sometimes done to obfuscate the crime and add complexity by adding additional countries with different laws. However, since all the RAM and processes can be captured, it should be possible to printscreen the whole trail of tunneling, and document the crimes being actively perpetrated live. This is also interesting if one wishes to check the memory of a corporate computer's memory without alerting the user. This could potentially avoid the need to create an unnecessary scene and reduce workplace violence. An allegation of hacking a computer while sitting in the office could be checked on the network by seeing the active memory of the netbook, laptop, or desktop. Many investigators say that in the future, it will not always be necessary to knock and enter an office. The option of remote acquisition may be the desired action. It is easier and reduces the potential for violence.

If there exists an employee who is engaged in an inappropriate chatroom conversion with a minor at work, this could be costly and embarrassing for the company. Such an allegation could be easily checked if the person is using the company computer on the network. Catching a child predator and having it handled quietly with law enforcement arresting the culprit is better than a public investigation where law enforcement is brought in and perhaps followed by a TV crew. Paraben's Shuttle or Shuttle Pro would be a good choice of tools for law enforcement to use to investigate a laptop, netbook, or desktop computer that is connected to the network and allegedly doing this type of activity. The ability to do stealthy investigations while not confronting the person under suspicion cannot be overestimated.

This Shuttle program also shows any device drivers that are being used. This is important because a person may immediately kill processes but print documents at a remote printer for an accomplice to obtain. Paraben's Shuttle allows the active device drivers to be seen and the printscreen feature can document it. This tool could potentially be used covertly by the AR to catch a thief pilfering trade secrets and then printing them on a remote printer for an accomplice. An AR could then stealthily dispatch an incident response team to the printer site and at the hacker station and catch everyone in the act. Since the evidence could be remotely captured, the chance for destruction of evidence is reduced. Law enforcement could be called at the same time and be there to backup the incident response team. The Shuttle tool could be used as part of the incident response policy in order to preserve evidence and reduce violence.

If the Shuttle tool is configured correctly, any type of disk on a network can become a local drive. An acquisition or forensic image could be done once the drive is considered local. Once an acquisition is done, email archives could be examined. However, if an email acquisition is done, the examiner better have had the proper authority because a lawsuit will most likely occur. If one is going to do live file remote acquisitions, then one should use tools such as F-Response or Paraben's Shuttle Pro. It is best to use tools that are tested for remote acquisition and do not possibly corrupt files while in transport.

BackTrack 5 FORENSIC TOOLS

Backtrack 5 was released in May 2011 and has many tools for the computer forensic professional and the network security professionals. Some tools could also possibly be misused for hacking. There are two programs in it that are particularly useful for finding rootkits. The programs are RkHunter and ChRootkit.

RkHunter: BACKTRACK FORENSICS TOOLS

There is a tool called RkHunter and it is located within the Backtrack 5 set of forensic tools. This type of tool is similar ChRootKit in that it looks for rootkits. This type of tool might be used on a netbook or laptop with a Linux- or Unix-type operating system installed. The digital evidence examiner might want to see if there are rootkits, if it seems that the machine in question was hacked. A person accused of intellectual property theft who seems unaware of anything regarding the accusation might be the victim of hacking and intellectual property theft. RkHunter could be used as a tool to look for rootkits by running it against the system with the –sk option so that one does not have to keep pressing a key to continue.

Mork.Pl

The Mork.Pl is a program in Perl that collects the database files for the older versions of Firefox and lets people see the Internet browser history. The program is good for looking at Mork files that were used in Mozilla. Many people have old computers around home and those have old browsers. The tool could easily be considered relevant in 2012 and beyond.

HexEdit

The hexEdit is very similar to a program that many computer forensic professionals encountered in the 1990s in a suite of programs called Norton Utilities. Version 2.0 had a hex editor that allowed one to look at any address on a storage device. One could examine the masterboot record of a drive or go inside a file of a picture to view the metadata. This type of tool allows one to quickly see some of the EXIF metadata embedded in pictures that has to do with the camera that took the photo. The ASCII characters and hexadecimal equivalent stored in each byte can be viewed. The program can also be used to replace sigmas in file allocation tables to manually restore nearly any file in a FAT file system. HexEdit is also a good tool to examine system areas and hidden partitions on the storage device where people may hide data.

ExifTool

This is a tool that allows the digital evidence investigators to examine the metadata in images, audio files, and video. The tool is more focused on metadata whereas HexEdit can go to any part of the file or storage device. ExifTool would be good to use if there was a strange audio file and the examiner wanted to know what type of program and version was used.

Evtparse.pl

The program is called Evt Parse because it parses through the .evt files to create a timeline for the investigator. The tool would be good to utilize after a program such as WebJob undid the tarballs into system and log files. This program takes the logs from the event manager and then processes it into a timeline that the investigators can look at and try to piece together the time shortly before and after the network penetration. The event files are usually stored in files with the .evt extension.

MISSIDENTIFY

This is a very useful program to uncover executable files that have been renamed with an extension such as .doc in order to avoid detection. File renaming has been known for some time. Pedophiles have been known to rename files such as 16yo.gif as 16yo.doc so someone thinks it is a document and not a picture. Access Data's FTK locates the pictures being renamed by analyzing the file format against the file extension. Missidentify is looking for those programs that run which have been renamed so they stay covert.

RegLookup

This tool could be very useful if you had a powerful laptop running Windows NT and wanted to see how the registry was configured and what values are in it. The output is made to be easy to read and it is more organized than a printout of raw values.

ReadPst

This utility helps people convert the Microsoft Outlook pst files to mbox files. There are tools that are said to make mbox files easy to examine. The program is run from a command line.

Pref.pl

This Perl script can be considered very useful because it parses what is in the prefetch files and directories of Windows XP and Windows Vista. The resulting output may be placed in a format of comma-separated values (.csv) which can be looked at in programs such as Microsoft Excel.

Ptk

This seems to be a very useful program that is browser based and allows one to look at all kinds of files to examine them and see the metadata. It does need to be configured to use a MySQL database. Some people compare it to the Sleuth Kit.

StegDetect

This utility looks for signatures of well-known steganography programs in the jpg files on the media in question. It can be used with a directory of pictures to flag down which ones need data carving to possibly remove a document from.

VINETTO

This tool runs on the command line of Linux and can be used to recover thumbnails and the metadata that is found on the thumbs.db file. The thumbnails are not high-resolution but can show the general image of a much higher-resolution picture that was deleted or is still on the system. Vinetto can be used on systems that have Microsoft Windows 98, Windows 2000, Windows XP, and 2003 Server.

FatBack

This tool seems to be good for recovering files from FAT file systems from netbook hard drives, laptop hard drives, digital cameras, USB flash drives, and floppy disks. The tool appears to be good for both FAT 16 and FAT 32 file systems. FatBack is considered to be a data carving tool.

FOREMOST

This program seems to work well in certain circumstances with forensic image files. It may be used to carve out pictures from unallocated space. Data Lifter 2 is a program that can also recover jpg pictures from unallocated space by collecting the header and footer of a picture and reuniting the picture.

RecoverJPG

This is a program that helps one to take either a forensic image or a partition and recover the JPG pictures. It is a data carving tool that works similarly to Data Lifter 2, Foremost, and Adroit. It may be an important tool for use in child pornography investigations.

SAFECOPY

The safecopy program could be considered by many to be very useful in the digital forensic examination of any device that got damaged. Many of the data carving or data recovery tools stop when they hit bad clusters or bad sectors. Safecopy skips that bad spot and moves on. It will try to go from any point to any designated stopping point.

SCALPEL

This program is included in Backtrack 5.0 and allows one to carve out a particular type of file. There is a database of file header and footer file signatures that it can use. This is another useful data carving utility for investigators.

SCROUNGE-NTFS

If someone had a laptop that was configured for NTFS instead of the traditional FAT file system, then Scrounge-NTFS might be good for recovering partition information and the file system tree.

TestDisk

This is a program that may be very useful at recovering the partitions on the disk. It can also assist at making a disk bootable again. It can often be very useful for rebuilding partition tables and could help someone locate a hidden partition. It might assist in fixing the MFT or master file table. Many people also see it as another file recovery tool. It is a good tool for both the NTFS file system and the FAT 32 system.

HashDeep

This is an important program for performing hashes on a set of files. Then those file hashes can be compared to a set of hashes known to be hacking tools or child pornography. This can basically be used to audit a set of files for known contraband. If there was an intellectual property theft case and a flash drive, the intellectual property file (IPF) could be hashed and then Hashdeep would hash every file on that drive and compare it for the IPF.

MD5DEEP

This is a program to create MD5 message digests on files and directories and compare them to other files or a list of MD5 hashes.

TigerDeep

TigerDeep is a program that compares tiger digests in the same way that message deep is used for message digests.

ddRESCUE

This is a program that can be useful for imaging a hard drive, SD card, or flash drive. The good part is that it tries to recover the bad areas of the drive unlike dd which just puts in zeroes for those areas. The tool could be considered important for any digital evidence examiner to have.

EWFacquire

This is another tool for forensically imaging a drive or partition. This differs from the other tools in that the results are placed in an EWF file. Some computer forensic personnel have tools that work with EWF files so this is a good tool for them.

DriftNet

This program seems really good for saving time because it can help sift through network traffic to find pictures. This might be a useful tool in an investigation where someone with a PDA, smartphone, laptop, or netbook is allegedly using the company network to pass child pornography pictures. It could also be used to check for images that might be passed with documents embedded in them. The collected pictures might be run against StegDetect for possible embedded documents.

TcpReplay

This program can be used to take captured packets and then replay them on a live network. This is very good to see how an attack might have been carried out.

WireShark

This is a famous network security and investigation tool that can be used to monitor traffic on the network and then possibly examine it. It allows the captured traffic to be looked at later. It is also possible to capture VOIP packets and then use a tool to replay it. Wireshark is considered the premier network sniffing tool. There are conferences devoted to the tool.

CmosPWD

This seems to be a useful tool for helping crack the BIOS password of a laptop, desktop, or netbook. Sometimes people use a BIOS password so that the machine will not fully bootup without the password. It is difficult to bypass the BIOS password because one usually has to open up the computer, change a jumper on the motherboard, and remove one or two batteries. Laptops and netbooks have very difficult screws to remove and sometimes require special screwdrivers with specialized star tips. There is very little space in netbooks too. It is better to use a tool such as CmosPWD to crack any type of password and get to the system.

fCrackZip

This is a program to crack passwords from password-protected zip files. There are times when password-protected zip files are found on a laptop or netbook system in question and they need to be examined since many files of interest could be in the zip file. The fCrackZip is a useful tool for quickly breaking the passwords with a brute force attack and dictionary attack.

SAMDUMP

Samdump is a program that netbook and laptop examiners can use to remove the password hashes out of SAM files. When people create passwords in Microsoft Windows-based systems, the hashes are stored in SAM files. Each local user who has a registry key has a SAM file for his or her passwords. Samdump receives the exported hash from the SAM files and puts them in a file called Sam.file. Then one can use another tool to try to get the password from the hash.

PDF-ID

This is a useful program for getting some of the header information from the pdf file.

PDF-PARSER

Hackers may use pdf files to trigger malware in a person's system. PDF-ID is an investigation tool that may be used to examine some of the content within a file as part of an investigation to see what might have triggered one or more pieces of malware. When some characters are discovered that appear to have no meaning, then tools such as ASCIIhexDecode can be used. Wepawet and FlateDecode can also be used to try to investigate malware in PDFs. Wepawet is usually used on files that were on a website.

PeepDF

This program helps the investigator go deep into the PDF to the byte level in order to investigate the possibility of malware or code that could be used to trigger malware.

PDFBOOK.PY

This is a little program that takes all the incoherent data that some people might refer to as "gibberish" from a process dump and removes the Facebook information. It could be a useful tool in finding out about Facebook page data for giving clues to a person's activities for use in missing persons cases.

PDGMAIL

This is a little program that takes all the incoherent data from a process dump and removes the Gmail information. It could be a useful tool in finding out Gmail communications between parties who are under investigation.

VOLATILITY

This is a program that is used to analyze memory dumps from Linux or Windows. The computer's RAM, also known as the memory, is the most volatile evidence and should be collected first. It is often good to use a tool such Volatility first to collect the RAM and then use a tool such as pdfbook.py to look for Facebook information. Perhaps there is the page of someone that the missing person was last looking at. The results of Volatility can also be combined with pdgmail to obtain Gmail communications. The good thing about Backtrack 5 is that the tools are in the same package so that various specialized utilities can be used sequentially to help solve a case.

PARABEN'S WINDOWS BREAKER

The Windows Breaker program is made by Paraben and greatly assists a person to enter a Microsoft Windows Operating System account that they do not have the password for. The program comes installed on a USB flash drive. If one can bootup to a flash drive, then one can use the tool. This may mean that one has to first bootup the computer, push a key such as F2, and then go into the bios to change the device bootup order. In the old days of computing such as the late 1980s and early 1990s, the bios checked the floppy for a bootable operating system first and if none was found, then the hard drive was checked. In 2012 we have numerous options such as the floppy disk, USB flash drive, CD ROM, and then the hard drive.

The program is very important for digital evidence examiners who need to get in a netbook or laptop that has the Microsoft Windows operating system installed but the account password is lost. It is also worth mentioning that the action of bypassing passwords should be done in situations where law enforcement has a search warrant or in an authorized investigation in a corporation where there is no expectation of privacy on the device in question. That means there should be some signed policy documents in human resources allowing the investigator the right to look at anything on the computer. In a criminal investigation, there should be a search warrant. Windows Breaker might also be used for a warrantless search by customs and border patrol at the border because of the Fourth Amendment exception. That exception allows all containers like luggage, and all digital containers like hard drives to be examined when entering the country.

LogicCube FORENSIC TALON ENHANCED

The LogicCube Forensic Talon may enable the digital investigator to forensically image hard drives. It is considered a data capture system. It could be of great use to private investigators, corporate investigators, military police, law enforcement officers, and computer forensic students who need to make a copy of a hard drive in the lab. The unit is reported to image hard drives at speeds of 7 GB per minute and then verify it [27]. Older LogicCube MD5 devices worked at speeds of up to 2 GB per minute. The device allows the suspect's drive to be captured into segments of 650 MB, 2 GB, and 4 GB. The vendor recommends putting the segments on CD or DVD for further analysis [27]. The investigator can choose to use either MD5 hashing or SHA-256 hashing for the DD files that are captured.

The system can be used with adapters from LogicCube that allow the 1.8-inch and 2.5-inch hard drives from laptops, notebooks, and subnotebooks to be connected and imaged. The digital evidence examiner may also purchase an adapter for compact flash drive imaging. There is support for the IDE, UDMA, and SATA drives. LogicCube salespeople can often be found in the vendor areas at many of the computer forensic shows and computer forensic conferences. It is a good idea to meet them, see a demonstration of the product, and get marketing literature to show your agency's management. It is also good that LogicCube has a phone number and email and that technical support is possible. Many people who are certified computer examiners use LogicCube products, so it is possible to pose questions to the CCE listserver and get an answer. It is important to be able to find a community of users to ask questions, seek advice, and have some camaraderie with.

The Talon also has a port on it for use with a cable. The cable can plug into the USB port of the desktop computer and there is a way to image the computer without removing the hard drive. This ability is extremely important because some people without A+ Certification and are not good with computer hardware may break the pins or bend them when removing an IDE drive from the computer. There is also another opportunity to bend pins when connecting the suspect's drive to the Talon. If one can leave the drive intact and image the hard drive without booting up the system's hard drive, then that is the best methodology. Perhaps it might be able to use the same methodology with a netbook or notebook. Both netbooks and notebooks are small and the drives are difficult to remove. If there is a way of imaging the drive without removing the drive from the unit, then that is the way to go. Desktops often have complicated latches and screws that need special screw drivers and if one can avoid opening them, then that is good. Many times when one opens the desktop, there is a hazard of electric shock or damaging a component with a static electricity discharge. It can also be difficult to get a desktop's metal cover back on and there is a risk of getting cut on the metal. The bottom line is that it is best if one can forensically image the internal storage device of any computing device without opening it.

The Talon comes in a ruggedized case with room for cables and a small printer. The case is airtight and water tight and is good for black bag operations. It is possible to use the Talon in a corporate after hour's investigation that leaves no mark on the suspect's computer. There is also a warranty that includes 1 year parts and labor. There is also an extended warranty available to push that warranty ahead two additional years and giving a total of 3 years. I believe in extended warranties because repair money may not be available later.

ADVANCED EFS DATA RECOVERY

Elcomsoft has a program called Advanced Data Recovery that costs $149 and the professional edition costs $299. This tool has significant documentation in multilanguages and may be useful for digital evidence investigators who must recover decrypted files from a notebook computer, desktop, or laptop that utilized Microsoft's Encrypting File System (EFS). This tool can be used with Windows 7, Windows Server 2008, Window 2003, Windows XP, and Windows 2000 for occasions where files were protected with encryption. This tool is reported to be powerful enough that recovery is possible for occasions when the hard disk is not bootable, the system suffered some damage, and the keys for encryption have been compromised [28].

Sometimes a hacker will enter a network and compromise machines by misconfiguring data, deleting user profiles, and possibly transferring users to another domain so that account could provide compromise to another domain later. It is reported that Advanced EFS Data Recovery recovers may still be useful for recovering encrypted file system data in such circumstances [28]. It seems that the digital evidence investigator would be advised to spend the extra money for the professional edition because the tool offers more time saving features. Consider the following quote from the vendor: "The Professional edition locates master and private keys in the deleted files as well, scanning the disk sector by sector and using patterns to locate the keys, allowing the recovery of re-formatted disks and overwritten Windows installations" [28].

A criminal who is using an EFS may have made his or her computer unusable because of a mistake such as upgrading the system without taking care that the system was using EFS. It is easy to forget that the EFS system was used since it seems transparent to the user [28]. The inaccessibility of the computer could be due to user incompetence, hacking from an outside source, or deliberate sabotage to the system to avoid examination if the police arrive.

Once the encryption is broken and files can be examined, then the real tasks of investigation can happen. Perhaps mailboxes of backed-up email could be recovered. Then other tools such as Vound Software's Intella Team Manager could be used to divide the work on the storage area network with other investigators who could use Team Reviewer in order to individually go over work and annotate it. The investigator's machine should have Windows 7 (32 bit or 64 bit architecture) or even an old computer with Windows 98. However, newer machines with faster processors should get results sooner than classic machines, but in these challenged times, it is good to know that the classic Windows 98 computer is sufficient for the job as long as administrator privileges are needed. The system requires only 2 MB free for the program. This is less than one high-resolution photograph in 2012. There is also a newsletter for the tool with tips and what some may consider to be called the best practices. There are online blogs about the tool where people can discuss problems, solutions, or new capabilities that they will ask the software vendor to include. There is a strong community of digital forensic professionals online and at conferences.

PsLoggedOn (INCIDENT RESPONSE TOOL)

The PsLoggedOn tool was written by Mark Russinovich and is available free online at the following URL: http://technet.microsoft.com/en-us/sysinternals/bb897545.aspx and the downloadable file is only 1.6 MB zipped and the website says that the publish date is April 28, 2010. This tool would probably be a good addition to the incident response toolkit of any incident response team, especially one that responds to system breaches. I could envision this program in use with WebJob to document and collect evidence to who may still be logged on the system and hacking a system or causing malevolent activity. The limitations of the tool are that you cannot see the resources of a remote machine which means that you can only see the next system logged on. Perhaps someone is controlling a computer that is controlling a computer that hacked in your network. The PsLoggedOn program carefully examines the hives of the registry and then checks HKEY_USERS for keys that have a security identifier (SID) and people with usernames who may be logged on your computer through the network. The NetSessionEnum API may be necessary to get the identity of the other person on shared resources such as a shared hard drive [29].

WebJob (INCIDENT RESPONSE TOOL)

There is a program called WebJob and it was written by Klayton Monroe. It is available at the URL http://webjob.sourceforge.net/WebJob/index.shtml. It could be considered to be a very good tool for allowing known good investigation tools to be run safely in a compromised hacked environment. It was created for assisting incident response teams who are performing an investigation on a single or large group of networked computers. WebJob might be good for laptops running Windows 2000 or Linux that got hacked and compromised. The tool was written in the C programming language and has run in many popular operating systems including AIX. WebJob can often be run from a floppy disk on old systems or a CD ROM on newer ones. According to the WebJob Project, "WebJob can be configured such that it is minimally invasive to the target system. This is important when trying to collect evidence of an attack on live systems" [30]. The tool is also good for massive searching with credit card numbers, hash numbers of hacking tools, and social security numbers [30]. It also needs just one port, 443, for outputting data to the WebJob server for collection and could be useful in creating a virtual evidence locker known as a VEL.

Once a hacker compromises a system, he or she may take a group of files and put them in a big file known as a tar ball. This is done to hamper investigation and frustrate investigators. The term "tar ball" is often associated with oil spills and is a messy object that people do not want to touch. Hackers often use tools that take system files, logs, and other objects and create a mess of files of fused files known as tar balls. WebJob is reported to include a tool known as PaD for taking tar balls and making them config files again [30].

Some investigators may prefer to run the minimally invasive WebJob program from the CD and collect data if the attack is still in play. Perhaps some evidence about the attacker or the connection path to the attacker's computer may be gathered. This may be important to successfully gather evidence and successfully prosecute the case. It is also possible that results may be put to a standard output. If the attack has been done, perhaps it is better to forensically image the system and then analyze it in an environment such as Encase. Later, it might be worth it to put the forensic image in a virtual environment and run programs on it to collect evidence and run tests. An experienced incident response team member who has worked on hacking cases could be the one to consult about this. It would also be worth to contact FIRST, the Forum for Incident Response Teams, and see about getting some advice or a consultant to help with a difficult hacking case. Sometimes it is necessary to seek help from experienced network forensic professionals on difficult cases. The International Association of Computer Investigative Specialists (IACIS) would be a good contact too.

AIR: AUTOMATED IMAGE AND RESTORE

There are open source tools designed and distributed by people who feel that computer security, computer forensic, and network security tools should be free. One such tool is called AIR and it was made by Steve Gibson and Nanni Bassetti. Steve Gibson has a trusted name in computing and his utility SpinRite has been used by computing departments worldwide since the 1990s. The tool has some similarities with Paraben's Forensic Replicator and both are very good for forensic preparation of digital media. Partitions or drives can be zeroed out by pointing and clicking on some buttons on menus that are found on an easy-to-use graphical user interface that comes with AIR. There is also a logging capability for both dates and times so that once a forensic acquisition was made, the start and stop times are logged. There is also a hashing function so that the examiner can verify that the source drive or partition has an exact image or clone on the target drive or digital media. There are five hashing functions which are: MD5, SHA1, SHA256, SHA384, and SHA512. Having a variety of hashing functions that are complex can increase confidence in the integrity of the data.

The program is also good for old and new computers in the home or workplace because there is autodetection for the SCSI drive and the newer IDE and EIDE drives. Sometimes CD-ROMs are hard to detect but AIR has autodetect for both tape drives and CD-ROMs. This is important because there may be some important data on some proprietary tapes and this can help in recovering the data and segmenting it into files that can later be reassembled and examined in a program such as Encase. The program also contains features for remote acquisition of a drive over a TCP/IP network. This can be done with the command lines of netcat and cryptcat. Netstat can be used to show what is connected. The logging is important when one uses command line tools such as netcat, netstat, and crypcat because some have little switches such as netstat -a and it is important to be able to show the court what was done later. It seems that logging features are more reliable than hand-written notes since they are generated by the system. The remote acquisition of files or entire tapes or drives by tools such as F-Response, Paraben's Shuttle Pro, and AIR are becoming more common as incident response teams are becoming smaller. Remote acquisition also allows for the avoidance of possible violence with a suspect when seizing evidence since they may not even be aware of the process.

It is also good that the program supports dd or dc3dd to create images files. Three dd files would be enough for an acquisition of an old laptop with a 6.4 GB drive such as the Toshiba Satellite 4080XCDT laptop with Windows 98 SE. The dc3dd version 7.0 is not currently supported at the time of writing in January 2012. The dd files are also good because they are so universally supported by a variety of digital forensic tools and can even be easily reassembled and examined. The dd files can also be processed and reassembled into a virtual machine and run in an examination's desktop Window. This is a very good method of previewing a machine because the examiner can point and click on icons, look at files, and get the feel and bird's eye view of all the programs and data available to the suspect. Perhaps even a steganography program may be noticed.

RUNTIME SOFTWARE

Runtime Software has a set of tools that could be of great use to anyone doing computer forensics or network forensics. They could also be of use to anyone who is doing network security and IT functions to backup drives and restore them later. The URL for their products is http://www.runtime.org/data-recovery-products.htm and it would probably be good to have a look there. The following are discussions of their tools, their requirements, and how they might be of use to mobile device investigators.

DriveLook VERSION 1

Runtime Software has a free tool for download that is 1.37 MB. It downloads very quickly with a broadband connection. It is a disk investigation tool for looking on a netbook hard drive. It is also possible to search strings of text. The tool is also good for searching forensic images for a word and then seeing the location of it. Every word on the hard drive can be indexed which is good for evidentiary purposes. One can also use a physical drive as input but if it is partitioned into multiple drives, logical drives can be checked. DriveLook 1 should run on the following operating systems: Windows 7, Windows XP, ME, or Windows 95 or 98. It is good that the tool runs on old or new system but 32 MB of RAM is needed. Lifetime software updates are possible too.

One thing that is very interesting is that DriveLook can be used for remote access for drives connected to the network. This is important for searches on remotely connected drives for contraband such as hacking tools or prima facie evidence. Many forensic tools are now offering a remote connection for viewing and collecting evidence. This is very important for implementing an investigation and reducing the possibility of workplace violence in a confrontation. Many people feel that the future of forensics is in the usage of remote connection and collection tools for gathering small amounts of data to prove a workplace policy violation. If more is found, then a regular investigation can be done once everything is turned over to the police.

DriveImage XML V2.30

This seems to be a good tool from RunTime Software for imaging a drive or making a drive to drive copy. It also seems useful for backing up the forensic computer in case it goes bad and the programs need to be reinstalled. This tool might be very good in a high-security computer forensics lab because after each investigation, the drive is wiped, and then the new image is restored with DriveImage. Using this wiping methodology, any electronic residue of malware or pornography is gone and the machine is restored to a clean image. The private edition is free but the commercial edition costs money. However, it is often better to purchase the commercial edition because of both the increased capability and the updates. It is also good to purchase pay versions of the software because technical help is also available. An answered question may save lots of time in an investigation. The program could be very useful to investigate a netbook on the network.

GetDataBack FOR NTFS V4.25

This tool has a pay version that costs $79 and could be useful for recovering data from a computer where the suspect has a computer that has an NTFS system that lost the data. The data could have been corrupted from a power surge, a virus, or perhaps the suspect formatted to try to destroy evidence. This tool can sometimes recover the files even though the Master File Table, root directory, or partition table is destroyed. There is an online ordering form for purchasing the tool.

GetDataBack FOR FAT V4.25

This tool has a pay version that costs $69 and seems useful for recovering data from a computer where the suspect has a laptop, desktop, or netbook computer that has a FAT system that lost the data. The data could have been corrupted from a power surge, a virus, or perhaps the suspect formatted to try to destroy evidence. This tool can sometimes recover the files even though the FAT was destroyed. There is an online ordering form for purchasing the tool.

CAPTAIN NEMO PRO V5.05

The program costs $90 and includes lifetime updates. It can be purchased from the Runtime Software Company. The system requirements are Windows 95, 98, ME, XP, 2000, 2008, Vista, or Windows 7. The program only requires a Pentium Processor and 512 MB RAM. The downloaded file is 2.9 MB. The program can be used to image a Novell, NTFS, Linux EXT2FS, EXT3FS, or XFS file system drive. Captain Nemo Pro V5.05 can be used as a file viewer and to look at the metadata concerning file size and creation. The tool is also useful for mounting images from dd files to a drive.

DISK DIGGER V1.0

This tool is available from the Runtime Software company and is only available as part of DiskExplorer Bundle or the RunTime Bundle. This tool seems very useful for navigating and inspecting any file system or on-disk structure. The file structures are displayed in an Internet Explorer window. The program has scripting features which may be customized for showing data. The scripts are for supporting various file system structures such as XFS, FAT 12, FAT 16, FAT 32, EXT2FS, and EXT3FS. The downloadable file is 1.66 MB.

ShadowCopy V2.0

This tool is available from the Runtime Software Company. It is free and the download file is 0.92 MB. The program requires Windows 7, Vista, 2003, or Windows XP. The program is good for copying files onto other media. Sometimes files are considered locked and cannot be copied. This tool seems good for them too.

RemoteByMail V1.01

This tool is available from the Runtime Software Company. It is free and the download file is 1.1 MB. The program requires Windows 7, 98, ME, NT, 2000, Vista, 2003, or Windows XP. The program seems very good for remotely accessing a file and having emailed it to oneself. This might be good for the computer forensic professional who forgot a report and needs to access it from home.

PYTHIA V1.02

This tool is available from the Runtime Software Company. It is free and the download file is 1.39 MB. The program requires Windows 95, 98, ME, NT, 2000, Vista, Windows XP, or Windows 7. The program requires 32 MB RAM. It seems good for looking for hidden patterns in data from various file formats. Perhaps one can do some research and see if it is possible to train the software to look for patterns indicating steganography.

ULTIMATE TOOLKIT

The Ultimate Toolkit (UTK) is a toolkit for computer forensic professionals made by Access Data. It includes many tools that are necessary for acquiring, preserving, analyzing, and reporting on digital data. The first thing one has to do before imaging the suspect's hard drive is prepare the forensic media to receive the image. WipeDriveTM is a tool for making a certain number of passes over a hard drive or flash drive to rewrite every byte. This is necessary because there could have been malware that got on the drive in the factory. The malware could be hidden or there might have been a rootkit that is difficult to detect. Wiping the drive to Department of Defense standards is necessary. Once it is wiped, it is still good to run antivirus and antispyware against it in case a lawyer asks.

Before a company considers purchasing the UTK, they need to see if their hardware supports it. The incident response team and examiners need Microsoft Windows 2000 or Microsoft XP. A newer operating system such as Windows 7 might be fine but it is best to call Access Data's technical support number in Utah to find out if the incident response team's operating system can run that UTK. The processor that will be running the UTK needs to be AMD compatible or a Pentium IV with a speed of at least 2 GHz. It may be possible to run the UTK on a machine with an older slower processor but too much time may be wasted on trying to search for keywords and checking hash files on the suspect's computer with hacker tools and child pornography. Manpower is often scarce and caseloads are high in investigative agencies. The best hardware is needed to maximize time so sufficient evidence can be gathered and the guilty are not free because thorough investigations cannot be done in a short time before trial. At least 2 GB of RAM is needed and the CD-ROM should be 4X speed. That should not be a problem since many mom and pop computer stores sell 16X CD-ROMs. The screen that the investigators will be using needs to support a minimum of XGA with a 1024 × 768 pixel resolution. This is because the screen has many menus and functions and is very busy. If the resolution is not there, it is difficult to see everything on one screen. There needs to be at least 1.623 GB of free space on the hard disk for program files. There is a USB security device or dongle shipped with FTK, so there needs to be a USB port. It is better if it is at least USB 2.0 in case the examiner uses a hub and a flash drive. Before a company considers purchasing the UTK, they need to check their budget too.

The UTK is from Access Data. Many people say Access Data has great support where people can call technical support, ask questions by email, and get news from the website. Access Data provides 1 year of free maintenance and support after purchase on many products. There is also a community of people who are certified computer investigators and use the UTK that one can ask in the list server. One can also join the IACSP and HTCIA and speak to people who this product and find people to ask questions about the Windows registry. The UTK also contains a Registry Viewer. The registry often holds the clues to the identity of a hacker or person who was connected to that machine. It may also be used in conjunction with a log file to tell the investigator about a specific flash drive or piece of equipment that was connected. It would behoove the investigator to take a class on the registry. The website also says that the Registry Viewer needs a USB dongle too. There is also a report-generating tool for marked evidence from places such as the "Protected Storage System Provider Key" which holds email passwords and settings. The same area also holds Internet passwords and settings to webmail programs, online vendors, or anything else people use frequently and store things in.

The UTK also contains the PRTK or Password Recovery Toolkit. This is important for breaking passwords and getting access to files. Brute force attacks and dictionary attacks that use large prefixes and suffixes need sufficient processing power to get the job done in a sufficient time. The perpetrator does not need the passwords to be unbreakable but only be unbreakable for a period of time during an investigation so that there is no opportunity to collect the needed evidence to establish guilt. That is why it is important to have a processor with a fast speed. The UTK also includes a 50 client Distributed Network Attack license.

eDISCOVERY SOFTWARE

Access Data makes the eDiscovery Software which is useful for preparing digital evidence for trial. Women in eDiscovery is a group whose conferences are listed on the Access Data website. Attending a chapter meeting or going to a conference might be a good way to learn more about using the software and seeing how it supports eDiscovery. Once the data is collected by the computer forensic specialists from a laptop, notebook, or desktop, it needs to be reviewed with document readers, paralegals, and attorneys who bookmark evidence and comment on certain documents and email. Without an eDiscovery software tool, it is too time consuming and too complex to organize all the data and make it in a format that courts want.

In the early days of computing and email in the 1990s, hard drives were less than 1 GB and people did not have the volume of email that there is today. It was easy for computer forensics investigators to image a drive, collate the evidence, and copy and paste some evidence in a report along with commentary and that was it. In 2012, hard drives are 1 TB and larger. That is one thousand times larger than before. Many people save everything and no longer delete unimportant emails to make room for new ones. People have more spam, junk email, and other things to look through. There are often large numbers of people involved in wrongdoings such as the Enron case and complex graphs need to be created with lots of players. It is too much for one investigator with a high caseload. County prosecutor's offices are needed to help prosecute cases and use eDiscovery tools. It is my opinion from attending HTCIA and other Homeland Security conferences that many small towns do not have the resources of trained people and eDiscovery tools to handle large complex cases such as those involving drug cartels. Many investigators will tell you that they believe there is a trend of drug cartels and gangs moving away from the well-resourced cities to the suburbs or small towns.

The eDiscovery software has wizards that simplify the process from litigation that is on hold, to the time reports are ready for court. Since the preparation of evidence and going to court is a repeatable fairly standard process, eDiscovery software provides the framework needed for all the evidence, preparation, annotation, and steps needed to go to court. The software also provides templates to add information. Templates appear to help organize the complex process of organizing evidence for forms and the steps to go to court. The software may also help a manager such as a lawyer mark evidence for review to give to readers and paralegals. Newer versions of eDiscovery software can also be obtained as reporting requirements change and new formats are needed. The legal industry is like any other industry in that there are changes, new regulations, and new ways of doing things. Software has to keep up with those changes. That is why it is important to take classes on eDiscovery with Access Data, keep the software up to date, and belong to some type of fellowship organization such as Women in eDiscovery. It is important that one can discuss problems that arise in this industry and discuss ways to create solutions to address them.

ChRootKit

This program called ChRootKit was written by Nelson Murilo and can be obtained free from the URL http://www.net-security.org/software.php?id=210 and it is important to see if there are rootkits on the netbook or laptop. If there was an intellectual property theft case and it seemed as if someone hacked in, it might be because of a rootkit. Rootkits are programs that allow people access to the computer and are very often difficult to detect. Worms are programs that crawl through the network and use up resources. Some of the famous ones are Ramen Worm, Romanian Rootkit, Monkit, LPD Worm, lrk3, lrk4, lrk5, Omega worm, Hidrootkit, and ZK Rootkit. There is a good book about rootkits called *Rootkits, Subverting the Internet Kernel* by Hoglund and Butler that really explains what they are and what they do [31].

KNOPPIX

The Knoppix CD is very important because it is a bootable GNU/Linux distribution that allows people to bootup a system and then preview it without making changes to it. It allows for the previewing of partitions and files and directories. It allows the examiner to look at the dates of files and file sizes. It is a good teaching tool for students who are starting computer forensics and need to get practice using some basic tools.

KNOPPIX SECURITY TOOLS DISTRIBUTION

The Knoppix Security Tools Distribution (STD) has many tools on it that are of great use to the digital investigator. The bview is a binary viewer. Sometimes is necessary to look at binary to see what value is in a particular word in the registry. A certain value for a port could mean that it is read-only or that the port was disabled and could exonerate someone who was accused of taking something by a device he or she connected to a USB port. bsed is the binary stream editor. Sleuth Kit version 1.66 is here and has many tools that are discussed in another place in the book. The fatback undeletes FAT files which is important for data recovery. There is also memfetch which is good for causing the operating system to do a memory dump. Md5deep is a program for creating md5 hashes on multiple files and directories.

Pasco is also included. This is a fascinating tool because it allows the computer forensic investigator to go into the index.dat file for Internet Explorer. Index.dat holds all the URLs that one visited. This is useful for an investigation in a corporation where one is accused of shopping online. There is also a program called "Wipe" which is good for wiping a USB flash drive to forensic standards and preparing it for receiving an image.

There are some very important tools such as TestDisk which may be used to recover partitions. Some criminals may try to delete a partition to destroy evidence in a partition. Sometimes in malware investigations, digital investigators want to see the browser's cookies. Galleta may be a good program for examining cookies. Ftimes is a tool for learning about the system baseline. Photorec is a program to extract digital images from digital cameras. If one was doing a malware investigation, then hdb might be a useful tool because it is a java decompiler. Fenris is another tool that is often useful for code debugging, tracing, decompiling, and doing some type of reverse engineering process. Rifiuti is a tool for looking at the items in a Windows Recycle Bin. Foremost is another tool for finding a specific type of file from an image. Foremost is good for extracting a JPEG or GIF. Hashdig is a tool for looking through hash databases. It might be useful for obtaining the hashes from SAM files in Windows. Once the hashes are obtained, another tool outside of this toolbox will be needed for converting them to a password.

Readdbx is a good tool for changing the Outlook Express files to the mbox format which many digital evidence tools can process. The readoe tool is often useful for changing the directories from Microsoft Outlook Express to an mbox format. It seems that the Knoppix STD tools are very good incident response tools to collect an image, extract pictures, collect email, collect hash files, and items from the Recycle Bin. There are some tools that allow the investigator to look at files at the binary level too. A lot of partition and file recovery as well as basic examination can be done with these tools. Then the results can be given to a law office that performs eDiscovery and they can use large-scale tools such as Intella Team Manager and Reviewer to share the evidence among people in the office. The incident response tools are important for any digital investigator to have and these tools are extremely low cost compared to other tools. Many of these tools are also in other kits and are frequently used by digital forensic investigators.

PASSWORD RECOVERY TOOLKIT

The PRTK is made and supported by Access Data. There is phone support as well as email support. The program is often used by law enforcement and corporate investigators so it is well known and should pass the Frye test in any state court in the United States. The PRTK is good for recovering passwords from applications that are used all over the world such as Microsoft Word and Microsoft Excel. There are over 100 applications that it can recover the passwords from. Law enforcement can also use the PRTK as a tool for breaking passwords when examining a netbook with password-protected files, and no password is known. The tool is also useful for people who just forgot the password of an important file.

One of the important things is that the tool has a security code associated with it so that unauthorized parties cannot use the tool. The present version can analyze many files concurrently. This is important because multitasking saves time. There are too many files to examine in a serial fashion. Many people use characters from Unicode and various alphabet sets to increase the security of their documents. It is common to hear of Cyrillic and Arabic letters being used in passwords in 2012. With more diversity in our society and nearly everyone having a computer, multilanguage fonts sets are encountered in netbooks, laptops, and desktops. PRTK can use multilanguage passwords.

People may also choose longer passwords to increase security. They may also use upper and lower case letters and numbers to increase complexity. Access Data's PRTK webpage says that it is possible to recover passwords regardless of length [32]. The time that is needed to recover a password that is very long may be impractical on a regular examination machine. A desktop computer with a very fast processor and graphics cards with multiple CPUs would greatly speed up the time to break the passwords and move the investigation along. PRTK needs a minimum of 2 GB of RAM to run and the processor should be at least an Intel Pentium 3 or a P4 or an AMD Athlon. The operating system may be Windows 2000 or Windows XP. There needs to be at least 100 MB for the program to reside in on the hard drive.

DISTRIBUTED NETWORK ATTACK

This is a tool from Access Data that can also be used for recovering passwords. It utilizes the concept of parallel computing which may also be observed in some form at the supercomputer labs and makes it something practical that the average computer forensic lab can utilize. There is software known as DNA Server that is made available to a network that certain workstations involved in the password breaking activity can access. Then there is DNA client or worker which can access the server. There is also a DNA Manager which coordinates the attack and sends elements of the keys to different workstations. The DNA clients work on the keys in the background process. It seems that if there is wide area network, the server and clients can be anywhere in the world. This could be very good for very large countries with centralized labs and remote outposts so that computers can work to solve a decryption or password-breaking problem across large distances. The bandwidth of the network needs to be a minimum of 100 Mb and

1 Gb is best. The server needs a 100 GB space free on the hard drive. It seems logical to check one's network topology first with Access Data before purchasing.

The worker machines need to have a minimum operating system of Windows 2000 or Windows XP. If the computer is a Mac, Macintosh OSX 10.39 or 10.4.x is needed. Linux Red Hat and Fedora Core 4 are acceptable too. Solaris is fine too. The processor needs to be Sparc, Power PC G4 or G5, or Intel Pentium III, P4, or AMD Athlon. There needs to be 40 GB free on the storage device and 1 GB RAM.

There is an opportunity to add or create customized dictionaries and the password attack can be set up for alphabets and dictionaries of specific languages. There are also graphs that give statistics on progress. DNA Server can recover passwords from MS Office XP, WinZip, and PKzip. It can recover passwords from PGP 4, 5, 6, and RAR up to version 2.9.

ISO BUSTER

There are occasions where a netbook or laptop is seized and all the media with it is also seized. There could be a stack of DVDs, CDs, or other optical media. A detective once told a story in a computer forensic class of how he once had probable cause and a search warrant for someone possessing and distributing child pornography. The police did a knock and enter and spoke to the person. They spoke to the man at length and found him arrogant and boasting that they would not find anything. The laptop and nearby media was previewed with a forensic tool and nothing was found. There was a small hole in the wall near the top of the ceiling behind the computer. The man's demeanor became very scared when they discussed the hole in the wall. The police then removed part of the sheetrock there and found a stack of CDs in cellophane with a magnet taped to it. The story related in class included a detail about the man using a fishing pole with a magnet and letting it fall down the hole. Then he reeled up the child pornography in the cellophane with the magnet on top. Once it was found, the person's body language was despair and he knew that he was caught. Once caught, it is said that some sex offenders seem as though they know they are guilty and want to plea bargain to minimize publicity and jail time.

If those DVDs, CDs, Blue Ray Disks, and optical media or whatever they were, got scratched or damaged behind the wall, a tool such as ISO Buster may be used to possibly recover, depending on the damage, of any of the data from the media. Some people will wonder what the difference is between a DVD and a CD. The DVD uses a more densely populated data format so that many more times the data can be stored on a DVD than a CD. The term "RW" means read and write. The acronym ROM means read-only memory. It is written once and then can only be read afterwards. The term "HD" means high-density or high-definition. BD ROM is Blue Ray Disk. The DVD VR is a DVD video recording. The vendor for ISO says that video or pictures from the following formats may be recovered: "CD-i, VCD, SVCD, SACD, CD-ROM, CD-ROM XA, CD-R, CD-RW, CD-MRW, …, DVD-ROM, DVCD, DVD-RAM, DVD-R, DVD-RW, DVD + R, DVD + RW, DVD + MRW, DVD + R Dual Layer, DVD-R Dual Layer, DVD + RW Dual Layer, DVD + VR, DVD + VRW, DVD-VR, DVD-VRW, DVD-VM, DVD-VFR, BD-ROM, BD-R, BD-R DL, BD-RE, BD-RE DL, BD-R SRM, BD-R RRM, BD-R SRM + POW, BD-R SRM-POW, BDAV, BDMV HD DVD-ROM, HD DVD-R, HD DVD-R DL, HD DVD-RW, HD DVD-RW DL, HD DVD-RAM, HD DVD-Video, …" [33]. The software can be downloaded and only costs $29.95. There are so many acronyms that it is best to go to an electronics store where digital media is sold to find out what each acronym means. An electronics store clerk could probably be a good source of learning the capacity of it, and what standard it uses.

There are many ways that the movies, videos, or pictures could have been made. Some could have been made with a drop and drag application such as "Roxio Direct CD, Roxio Drag-to-Disc, Ahead/Nero InCD, Prassi/Veritas/Sonic DLA, VOB/Pinnacle Instant-Write, CeQuadrat Packet CD, NTI FileCD, and BHA B's CLiP" [2]. Since digital media is so prevalent in the twenty-first century, there are many ways to create the video. Sometimes there are orphaned files or deleted files that can be recovered with this tool too. If someone had tried to format the DVD or CDs, the ISO Buster can often recover video or pictures from that too. The tool ISO Buster can also be used to recover E01 files which means if a Guidance Software Encase image file was damaged, it might be possible to recover that too. ISO Buster seems like a very useful tool that should be in the computer forensic examiner's toolbox.

IRT CONDUCTING RESEARCH IN THE LAB: IRB–NON-IRB RESEARCH

If the research in the lab uses people's opinions and requires their interaction, then a proposal must be submitted for the Institutional Review Board (IRB) for review. If the research consists of an examiner doing an activity on an isolated computer in the lab with no associated risks, then it is generally considered expedited research and does not need a full IRB review. It is very important to seek IRB approval because it lowers the risks associated with research. People who participate in research often fill out and place their signature on an informed consent form which makes the risks of the research known to them and waives their right to sue. The Office for Human Research Protection (OHRP) can provide more information on the subject [34].

In an academic lab, it would be advisable to use simulated evidence for teaching purposes because many real-life examples of evidence, even from public cases whose evidence has been made available, may be disturbing and may expose students to examples of human behavior that surpass what is seen on television. Real evidence could also expose students to authority figures violating public trust and cause them to seek counseling.

OSForensics HAS A FILE AND METADATA VIEWER

The metadata is information about the file. Some digital images include the camera brand, model, and occasionally the serial number. Sometimes metadata includes the GPS coordinates as well as information about when the file was created and last modified. Some people also use features of Microsoft Word and image tools such as Picasa 3 to remove metadata from files. OSForensics has a section where metadata can be viewed and the contents of the file down to the byte level can be viewed. This is an important feature.

OSForensics BASIC FILE RECOVERY

There is an option in OSForensics for basic file recovery. An examiner can save processing time and searching time by selecting documents only or pictures only. The results are displayed in a window with the relative quality of recovered file. Files with a number close to 100 are considered intact. Files may get corrupted from bad sectors on the storage device or clusters may have been written over. It is important to recover the files before much activity has occurred on the suspect's computer. The recovered files can be viewed and OSForensics can give hashes in MD5, SHA-1, or SHA-256.

OSForensics RECENT ACTIVITY

If the examiner needs to show recent activities of the suspect, such as see searches, websites visited, and documents accessed, then the recent activity option should be utilized. It is important to show a timeline of activities from signing in to doing misdeeds to establish guilt or innocence. However, a person may say that someone using my username and password did the activity from that machine at that time but it was not me. Quite often, labs, schools, and corporate environments are now using IP cameras with digital video recorders to archive activities at certain computers at certain times. Archived video in conjunction with computer activity logs is very strong evidence and cannot easily be repudiated.

SURVEY OF TOOLS: HARD DRIVE IMAGER/FORENSIC PREPARATION

One of the tools that are very useful for forensic preparation of hard disk media is called Disk Jockey Pro. The device allows one to connect a hard drive and then one can do a quick wipe of the data. However, it is possible that one pass of data wiping may not be sufficient for very old hard drives. Very old hard drives had wide spaces between the concentric rings and some sources have said that it is possible that there might be some residue of data that an electron microscope might pick up. That is why there was a Department of Defense standard of wiping digital media with three passes over each byte. The Disk Jockey Pro also has a feature to allow one to do this set of three passes over each byte. This version of the Disk Jockey Pro is great for information technology departments who must quickly wipe drives and copy drives or even combine two small hard drives into one large hard drive. However, there is a Disk Jockey Pro Forensic Version that is more suitable for wiping drives and making forensic images of drives. That will be addressed in the next paragraph.

The National Institute of Standards and Technology (NIST) have a report about the Disk Jockey Pro Forensic Edition version 1.2 [35]. The tool was tested in a very thorough systematic method for its ability to wipe a hard drive as part of the forensic media preparation process. Various sectors are examined and the report is available to read on online. The Disk Jockey Pro Forensic Edition has a feature to do a Department of Defense standard wipe that passes over each byte seven times [36]. This device can also be used to copy the Drive Configuration Overlay (DCO) of the hard drive. The operating system and bios cannot usually access the Host Protected Area (HPA) and DCO areas of a hard disk drive which makes it a perfect place to hide very small amounts of information. The fact that the Disk Jockey Pro Forensic Edition can copy the DCO and HPA makes it an important tool because those areas can be searched later by a tool such as X-Ways Forensics. The DCO was introduced on the ATA-6 standard while the HPA, an area that holds the maximum address of the device, was introduced during the ATA-4 standard. The DCO can be written to and then the location holding the numbers of sectors could be understood so that an 80 GB drive can be made to appear to be 40 GB, thus creating a hidden partition of 40 GB. It is also possible that some low-cost nonforensic imaging tools might capture this hidden part of the disk. That is why it is good to read the Forensic Tools section on Forensic Imaging on the NIST website.

There is also a write-protect feature that has been tested when the hard drive is connected correctly in the proper place and never allows the suspect's hard drive to be written over. The device also allows for a write-protected USB 2.0 port so that cell phones, PDAs, digital cameras, and hard drives can be connected and previewed. This would be another simple option for creating a USB read-only port instead of doing something complex such as changing some bytes to a binary value in the registry. The device also has the ability to accept 3.5-inch and 2.5-inch cables. This is important for imaging the palmtop, subnotebook, notebook, and laptop computers. The reason for removing a hard drive from the computing device is that it is often many times faster than booting up with an imaging tool and trying to copy everything through a parallel port or an old USB port. Small computing devices may not have a CD ROM and high-speed port that would normally allow one to bootup with a program such as FTK imager and then copy the drive. One can also use Disk Jockey Pro Forensic and copy one hard drive to another hard drive, including the DCO and HPA areas and then verify both drives are the same.

The Disk Jockey Pro Forensic unit also has a feature that lets one check to see if there are bad sectors on the hard disk. This is very good to know. If the drive has many bad sectors and it is an 80 GB hard drive, for example, then it is not a good idea to try to copy another 80 GB hard drive to it because it will not be able to hold a forensic image of the other drive. The LogicCube MD5 also has a feature that allows for the testing of a drive for bad sectors. One of the excellent features of Disk Jockey Pro Forensic is that its documentation tells you what operating systems it works with and what ones it does not. "Windows 95/Windows 3.x/Windows NT/Mac OS 9.x and earlier are not supported and computers that were sold prior to the release of Windows 98 that have USB ports may not operate properly" [37]. The device is reported to work well with Windows 98 SE, Windows 2000, Windows Me, Windows XP, and Mac OS X version X or greater [37]. There are many signed testimonials to how well the device works with estimates of how much time the device saved the computing professionals [38].

ImageMASSter SOLO-IV IT

Another device made by the same company is the ImageMASSter Solo-IV IT which allows for forensic preparation of media such as hard drives by wiping hard drives that have an IDE, SATA, or SAS interfaces. The device also allows for the forensic imaging of digital media cards such as the SD, Micro SD, and Compact Flash cards. There are also cables and adapters so that the small computing drives that are 1.8 or 2.5 inches can be imaged. Some 1.8-inch drives are ZIF drives and hold as much as 128 GB. There is safety mechanisms built in so that the suspect's media will not be written to. Some ZIF drives use the PATA interface so the vendor might have to be contacted because the documentation says that there are some cables available for proprietary hard drive and laptop drive interfaces. The device also supports forensically wiping two drives simultaneously and then copying two suspect drives to two forensically wiped drives. This is done in a parallel fashion and the duplication can be verified with either SHA-1 or SHA-2 hashes. The hardware has also been used by many computer forensic professionals so its use for investigations should not be a problem. It should pass the Frye test without a problem.

One of the security features built into the device is for the encryption of forensic images. As it images, one can encrypt it with the AES 192 or AES 256 algorithms. This may seem contrary to what one thinks about preserving evidence because someone is taking an image and putting it in a completely different format. Perhaps one might even question if information is being spoiled. However, consider the high-profile case where a forensic image is made of a suspect's computer who is believed to have stolen tens of thousands of credit card numbers, date of births, and social security numbers. If the forensic image is stolen from the examiner during an armed robbery, for example, the thieves could potentially make millions of dollars from identity theft. For better or worse, the computer forensic professional making the image and later examining it is considered the custodian of that information. That is why many examiners take the precaution of encrypting the data [39]. There is digital examiner insurance of amounts up to one million U.S. dollars but such a loss could result in the tens of millions of dollars. Some examiners also carry omission insurance but even that would not cover such a loss. Even if the encrypted forensic image was stolen, the breaking of AES 192 or AES 256 is impractical outside of a theoretical setting according to Jack Germain, a technical writer for *TechNewsWorld* [40].

The ImageMASSter Solo-IV IT is also reported to have a Windows XP-embedded operating system. This is good because high-end mobile devices might be using a Microsoft-embedded operating system such as Windows CE 5.0 and then previewing the files on the screen might be convenient with this device. The built-in screen on the device also works with a touch pen which is included with the device. This is important if one is working in Taiwan, Indonesia, Vietnam, or other hot and humid environments where one's fingers might be sweaty and cannot effectively use a touch screen or touchpad. The device also operates on 100–240 V which means that the device can be operated in Oceana, South East Asia, the Americas, or the United Kingdom. However, it is still important to see what plugs come with the device and consider where one will work. Some power outlets in places such as the United Kingdom have power connects that include metal tabs that are parallel while one is horizontal. The time to consider such operations details are not in the field, but before one leaves for overseas work. There are often power adapter kits at large international airports such as Liberty International in Newark, New Jersey. Some devices that do not operate with a variable range of voltages means that computer forensic professionals also need to purchase adapters, some of which can get hot if the device attached consumes much power.

The device also allows for imaging in a raw format or to an E01 format that Guidance Software's Encase would use. It also allows for the standard Linux DD file or segments depending on the image size. This is exceptionally good because one could take the captured image in the E01 file format and then examine it with Encase. This would allow one to use a premium software forensic tool with enormous support for users. The DD files would be great for those who wish to use a utility to change the file format and prepare it for running in a virtual machine window on an examination machine.

One of the interesting things about the device is that it is reported in the specification sheets to operate anywhere between 5 and 55°C. That means that the device could be operated in the summer of a developing nation, and stored there without a problem. There are sometimes requests for short-term computer forensics work in Vietnam on the CCE list server. It is important to consider the temperature and humidity of the place where the device will be used since air-conditioned environments are not always possible to work in. The device was also listed as only consuming 9 W which is also important. In some developing nations it is not always possible to get power and one may have to use an uninterruptible power supply if the power goes out. The specification sheet also says that it operates in 20–60% noncondensing humidity [41]. This is also important if one will be working in humid places such as India, Indonesia, or Vietnam. Digital forensics is growing worldwide as 60% of the earth's population has cell phones, many of which have external media cards. By 2013, some computer scientists estimate that billions of people will have a desktop, laptop, notebook, palmtop, or netbook computer. Statistics show that nine hundred million people on the planet have a computer which means that there is a lot of potential digital forensics everywhere [42].

The weight of the device is only 5.35 pounds and the overall dimensions are 10.6″W × 3.8″H × 7.7″D (270 × 98 × 194 mm). If one is going to take the equipment overseas, it is important to consider the size and weight of everything that will be shipped. It may also be advisable to get an ATA Carnet so that one may be able to go through customs quite easily with the equipment and not have problems with customs and border patrol in the foreign country or on the way home. The device is listed at $2800 which is reasonable considering all the work that the device can be used for.

DriveSpy

DriveSpy can be used for forensic media preparation since it can wipe a drive or partition. This is a tool that can be used to do the most important task in computer forensics, which is to make a disk to disk forensic duplicate. A partition can also be forensically imaged and compressed with the tool. Drives or partitions that are imaged can also be hashed with an MD5 hash. This is important for the chain of custody form and to show that nothing was tampered with later. If one only wanted to copy a few picture files, those files could just be copied and given an MD5 hash. DriveSpy is also reported by Digital Intelligence to be able to process drives that are duplicates even when the sector translation is different and if the drive geometry is different. Once a forensic copy exists, an examiner can look at the physical addresses of SD cards, hard drives, floppy disks, and flash drives with DOS and non-DOS partitions. The hex viewer can be used to look at sectors and clusters. This is important because small files that do not use all the space in the cluster have what is known as file slack. This file slack, often called slack space, is viewable too. Interrupt 13 hex is used to access the drive. The slack space can be gathered and put in a file. This file may be used with other tools later to create a dictionary for trying to break passwords.

Large hard drives that exceed 8.4 GB can be viewed. Hidden partitions could also be viewed to see if they are empty or may hold covert data. There is also an option to log the investigation at the keystroke-by-keystroke level for high-profile cases. There is also scripting available to help automate routine tasks that require frequent examiner attention. A drive can be copied to another drive or just a range of sectors can be copied from drive to drive. Then, any erased files can be copied to a work area without changing the modification date and showing the file was accessed. Certain types of files could be selected and copied. Perhaps the investigation is only concerning system files and hidden files to see if there was a type of system compromise.

DriveSpy is also good for viewing unallocated space. That means areas that are not being used by the file allocation table for file storage. That unallocated space can be gathered by DriveSpy and put in a file. Old flash drives as well as new ones could be examined since the tool supports FAT 12, FAT 16, and FAT 32 partitions. DriveSpy also supports FAT 16x (FAT extended) and FAT32x (FAT 32 extended) partitions. FAT 16x is often said to be like FAT 16 but the difference is that the extended one can support up to 4 GB. DriveSpy can also process those long file names associated with Windows 98 and above. If one was looking for all JPG image files, DriveSpy can be used to select files based on the file extension but it can also select files based on internal header information. This is important in case someone renames a picture to look like a Microsoft Word document. DriveSpy is also good for use in looking at the cluster chains in the partition's file allocation table. This could be useful in recovering a file that is quite large and fragmented.

The company that sells and supports DriveSpy is located in New Berlin, Wisconsin. Law enforcement licenses for the program cost $199 95. There is email support, phone support, and the website tells you how many forensic experts are online at the present time. Digital Intelligence also offers classes for their hardware and software products which is very important. DriveSpy is often bundled with computer forensic textbooks so that students have a real tool to practice digital forensics with.

PDBlock

This is a program by Digital Intelligence that acts as a write blocker and prevents writing to a drive. It is also usable on systems with FAT 32 extended systems. There is ample support and documentation for the program and it has been rigorously tested. There are reports for download that show testing results. The documentation is portable and in pdf format. This is important for high-level cases that may go to trial. The program is not expensive at $29.95 for law enforcement personnel. The program is a DOS application and it does not run in Microsoft Windows.

PDWipe

This is another Digital Intelligence program that is used for forensic media preparation. When the investigation is over, PDWipe could also be used for purging sensitive information from hard drives or flash drives. The Digital Intelligence website reports that "this wiping algorithm exceeds that specified by the DOD 5220.22-M specification for both 'clearing' and 'purging' of sensitive information on Hard Drives" [43]. The law enforcement license is for one machine at a time and only for use by the licensed investigator who is the law enforcement professional. This is good because the licensed person may have a copy on the desktop at work and a laptop in the field but would only be using one machine at a time. The law enforcement license is $19.95.

PART

Sometimes it is important to be able to unhide a partition. Part, a program sold by Digital Intelligence, is good for unhiding or hiding a partition. The website says it has mechanisms similar to Partition Magic and Ghost.

DIGITAL INTELLIGENCE CONSULTING

It is also important that Digital Intelligence offers consulting services. It is possible that a corporation starts an investigation that becomes too complicated for them and they need consulting services. Perhaps the equipment that they need is far beyond their budget, so a consultant with his or her own equipment is needed. It is also good that Digital Intelligence has a good reputation in the field of digital forensics.

DIGITAL INTELLIGENCE TRAINING

There is often interactive training for law enforcement anywhere in the world. The remote forensic examiner can see and hear the professor running the class by a video feed. He or she may have questions and can ask them via Voice over IP. Since the remote student preregistered, a workbook and CD would be provided. At times, the remote student can also take control over the forensic workstation to practice concepts. This seems to be a good option for digital forensic examiners who may be deployed in Somalia or Afghanistan and cannot get to a class.

CHAIN OF CUSTODY

If one goes on a tour and visits the New Jersey Regional Computer Forensics Lab (NJRCFL) in Hamilton, New Jersey, one may get to see the evidentiary process for seized digital devices that could include phones, netbooks, and desktop computers. A case manager first interviews the law enforcement officer about a case. Many questions are asked, documents may be reviewed, and a priority is assigned to the case. Then the law enforcement officer who is part of that investigation and may have been part of the incident response team may bring the evidence to the NJRCFL. The evidence will be checked in and bundled. A bar code will be placed on the bundle. The evidence area has steel racks holding bubble wrapped computers with bar-coded evidence tags on them. If a bar code is scanned with a hand scanner, that piece of evidence will show its whole history from the time that it was seized, who transferred it, the case number, if it was examined yet, and any other pertinent details. This is an example of the chain of custody, also known as the chain of evidence.

There is usually a chain of evidence form. Nelson, Enfinger, Phillips, and Steuart give an excellent detailed discussion of the chain of evidence form in their book *Guide to Computer Forensics and Investigations* [44]. A chain of custody form may include a variety of information but may look different depending on how the organization decides what the layout is and what information should or should not be included. The search warrant, case number, and time and date of seizure may be on a law enforcement chain of custody. The investigating agency, nature of case, the place where the evidence was seized and tagged, as well as the person who seized the evidence may be listed on the form. Associated photograph numbers may be on the form too. The serial number, vendor name, evidence description, and any processing of evidence would be on the form. Any transfer of evidence will be noted too and perhaps signatures accepting transfer and responsibility would also be present.

The chain of evidence form is basically a history of that piece of evidence that tells who had that evidence and in what circumstances from the time it was seized until the time it appears as an exhibit in court. The chain of evidence form, also known as a chain of custody form, also gives the court and its officers a sense of assurance to the quality of that piece of evidence. It shows that it was under lock and key and not tampered with during the entire journey of the justice process. In Bergen County, New Jersey, evidence associated with a murder case is held for 100 or more years. Can you imagine that the netbook, laptop, or desktop computer of a serial murderer may be stored for over 100 years? The burden of storage for law enforcement is unbelievable.

INCIDENT RESPONSE POLICY

Incident response policies need to have a protocol of how to define an event and how to respond to it. If there is reported malware on a computer, the incident response policy might be to dispatch a team to respond and preserve the evidence. Preservation may be done by running a tool that collects the most volatile evidence first such as RAM, and then acquires both allocated and unallocated space on the person's workstation. Preservation may also mean running a NETSTAT—a command from the DOS prompt or command prompt—and taking a picture of all the connections to the machine. Perhaps an IP address can be seen if the connection is live and a port that the malware is using can be documented. Containing the spread of the malware might be as simple as releasing the connection to the network or physically removing a network cable. Eradicating the malware might be as simple as wiping the person's hard drive and RAM and then restoring a standard workstation image of a hard drive on that computer.

Thresholds need to be defined. An incident definition chart that explains incidents and thresholds should exist so that if certain conditions are present and thresholds are exceeded, certain law enforcement agencies be called. Perhaps if *X* dollars are stolen, the AR contacts the local municipal police. If *Y* dollars are stolen or a certain breach of data occurred, the FBI would be called by the AR. There should be a chart that informs the incident response team that it is a criminal act of a certain level and that the AR can make a judgment on that and call law enforcement. The public information officer, management, and CIO might all have a defined role in the incident response policy.

The incident response policy should be an up-to-date "living document" that incident response team members understand. It should act as a guide of how to respond to an event and what tools to use. It should cover how to respond, preserve evidence, contain evidence, eradicate a threat, and restore the system. This living document should be practiced in exercises so incident response teams know how to deal with an event. The policy may also define and set the scope of what may be asked or not asked of the suspect. The policy is also a document that tells each IRT member their role in an event and defines the scope of their role. This should alleviate some of the stress and if IRT members know what to do and what not to do, it should help limit improper responses and help limit lawsuits from the suspect.

The IRT policy might state that each member of the incident response team should use freeware that is known not to contain malware or to use licensed software so that evidence is not tainted and later suppressed by a judge. The policy might also say the IRT members use forensic techniques and tools accepted by their peers so that they may pass the Frye test. The policy might have rules and guidelines for reducing verbal conflict and workplace violence with the suspect. The policy might also discuss how and what to label and photograph as well as how to fill out the chain of custody form. The IRT policy might discuss when to use Faraday bags to limit the potential for wireless tampering during the event.

POLICY OF REMOTE COLLECTION OF DATA BY INCIDENT RESPONSE TEAMS

One can often find informal conversations at conferences such as the High Tech Crimes Investigators Conference where corporate investigators have spoken about the increase of workplace violence and the danger of encountering an employee during the digital evidence collection process. Sometimes corporate investigators will discuss the "black bag" operations that they have been a part of during their career. They say it starts with a security guard who is instructed to let a digital forensics professional in the building from a side entrance after hours to image a desktop, laptop, notebook, or PDA. The security guard has rounds to make and most likely cannot spend much time with the corporate investigator. These investigators or digital forensic specialists may be alone, unaccompanied for hours, while a forensic image of the device is being made. People who relay such experiences have privately discussed fears of employees returning after hours. The same people also privately discuss that they may not have the money to invest in newer equipment and that forensic images take longer to create with old LogicCubes and increasing larger-capacity hard drives.

Such stories cannot stay secret long and do find their way back to vendors who create digital forensic products for corporate, military, law enforcement, and private investigation communities. There has been a desire to remotely access a person's workstation and collect an image of the suspect's computing device for some time. Some say this has only been possible on a large scale since the prevalence of optical fiber and broadband Internet to offices and homes around the world. A person who has a 1 terabyte hard drive and a dialup connection to the Internet would be safe from remote collection of data.

Remote acquisition of a hard drive decreases the commuting time to and from a site where the device in question is located. It could be considered a green technology and such adaption of such technology might allow companies some tax break. It also decreases the possibility of physical harm to the incident response team member who must collect the evidence. Remote acquisition can also allow the person to perform more data collections since he or she could be a secure stationary location. This type of remote acquisition could also open up digital forensics to people with disabilities, thus providing them with an interesting high-paying career.

Interrogating a networked computer for evidence and then obtaining the needed evidence remotely without shutting down the computer has been a desired outcome for some time. Frank Adelstein and Matthew Stillerman filed a patent in 2009 for "Remote Collection of Computer Forensic Evidence" [45]. The first sentence in the their patent abstract says, "The invention is directed to techniques for allowing a user to remotely interrogate a target computing device in order to collect and analyze computer evidence that may be stored on the target computing device" [45].

F-Response has a website that states, "F-Response is an easy to use, vendor neutral, patented software utility that enables an investigator to conduct live forensics, Data Recovery, and eDiscovery over an IP network using their tool(s) of choice" [46]. It is actually F-Response Consultant + Covert supports that can be a major component of remote forensics. The F-Response Consultant + Covert tool has to be setup, properly credentialed, and have the license dongle. Then it can be dispatched to certain IP addresses. Once a connection is made remotely, a tool such as Access Data's Forensic Toolkit or X-Ways Forensics can be used to collect the evidence remotely on a remote computer's hard disk and memory [47].

The tool includes support for Windows 2000, XP, 2003, Vista, 2008, and 7, 32, and 64 bit physical memory only supported on 32 bit and 64 bit Windows [48]. Various versions of Apple OSX, Linux, Solaris, IBM AIX, and FreeBSD are supported. It is best to check the website for more exact details.

EXAMPLE OF AN INCIDENT INVOLVING NOTEBOOKS THAT THE IRT MAY BE DISPATCHED TO

The corporate IRT may be dispatched to investigate allegations of email spam and harassment [49]. Email spam is large amounts of unsolicited email that jam up bandwidth and fill mailboxes. It is a problem because it can cause a person's productivity to be severely reduced because their mailbox is full and they cannot receive legitimate email or because there is so much mail to look through that it slows down productivity. If certain emails that look legitimate are opened, rootkits, keyloggers, and other forms of malware may be installed on the computer, thus allowing criminals to steal information or do corporate espionage. The IRT has to first investigate that an incident happened, the extent of the incident, and then preserve evidence, contain the problem, eradicate the harm, and restore the netbook to its former working self.

Many incident response teams in corporations will coordinate with IT security in order to install proactive security measures to reduce future incidents. Webmail interfaces with spam filters such as Postini have been one proactive countermeasure utilized in many universities and corporations in order to reduce spam and spam infected with malware from entering a person's email application on a netbook. Webmail filters can also be safely perused without downloading malware on the computer. If a legitimate email was mistakenly caught in the filter, it can be redirected to a person's email application.

TamperProof BAGS

The tamperproof bag is one tool that is used as part of a group of tools and procedures to protect the integrity of data. Sometimes in an investigation, a law firm wants the original hard disk sent to a digital forensics lab for testing. The computer forensic technician may go to the site and remove the drive. He or she may place it in a plastic box and pack it with bubble wrap or plastic bags that are inflated. Then the investigator would drill a hole in the box and place a seal through it. The whole thing may be put in a tamperproof bag. These bags contain a strip that can be peeled off at the top and then a gummy flap can be pressed to the plastic. The item is then sealed. There is usually a chain of evidence form on the bag. It is there that one can write if it is bag one of two if the evidence is part of a sequence. Then the case number can be written on the bag. The date, investigator's name, and any signatures of people who signed off on the package of evidence are displayed. There is usually a place for notes and special handling instructions about temperature or if it is fragile.

COMPARISON OF AntiStatic BAG, BOOSTER BAG, TamperProof BAG

The tamperproof bag may be sent by FEDEX to the lab or dropped off if it is close by. FEDEX is good because of their high security and their tracking. Many places use them as an accepted part of their chain of evidence. Basic Ltd. in Brooklyn, New York, sells Evidence Bags/Property Bags and they would be a good contact to ask about tamperproof bags. The tamperproof bag is different from an antistatic bag. The antistatic bag is good for protecting a device from static discharges. The damage to electronic chips due to static electricity discharge is a problem in dry environments where static electricity builds. The tamperproof bag is also different from the Faraday bag which protects the mobile device from signals which may alter it. All these bags are different from the booster bag which may be used by people to avoid security detection devices in a store. Booster bags may have an aluminum lining. Some people say that booster bags are essentially Faraday cages and may work like a Faraday bag. It would be good to see the results of research that involved the testing of various types of bags in different environments.

PACKING AND TRANSPORTING SEIZED EQUIPMENT

The incident response team may have to at some point in their existence engage in the transport of a large amount of equipment to a lab. In such circumstances, it might be worth getting a pallet, tagging evidence, and then using a digital camera to video tape all the cables and how they are connected to the equipment. Then one may pack bubble wrap on all the equipment and tape it. ULINE stretch wrap is a good item for securing and protecting items on a pallet. ULINE Opaque is good because the contents are not visible and it also offers some security from potential thieves. The black UVI Goodwrappers http://www.uline.com/BL_2952/Opaque-Stretch-Wrap are also reported to be good for some protection from ultraviolet rays from the sun. This type of opaque ULINE stretch wrap clings to itself and seems nearly impossible to penetrate without a knife or razor. Some of the wraps come in sizes of 30 inches by 100 feet. Rolls cost between thirty and forty U.S. dollars.

There are also choices to select either a manual or an automatic stretch wrap machine. It is not difficult to wrap up evidence with one of these machines. One starts by pushing a dolly with a pallet of equipment onto the ramp that comes with the stretch wrap machines. Then, if it is automatic, one pushes a button to start the motor. An outstretched arm puts ULINE wrap around the pallet. The motor rotates the pallet. The arm starts wrapping the pallet near the floor and continues wrapping as the arm rises until the entire pallet is wrapped. The automatic machines are good because they do not waste as much wrap as humans. The machines use standard U.S. household voltage and may use as much as 15 A. It is important to be prepared to seize any size load of equipment from a suspect or organized crime operation.

X-FIRE (INCIDENT MANAGEMENT SOFTWARE)

The X-Fire tool is made in Canada and seems very useful for the law enforcement and public safety communities. Agnovi is the company which is the maker of this investigative management software. The company was founded in 2001. The X-Fire case management software lets the AR track the incident or case from its initial reporting until the time it goes to court. The tool supports the investigative triangle that is often studied in computer forensics and in criminal justice. The software allows law enforcement to make customized reports which is important because states in the United States are very different and each has different ways of handling computer incidents. The tool has custom user interfaces which are good because corporate investigations are very different than law enforcement investigations. The details for the case and the information needed to be inputted are different depending on if it is a criminal case or a corporate policy violation. There are fields for the contact management subsystem. The home phone numbers, work phone numbers, home addresses, work addresses, and time of days that one is available could be customized to be part of the contact management subsystem. Functionality can be customized too. Perhaps an onscreen calendar or a calculator might be needed. The tool allows for customized reporting. There are so many types of regulatory bodies depending upon the industry that one is in. If one is in a health care setting, then HIPAA compliance is an issue. Certain information has to be safeguarded from certain people. Role-based access (RBA) could be important. The person is only allowed to see what he or she needs to based on his or her needs.

X-Fire allows for legacy integration. This is important because some offices might have a politically connected person still using an old program on the original IBM AT and a Lantastic network while others are on a different system. It is great when systems can be made to integrate the old with the new. X-Fire also allows for the import and export of data. It gives one the ability to bring data in quickly or export it, or to create a customized report about incidents such as computer hacking incidents at a school. This would be important for the university public safety departments being compliant with the Clery Act. The website for X-Fire also reports that it has a 24 h/7 days a week support. X-Fire also has mobile access which is important because an incident response team may want to add something to the system, look up a related case, or check on the status of an incident.

The X-Fire incident management system is also multilingual which is important. French and English are the languages of Canada. Spanish and English are important languages in the United States. Some tribal governments in the United States such as the Navajo Nation may want things in English and Navajo. The Navajo Nation covers four states and seventeen million acres. X-Fire also allows for group management. Different groups may need to access or input information regarding an incident in different ways. The system also supports multicurrency which is important since the software is used in Canada and the United States. Both countries have different currencies. The X-Fire incident management software runs on Windows, Linux, Mac, Unix, and can also be web based.

There are also subject profiles which might be customized by corporate computer forensic incident response teams. The teams may be defined with a variety of fields such as incident response team members, leader, documenter/photographer, technician, and victim. Each role might be defined with responsibilities and perhaps some personal details of the person that fills that role. Workflow can also be tracked so that if an incident seems to be delayed, one can see where the delay is. Perhaps the investigation cannot continue until the depositions are finished. Events can also be scheduled. Perhaps when a subpoena was served, or a deposition date or court appearance might be noted. In any case, software that helps organize incidents from start to close is very important. It is good that companies are vending incident management systems to a variety of law enforcement, public safety, and corporate environments in many countries.

POLICY INFRACTION: SELLING STOLEN ITEMS ON EBAY AT WORK

I was third party to a story that may or may not be true but illustrates an example of what started as a policy investigation and then progressed to a criminal investigation. A person who did uncover investigations for a municipal law enforcement agency was a student in one of my classes and told me a story but asked for anonymity. The law enforcement did not mention the arrested person's name or any specific details about the case but spoke of generalities. The undercover investigator spoke of someone who was arrested at their workplace for misappropriation of property. The person who was arrested originally sold a boat from his home which was in the same town as the investigator. The item was posted at the person's workplace. The online auctions sites were on a list of prohibited sites at the person's workplace because the person's job function did not include ecommerce. Anytime the employees at that company accessed an online auction site, the network administrator at that company received a notice and the person's username, IP address, and MAC address of the computer that was used to access the site was logged. Then the AR in that company was notified. The AR checked with HR and the suspect had signed an acceptable use policy for the computer and Internet which prohibited online auction sites and a plethora of other activities. The employee handbook also stated that the man's job did not have anything to do with ecommerce.

The AR dispatched an incident response team. The incident response team already had copies of policies that were signed and dated and filed with HR. These prohibited the aforementioned activity of the suspect. Then the incident response team confronted the person. The employee had a banner on the computer stating that the person had no expectation of privacy. The signed polices also stated that the person had no expectation of privacy. The story did not cover any dialogue about the initial confrontation between the incident response team and the suspect. The incident response team removed the hard drive and used a device such as a LogicCube to clone the hard drive.

Later, the hard drive was inspected and the activity showed that the person did in fact sell a boat on eBay. That would have caused the person to have a letter put in his or her file and face some type of sanctions. However, it also showed that the same boat was sold two more times. The first sale was legal but the second and third transactions were not and this became a misappropriation of property which is a criminal offense. Law enforcement was called and everything concerning the investigation was turned over to them. This is an example of the silver platter doctrine. Due process now applied and the incident response team had to stop investigating. If further investigation was needed, the law enforcement agency had to sign an affidavit, give probable cause, and ask a judge to issue a subpoena. If the incident response team kept investigating for the police without a subpoena, the suspect's lawyer may say that there existed a failure to respect the Fourth Amendment rights of his client.

There are many investigations in corporations that start out as policy investigations but then proceed to criminal investigations once a violation of the law is encountered. The incident response teams need to know exactly how to act, what to say, and the boundaries of their investigative powers. Corporations have a lot more power than law enforcement in the workplace because companies own the devices and networks in question and often have policies that are signed where employees have waived their rights to privacy. Some computers, netbooks, and company-owned computing devices also have banners that employees must click on each time they use the device that remind them that they have no expectation of privacy and that the device is owned by the company and may be inspected at the office.

NEED FOR INTERNET USAGE POLICIES

The Internet usage policy should discuss how the Internet will be used by the organization and the person using the device that is owned by the company. "According to the filings in Blake J Robbins v Lower Merion School District (PA) et al., the laptops issued to high-school students in the well-heeled Philly suburb have webcams that can be covertly activated by the schools' administrators, who have used this facility to spy on students and even their families. The issue came to light when the Robbins's child was disciplined for 'improper behavior in his home' and the Vice Principal used a photo taken by the webcam as evidence" [50].

Perhaps if the alleged event occurred, the motivation might have been that the school wanted to protect the students by making sure they did not reveal too much information online about themselves to potential pedophile. Perhaps if the alleged event occurred, perhaps the school wanted to make sure there was no cyberbullying. Could someone have gone beyond reasonable standards of conduct and spied on students? Perhaps students and their families perceive a sense of intrusion and a loss of expectation of privacy. In any case, it would be beneficial for the school to state their expectations for the use of the laptop and if or how students will or will not be monitored in their homes through the Internet. Perhaps the policy should have been developed with the parents and discussions about expectations of privacy and limitations for monitoring should have been signed off on by representatives of all interested parties. Perhaps then an information session should have occurred. Good policies that are communicated and agreed upon by all parties can reduce activity that results in perceived violations and investigations. No matter what the outcome of the investigation is, the community where the investigation occurred will have spent much money on legal representation in a time when resources are scarce.

NEED FOR COMPUTER USAGE POLICIES

There needs to be a computer usage policy that discusses the proper use of a computer so that there are no false expectations from management of how the employee in a corporation should use the computer. If a person has done with their work, they may feel it is ok to check personal email, check retirement accounts, play solitaire card games, and read the news. The computer usage policy goes hand in hand with the Internet usage policy and should discuss the proper rules for using the computer and the Internet. There need to be boundaries of when leisure can be done, for how long, and the type of leisure activities that can be done. Suppose a person works in a power company as a dispatcher or secretary and is legitimately allowed to play video games on the night shift if things are slow. Then suppose there are power outages and people without heat and power. A reporter at the company may have found out that while people were freezing and had no power, the secretary was playing video games. If this was reported, the public would be furious and the perception of that company would be unfavorable in the media.

In the case just stated, the person playing video games might have been on break and entitled to that time. If he or she was not playing video games, it would still not have helped the teams dispatched on the road to clean up any other wires or restore power. However, the media can report that a person in that company was doing a frivolous activity while others were depending on that company for critical services. This can give an unfavorable opinion though nothing wrong was done. Sometimes the policies need to address these potential situations and ban activities such as game playing or leisure on the computer to mitigate such circumstances and avoid bad press or public inquiries. Some policies may seem draconian to employees but they could be a proactive result of the general counsel, HR, the AR, and management addressing potential problems.

Policies also need to state that there is no expectation of privacy and that the incident response team may look at the computer on premises anytime they wish. If the computer is taken home, then the policy needs to set limits on the behavior of IT staff as well as employees. There were some school districts in the news that remotely spied on children at home using some type of computer or laptop and parents were outraged. There are lawsuits over this. In any case, there will be expensive court cases that will cost the community and the school boards hundreds of thousands if not millions of dollars depending how the case goes. Then there is the bad publicity and lack of trust in the back of parents' minds. Perhaps the school was only seeing if the person was accessing objectionable websites and the camera was enabled but the harm is done, confidence is lost, and lawsuits are filed. Good policies are like good contracts. They discuss rules and expectations for both sides and define clear sanctions for noncompliance.

NEED FOR TELEPHONE POLICIES AT WORK

The telephone policies in an organization must include the usual items that one would expect to have such as using the phone for business purposes and not making or receiving personal calls. Other considerations would be such cost-saving measures such as limiting long distance calls, especially to foreign countries. These policy items would be instituted in order to limit the costs to an organization as well as limit distractions to productivity.

There are other items that need to be in a telephone policy that people would not normally expect to be in there such as the connecting of unauthorized devices. The reason for this is that people may bring in a fax machine from home and send and receive faxes for either personal or work purposes. Once a device such as the fax is brought in, the activity on it cannot be easily monitored on a dialup line. This means that an employee could be doing any number of prohibited activities such as espionage and running their own unauthorized business within the company on company time. The employee could also be engaged in a number of illegal trafficking activities as well as sending or receiving illegal pornography.

Unauthorized thermal fax machines contain large numbers of documents on a ribbon which may leave the company or its employees vulnerable to identity theft, corporate espionage, or unfavorable publicity. A mobile device forensics student named Elly described in class how she had helped her employer recover an important lost document from the fax machine. Elly had removed the ribbon from the thermal fax, unrolled it to point on the roll where documents were faxed approximately three weeks earlier. Then Elly found the negative of the document and placed it on a Xerox machine. She revered the colors so black was white and then white became black. The document was reprinted with only some minor clarity or resolution lost. Elly told her classmates that telephone policies in corporations should not allow employees to connect, disconnect, or add devices to the telephone network without permission of the employer.

Sometimes people think that they are saving a small company money by dumpster diving and obtaining equipment such as fax machines. The fax machines are connected at the office and occasionally start fires or have short circuits that disable the phone line. When a phone line is disabled, everyone else on the line cannot use that line and productivity is lost. A service call may also need to be made which may cost an additional large sum of money. An employee who wishes to be a good Samaritan may wind up costing the company a thousand dollars in maintenance bills and lost production. The employee who brought the device in may become the subject of an investigation. The device in question could also have been stolen and dumped in the dumpster. The possession of stolen equipment could cause further investigation. It is better to prohibit outside equipment in the telephone policy and provide a mechanism for employees to ask for telephony equipment that they may need.

It is also important to state that approved office equipment such as a netbook cannot be used in an unauthorized manner on the telephone line if additional items such as USB modems are added. I was at a symposium and was told by someone who managed a small office that one of their employees brought in a netbook with a USB external modem and connected it to the phone lines. During lunch hour the employee was going into chatrooms and communicating with others, sometimes sending pictures. When others came by, the person hit the Alt Tab keys and went to another task. The subject of the investigation would not let the organization investigate his machine. The suspect said that there was nothing in the telephone policy that prohibited him connecting his netbook and dialing a local number to connect to the Internet. He said it was his private laptop that he bought and the organization had no right to inspect privately owned property. He also added that his activities were on lunch time and not company time. He also added that the internet service provider was local and that no unnecessary expense was incurred by the company.

I told the person that all these activities could either have been prevented or easily prosecuted if his organization had a policy that prevented the use of unauthorized devices on the telephone network and that the phone was only to be used for authorized telephone conversations with company customers during business hours. The telephone policy needs to say what devices can be connected such as telephones, fax, computers, netbooks, mobile computing devices, or modems. The policy needs to specify the activities that are supported by the phone and during what hours. The telephone policy needs to also specify what activities are prohibited on the network and what equipment is prohibited. The telephone policy may also work in tandem with the employee handbook or employee conduct code.

Many companies also use the telephone policy to discuss sanctions for noncompliance of the policy. The telephone policy is a document that communicates expected behavior, prohibited behavior, and sanctions for noncompliance. It should also state that devices that connect to the telephone network are subject to inspection and that there is no expectation of privacy. A well-written policy that is signed by employees gives the AR the authority to have an incident response team investigate the devices in question on the network while minimizing the fear of a lawsuit by the employee. There are many places online that offer sample telephone policies that can be downloaded and serve as a starting point if a company has no policy. It seems logical that the legal department, CIO, AR, telephone services, the IRT, and HR should all be on a committee that creates and updates the telephone policy.

NEED FOR A LAPTOP/NETBOOK POLICY

The group of people who create policies should have a policy regarding acceptable use and unacceptable use. One of the points of contention is games. I took a class on PC repair many years ago and an employee of a large organization relayed a story about his being investigated for playing video games at work. Some people who had mobile computing devices that had older versions of operating systems had games that were installed on them. Some of these games were Freecell, Solitaire, and Minesweeper. In one organization, a person was observed by another employee playing Solitaire on a laptop. An incident response team was dispatched and the person was confronted. Perhaps the suspect had an employee monitoring program such as Spectorsoft installed that took snapshots at various intervals. This revealed the employee's activities. The suspect did not deny that he played video games. He said that he was finished with his work and was improving his hand eye coordination with the track point. The track point is a knob that is located on the keyboard and performs the same cursor movements as a mouse. The suspect also said that the mobile computing device was provided with the track point, the game, and that there was no policy prohibiting his actions. Perhaps the union supported the subject of the investigation's claims. The matter was dropped.

If that organization was proactive and wished to anticipate and prevent such situations, they would have consulted and implemented advice from a book called *IT Ethics* by Stephen Northcutt that discusses the subject of games on computing devices at work as well as other topics such as shopping online at work [51]. Many technical employees would benefit from a mandatory training session that explains the policies of the workforce as well as the motivation and reasoning behind the rules. The organization may also wish to have an online quiz that records the person's comprehension of the organizational rules. This could be a useful tool if an investigation in the organization occurs later because it would show that the person knew an activity was prohibited but did it anyway. It is important to understand that policy is both a communication and prosecution tool for the organization. It should be clear, concise, complete, be signed, enforced equally among all employees, and filed in a safe place.

SOMETHING NEW: ELECTRICAL OUTLETS POLICY

It is important to have a policy that employees sign an "Electrical Outlet Policy." A person who is has had emergency management or risk management training would agree with this because limiting what is plugged in the outlet reduces the risk of overheating wires and fire. That is good for both safety and lowering insurance premiums. However, such a policy would prohibit employees of a corporation to use the electrical wiring within the wall for a covert area network that bypasses all safeguards of the organization and allows data to flow unnoticed to outlets outdoors.

Intellectual property theft is a big problem in 2010 and 2011. One of the ways it could be done is by having a person in a corporation bring in their netbook from home and then connect a USB hub to it. Then external USB CD-ROMs, 3.5-inch drives, and flash drives with company data could be put in the hub. A USB Ethernet adapter could connect to the hub. Then that could be connected to an Ethernet powerline adapter. Now information from flash drives, 3.5-inch disks, and CDs could be transmitted over the power lines in the building. Someone on the outside pretending to be a landscaper could plug the same equipment in an outlet outdoors. Then any type of digital files could be passed from the research and development area of the corporation to the outside and be completely undetected.

The signed and evenly and fairly enforced policy is the legal mechanism that says this is not allowed and violation of the policy would give the organization an easy road to prosecution. The information security department in the organization could prevent the electrical wiring from being able to carry covert data by filtering the power. Filtered power can be implemented with uninterruptible power supplies often called a UPS. These signed policies should be signed by employee and filed in human resources.

SILVER PLATTER DOCTRINE (TURNING OVER EVERYTHING TO THE POLICE)

When one takes the online class to become a certified computer examiner, the course materials tell the class that one day in the future, you may be performing a digital forensics examination for a client. Perhaps you are in the middle of a policy violation such as determining if an employee abused the system and child pornography is found. It is at that time you cease the investigation and take everything to the local law enforcement agency. That is an example of the silver platter doctrine. It is putting everything together and presenting it to the police for a criminal investigation. The moment that child pornography was found, the investigation went from a policy infraction investigation to a criminal one. Nelson et al. define the silver platter doctrine as "the policy of submitting to the police by an investigator who is not an agent of the court when a criminal act is uncovered" [52].

Once a corporate investigation turns to a criminal one, the Fourth Amendment applies and any future investigation at that company needs to follow a search warrant. If an employee of the company continues investigating the case after everything is turned over to the police, then he or she runs the risk of becoming an agent of law enforcement which is not a good thing in this case and exposes one to being sued. Law enforcement can work with a judge and have a subpoena issued so the corporate person can provide the needed extra information for the case which minimizes one's risk to potential civil liability [53].

A corporate investigator who is a good citizen would naturally want to help and keep investigating to help police but such efforts are misguided unless done within the framework of the U.S. Legal System. There is no reason why law enforcement cannot look through the material that they received through the silver platter doctrine and then ask the judge for a subpoena for more information. It is important for the AR to explain the boundaries of the incident response team members' responsibilities to law enforcement and to the corporation. It is easy to be a well-meaning person who thinks they are doing good deeds and then violates another person's Fourth Amendment rights. Incident response team members in a corporation should ask questions of the AR and the general counsel.

COUNTERING SENSITIVE INFORMATION LOSS FOR DIGITAL EVIDENCE INVESTIGATION LABS

A digital forensic lab has to be careful of how it discards paperwork since many people dumpster dive to recover equipment or try to get information from discarded papers to use for network intrusion. Many organizations dispose of documents by using a shredder that creates single strips. This may sound as an effective countermeasure to deter criminals or terrorists, who wish to steal paper documents. However, it is not effective. Here is a famous example of strip reconstruction. A group of carpet weavers were hired in 1979 to reconstruct CIA documents recovered from the American Embassy in Iran. Many businesses and government offices later used cross-cut shredders which took a standard 8.5 × 11-inch paper and created approximately 300 1-inch strips. Even this is not effective; I was able to shred a roadmap and piece together strips using the same skills as someone constructing a puzzle at about one strip per minute. A standard paper might take up to 5 h to reassemble. In order to protect information, documents are now cross-cut shred into very small squares of approximately 1/4 inch on each side. However, even this technique is not a sufficient countermeasure since shreds can be scanned and algorithms employed that utilize image-based document reconstruction techniques that are known commonly as unshredding. Many people suggest utilizing an information destruction service that is certified by the National Association for Information Destruction (NAID).

If you are in law enforcement and giving your computer to someone else in the department because you are getting a new computer, then it is advisable to first copy all of your data and emails to CDs. Then the IT department may reimage the drive so that it is factory fresh or they may delete all your personal data and run a data scrubber on it. The free space and all other deleted files may be wiped clean of all information with a program such as CCLEANER. CCLEANER can make up to seven passes over each byte thus surpassing even the Department of Defense Standard of three times. The important thing is not to dispose of sensitive information or accidently give it when recycling computers with coworkers. In any case, you need to follow the policies of your organization.

FINAL THOUGHTS ON INCIDENT RESPONSE TEAMS

It is important for any incident response team to use any forensic tool with a hash library of system files so that the team can easily find out if the operating system was compromised. Compromised files may have the correct file name and file size but have a different MD5 hash. Some of these files may have code that allows outsiders remote access to the system and deletes log entries to cover tracks. Programs such as Guidance Software's Encase and Access Data's Forensic Toolkit also have large databases of MD5 hashes of files associated with hacker tools and child pornography. This can save investigators a lot of time getting to *prima facie* evidence and tools indicating a compromised system. In any corporation, it is important to recognize when criminal activity occurred and then call law enforcement.

REFERENCES

1. Nelson, B., Phillips, A., Enfinger, F., Steuart, C. (2004). *Guide to Computer Forensics and Investigations*. Thomson Technology, Boston, MA, p. 649, ISBN 0-619-13120-9.
2. Bigelow, S. (1993). *Maintain and Repair Your Notebook, Palmtop, or Pen Computer (Save a Bundle Series)*. McGraw Hill Publishing, Texas, USA, ISBN 978-0830644544.
3. Keefe, E. (1996). *PC in Your Pocket!: Information When You Need It: Best of the HP Palmtop Paper*. Thaddeus Computing Company, Fairfield, IN, ISBN-13 978-0965218702.
4. Vaaca, J., Rudolph, K. (2011). *System Forensics, Investigation and Response*. Jones & Bartlett Learning, Boston, MA, p. 316, ISBN 978-0-7637-9134-6.
5. Vaaca, J., Rudolph, K. (2011). *System Forensics, Investigation and Response*. Jones & Bartlett Learning, Boston, MA, pp. 254–257, ISBN 978-0-7637-9134-6.
6. Cardone, E. (2011). Bomb found, dismantled at Victoria Drug House. *Victoria News*, Victoria, Canada, June 16, 2011, http://www.vicnews.com/news/124038559.html.
7. McNeil, D. (2005). M.R.I.'s strong magnets cited in accidents. *New York Times*, August 19, 2005, http://www.nytimes.com/2005/08/19/health/19magnet.html.

8. Virtual Reality Environments (December 8, 2011). *Virtually Better*. Retrieved December 16, 2011, from http://virtuallybetter.com/environments.html.
9. Resilience Training (September 27, 2011). *U.S. Army Medical Department*. Retrieved December 16, 2011, from https://www.resilience.army.mil/.
10. Rizzo, A. et al. (2011). Virtual reality goes to war: A brief review of the future of military behavioral healthcare. *Journal of Clinical Psychology Medical Settings*, 18, 176–187, DOI 10.1007/s10800-011-9247-2, http://projects.ict.usc.edu/vrcpat/PDF/Parsons_(JOCS)_Virtual%20Reality%20Goes%20to%20War.pdf.
11. Kingsley-Hughes, A. (2004). *The PC Doctor's Fix It Yourself Guide*. pp. 4–25, McGraw-Hill/Osborne, Emeryville, CA, ISBN 0-07-225553-6.
12. Mueller, S. (2008). *Upgrading and Repairing PCs*. 18th edition, pp. 1390–1400, Que Publishing, Indianapolis, ISBN 978-0-7897-3697-0.
13. Doherty, E. (2011). Teaching cell phone forensics and e-learning, *2011 International Conference on e-Learning, e-Business, Enterprise Information Systems, and e-Government (EEE'11)*, July 18–21, 2011, Las Vegas, NV, accepted as a Regular Research Paper (RRP), that is, publication in the *Proceedings and Oral Formal Presentation*.
14. Doherty, E. (2011). Teaching digital camera forensics in a virtual reality classroom, *The 2011 International Conference on Computer Graphics and Virtual Reality (CGVR'11)*, July 18–21, 2011, Las Vegas, NV, accepted as a Regular Research Paper (RRP), that is, publication in the *Proceedings and Oral Formal Presentation*.
15. Building a Bootable Version of OSForensics using WinPE (n.d.). In *PassMark Software*. Retrieved November 16, 2011, from http://www.osforensics.com/faqs-and-tutorials/booting-osforensics-with-winpe.html.
16. Carrier, B. (2005). *File System Forensic Analysis*. Addison and Wesley, Upper Saddle River, NJ, ISBN-13 978-0321268174.
17. The National Repository of Digital Forensic Intelligence. URL accessed February 12, 2012. Retrieved from www.nrdfi.net.
18. *DriveProphet*. URL accessed February 12, 2012. Retrieved from www.driveprophet.com.
19. Elcomsoft Internet Password Breaker. *Elcomsoft*. URL accessed February 4, 2012. Retrieved from http://www.elcomsoft.com/einpb.html.
20. Elcomsoft Wireless Security Auditor. *Elcomsoft*. URL accessed February 4, 2012. Retrieved from http://www.elcomsoft.com/ewsa.html.
21. Contact CIA. *Central Intelligence Agency*. URL accessed February 4, 2012. Retrieved from https://www.cia.gov/contact-cia/index.html.
22. Advanced Archive Password Recovery. *Elcomsoft*. URL accessed February 4, 2012. Retrieved from http://www.elcomsoft.com/archpr.html.
23. Advanced Office Password Breaker. *Elcomsoft*. URL accessed February 4, 2012. Retrieved from http://www.elcomsoft.com/aopb.html.
24. winfingerprint. *Sourceforge*. URL accessed February 13, 2012. Retrieved from http://sourceforge.net/projects/winfingerprint/.
25. Hide IP address free. *Proxy Way.com*. URL accessed February 7, 2012. Retrieved from http://www.proxyway.com/www/hiding-ip-address-free/hide-ip-address-tool.html.
26. Live Forensics on a Windows System: Using Windows Forensic Toolchest (WFT). *Fool Moon*. URL accessed February 5, 2012. Retrieved from http://www.foolmoon.net/downloads/Live_Forensics_Using_WFT.pdf

27. *Logicube*. URL accessed February 12, 2012. Retrieved from http://www.logicubeforensics.com/products/hd_duplication/talon.asp.
28. Advanced EFS Data Recovery. *Elcomsoft*. URL accessed February 5, 2012. Retrieved from http://elcomsoft.com/aefsdr.html.
29. Russinovich, M. (April 28, 2010). PsLoggedOn v1.34. In *Windows Sysinternals*. URL accessed January 30, 2012. Retrieved from http://technet.microsoft.com/en-us/sysinternals/bb897545.aspx.
30. WebJob. *Sourceforge*. URL accessed January 30, 2012. Retrieved from http://webjob.sourceforge.net/WebJob/index.shtml.
31. Hoglund, G., Butler, J. (2005). *Rootkits, Subverting the Internet Kernel*. Addison and Wesley, Upper Saddle River, NJ, pp. 1–50.
32. Password Recovery Toolkit. *AccessData*. URL accessed February 3, 2012. Retrieved from http://accessdata.com/products/computer-forensics/decryption#passwordrecoverytoolkit.
33. *IsoBuster*. URL accessed January 21, 2012. Retrieved from http://www.smart-projects.net/cdrecovery.php.
34. IRB Guidebook (n.d.). In *OHRPArchive*. Retrieved November 16, 2011, from http://www.hhs.gov/ohrp/archive/irb/irb_guidebook.htm.
35. Test Results for Forensic Media Preparation Tool: Disk Jockey PRO Forensic Edition (version 1.20) (October, 2010). In *NIJ Special Report*. URL accessed January 21, 2012. Retrieved from https://ncjrs.gov/pdffiles1/nij/231988.pdf.
36. Disk Jockey pro (n.d.). In *Diskology*. URL accessed January 21, 2012. Retrieved from http://www.diskology.com/djforensic.html.
37. Disk Jockey Specifications (for the nerd in you) (n.d.). In *Diskology*. URL accessed January 21, 2012, Retrieved from http://www.diskology.com/djspecs.html.
38. Testimonials (n.d.). In *Diskology*. URL accessed January 21, 2012. Retrieved from http://www.diskology.com/djtestimonials.html.
39. Hogfly. (July 16, 2009). Drive encryption. In *Forensic Incident Response*. URL accessed January 21, 2012. Retrieved from http://forensicir.blogspot.com/2009/07/drive-encryption.html.
40. Germain, J. (2009). Is AES Encryption Crackable?. URL accessed January 12, 2022. Retrieved from http://betanews.com/2009/11/05/is-aes-encryption-crackable/.
41. ImageMASSter Solo-IV IT. *Diskology*. URL accessed January 21, 2012. Retrieved from http://www.diskology.com/soloIVit.html.
42. How many computer users in the world? (n.d.). In *askville*. URL accessed on 21, 2012, from http://askville.amazon.com/computer-users-world/AnswerViewer.do?requestId=547827.
43. PDWIPE. *Digital Intelligence*. URL accessed February 2, 2012. Retrieved from http://www.digitalintelligence.com/software/disoftware/pdwipe/.
44. Nelson, B., Phillips, A., Enfinger, F., Steuart, C. (2004). *Guide to Computer Forensics Investigations*. Thomson Technology, Boston, MA, pp. 36–39, ISBN 0-619-13120-9.
45. U.S. Patent 20090150988, Inventors Frank N. Adelstein and Matthew Stillerman, publication date 6/11/2009.
46. What is F-Response? (n.d.). In *F-Response*. URL accessed December 17, 2011. Retrieved from http://www.f-response.com/.
47. Remote Analysis Capability for X-Ways Forensics: F-Response. In *X-Ways Software Technology AG*. URL accessed December 17, 2011. Retrieved from http://x-ways.net/forensics/f-response.html.

48. X-Ways Forensics: Integrated Computer Forensics Software. In *X-Ways Software Technology AG*. URL accessed February 24, 2012. Retrieved from http://www.x-ways.net/forensics/index-m.html.
49. Mandia, K., Prosise, C., Pepe, M. (2003). *Incident Response and Computer Forensics*. 2nd edition, McGraw-Hill, Emerysville, CA, p. 12, ISBN 978-0-07-222696-6.
50. Doctorow, C. (February 17, 2010). School used student laptop webcams to spy on them at school and home. In *boingboing*. URL accessed December 17, 2011. Retrieved from http://boingboing.net/2010/02/17/school-used-student.html.
51. Northcutt, S. (2004). *IT Ethics Handbook, Right and Wrong for IT Professionals*. Syngress Books, MA, USA, ISBN 1-931836-14-0.
52. Nelson, B., Phillips, A., Enfinger, F., Steuart, C. (2004). *Guide to Computer Forensics Investigations*. p. 25, Thomson Technology, Boston, MA, ISBN 0-619-13120-9.
53. Nelson, B., Phillips, A., Enfinger, F., Steuart, C. (2004). *Guide to Computer Forensics Investigations*. pp. 295–296, Thomson Technology, Boston, MA, ISBN 0-619-13120-9.

CHAPTER 6

Cell Phone Investigations by Police

POLICE FOUND A CELL PHONE AT THE CRIME SCENE

A reporter named Kevin Brady wrote an online article titled "Dropped Cell Phone Leads to Arrest of Valrico Walgreens Robbery Suspect." A person who allegedly did a robbery drops the phone and takes some syringes. The police are called and they call the last person in the call log. Brady writes, "On finding the phone, deputies called the last numbers that the suspect had called and reached a TECO auditor, who was at the suspect's home, at 509 Holiday Terrace, conducting an electrical audit. The TECO auditor reportedly told deputies he saw Rodriquez running into the house, which was shortly after the robbery" [1]. This is interesting because it shows how a cell phone is useful in a police investigation.

GETTING A SEARCH WARRANT

It is first important to examine the history of the search warrant and discuss the ideas of where government gives law enforcement the authority to take a privately owned cell phone or mobile device and examine it. The U.S. citizens are often concerned about privacy and where government gets its authority. There are many movements today that are very intimidating to law enforcement, one of which is people who call themselves sovereign citizens. They challenge law enforcement's right to do anything to them. This chapter will discuss the legitimate derivation of the government's power to seize and search papers and items from people from antiquity to twenty-first-century America.

Before one goes through all the work to get a search warrant to look at a phone, a law enforcement officer might just ask the owner for permission to look at that phone. Then a search warrant is not needed. Suppose the robber of the cell phone passed out on the floor, the phone might hold clues to the next of kin, and this might be an exigent situation where the law enforcement officer needs to get in the phone to call 911 and talk to the next of kin. In exigent situations, a warrant is not need to go in the phone. If there was unread email or voicemail messages in the phone, then a communication data warrant would be needed, at least in New Jersey. The next section will address the communication data warrant.

COMMUNICATION DATA WARRANT

A Communication Data Warrant is what is commonly known as a CDW in New Jersey. Rubin Sinins is a NJ Certified Criminal Trial Attorney and wrote an article for the advocate's almanac, a New Jersey trial lawyer's blog. Rubin discussed a case called the State v. Finesmith, "The Court noted that, under the State's Wiretap Act, communication data warrants are different than wiretaps and require a lower standard for issuance" [2]. Suppose someone had some unopened email, then communication data warrant is needed. If the email has been read, it is treated differently than the email that is unread. Many people has varying opinions about the CDW depending on one's role in society. Citizens like it because it is another layer of protection to their right to privacy. Even though an investigator looked at someone's cell phone because they had a search warrant, in New Jersey, they would still need a CDW to get some recent email that was not downloaded yet and read. Police who perform mobile forensic investigations may see it as an additional burden of paperwork and a further constraint of their time when so much has to be done quickly because people have a right to a speedy trial. Some citizens may view the CDW as a retroactive wiretap that is easier to obtain than a real-time wiretap. However, it is not that easy because there is enough material on the subject that there exists a half-day class at the Morris County Police and Fire Academy on the subject.

There is a document called "Procedures to Be Followed in Handling Applications for Communications Data Warrants and Communications Order." It was a directive #9-99 that was issued by Richard J. Williams. It discusses that there exist special CDW judges to receive applications for CDWs and issue them. Specialization is always good for the public because it makes sure one has an expert who is familiar with the technologies of communication and the law. This is an excellent document to read if you are interested in the subject further.

Many people will ask how did the lack of privacy on phone calls occur? Is it something new after 9/11? When did the first occasion of this occur? Here is a history of intercepting telephone calls which starts with the earliest legislation which was created by the British government in Queen Victoria's time for India. Many of those ideas migrated to the United States much later and can be seen in the Patriot Act.

Many people think that people listening in on other people's calls is something relatively new but there is a joke among antique telephone enthusiasts that goes like this, "Does anyone know when the first wire tapper listened in on a call?" The punch line is, "shortly after the first telephone pole was put up." According to the *Random House College Dictionary* (1975), wiretap means "an act or the technique of tapping telephone or telegraph wires for evidence or other information" [3]. The Indian Telegraph Act of 1885 describes a telegraph in a very broad sense so that it includes telephones, faxes, and later was interpreted to include computer communications that used a modem. The legislation created rules of conduct and penalties for misconduct that regulated the behavior of the government, those who worked for the Department of Telecommunications, and people who used or abused the "telegraph" system. I have said in my classes that it is a genius piece of legislation that has stood the test of time. It can be read in its entirety online along with its 2003 amendments at the URL http://www.dot.gov.in/Acts/telegraphact.htm. The fascinating thing about the 1885 Indian Telegraph Act is that Section 24, which discusses "Unlawfully attempting to learning the contents of messages" [4]. John Ribeiro of IDG reports that the Indian Telegraph Act of 1885 still governs people possessing or using wiretapping equipment for use on telephone or cell phone calls in 2010 in India [5]. This Indian legislation has been studied by many people of many countries and is still looked at a model for governing telecommunication networks.

The Indian Telegraph Act of 1885, Section 5, states, "Power for Government to take possession of licensed telegraphs and to order interception of messages." This allows for the government to intercept message for reasons of public safety and what could be considered national security today. The *Spy Factory* is a documentary that says that the National Security Agency in the United States can monitor phone calls in certain circumstances for reasons of national security starting in the late twentieth century and beyond [6]. This is a similar function of national security and does not seem unreasonable. It seems that perhaps some type of government monitoring and interception of calls from bad actors that plan to harm the public is allowed. This seems reasonable because other types of spying on phone calls and communications could give people unfair advantage in business, campaigns, or for personal agendas. This idea is discussed in more detail in John Vacca's book *Computer Forensics, Crime Scene Investigation, Second Edition.*

GETTING A SEARCH WARRANT: HISTORY OF SEARCH AND SEIZURE

There have been many cell phone investigations by the police. Let us speak of a few different types of investigations that are common before getting to one that involves a murder. The reason for discussing these other investigations and laws is to give the reader some background in the laws that govern telecommunications which includes cell phones. This chapter will describe some of the ways that cell phones are being misused, some of the laws being violated by the public, and then the laws that constrain or assist law enforcement with regard to cell phones and investigations. Before we look at anything of this nature, it is first important to look at the history of search and seizure from ancient Roman times, to seventeenth-century England, to the founding fathers of the United States, and then to the American Constitution and Bill of Rights, and especially the Fourth Amendment. The chapter also surveys a variety of warrants and then looks at a case where a cell phone must be seized and examined, and a person arrested. It is not possible to discuss the idea of warrants and search and seizure in the twenty-first century in the United States without understanding the migration of ideas of the ancient Romans, seventeenth-century English, and the American founding fathers. These ideas of search and seizure, warrants, and arrests are still being discussed today in courts as people question the government's authority to look through their computers, search and seize data, and punish people for criminal activity.

If one reads the writings of the early fathers of the American Revolution, there is an emerging theme that the colonists do not like the British stopping them on the street and searching their personal papers or goods. There is also anger about knocking on the door, asking to come in, and then searching the house. However, it has been established since the time of 1603 by British Courts with Semayne's case that there is a need for safeguards to the public from unreasonable search and seizure but there is a genuine need for law enforcement to arrest someone in their home and seize certain items relevant to the crime. Then in *Entick vs. Carrington*, it was shown in 1665 that specific warrants for specific crimes, for specific evidence, with specific people and places was needed. An early document of the Congress of United States when it was held at the City of New York on Wednesday, March 4, 1789 has a six article about search and seizure. It says, "The right of people to be secure in their persons, houses, papers, and effects, against unreasonable searches and seizures, shall not be violated and no warrants shall issue, but upon probable cause, supported, supported by oath or affirmation, and particularly describing the places to be searched, and the persons or things to be seized." This paper was signed by Frederick Augustus Muhlenburg, Speaker of the House of Representatives and the Vice President of the United States John Adams. The curious thing is that it is item six but it was moved up because of its importance and became the Fourth Amendment. This item against unreasonable search and seizure is also repeated in many state constitutions such as Virginia because it is so important.

SEARCH WARRANTS IN THE UNITED STATES

People in the North American colonies on the East Coast probably had their first experience with search warrants under the British with the "Writs of Assistance." These were documents that were issued in the name of the King and allowed a representative to look for items that were taxable and had no tax stamp or contraband items. These writs were not specific as search warrants today but did give some specificity. From the history, we learn that a Plymouth Massachusetts lawyer named James Otis was publically against search warrant due to violation of people's rights [7].

Most people know that in the United States a policeman will speak to a judge and fill out a form and take an oath that what was said was true. The judge calls that an affidavit and it gets notarized. The judge will determine if there is probable cause and issue a search warrant for collecting evidence. If you wish to read about such topics in detail, then find LawBrain online. It is a fabulous online source to read about a variety of legal topics including affidavits, search warrants, and probable cause. LawBrain tells us that the U.S. Supreme Court does not require every search and seizure to require a search warrant [8]. There are warrantless searches such as in the cases of the Fourth Amendment exceptions discussed previously in this chapter. One such exception is the Border Patrol Exception. Since our days as a British colony, there have been border patrol searches of ships entering the country seeking contraband. This function is continued by tradition until today and everyone experiences it at the airport or at the port when returning from a cruise. The hard drive is considered a digital container, and so by similar logic, an SD card is a small container. Cell phones contain digital storage so these too would also most logically fall in that category.

AFFIDAVIT FOR A SEARCH WARRANT FOR SEIZING DIGITAL EVIDENCE

The United States Department of Justice provides a document that gives the reader an understanding what should be included in an affidavit necessary for a search and seizure of digital evidence [9]. The affidavit is from truthful things seen and heard but it can also have some reliable hearsay in it. For example, if one thought there was a cell phone with evidence of a murder; the first section would include the definitions of technical terms used in the affidavit about the cell phone and evidence.

Some of the definitions or descriptions could include pen register, JPG files, and call log. JPG or JPEG files are files that contain photographs and are in a specific formation as defined by the JPG standards groups. LawBrain says the following about pen registers: "The use of pen registers is governed by a 1986 federal statute, Pen Registers and Trap and Trace Devices (18 U.S.C.A. §§ 3121–3127). The statute also governs the use of trap devices which are used to identify the originating number from which the wire or electronic communications were transmitted. Neither device enables the listening or recording of the actual communication" [10]. The call log might be defined as simply as a file showing phone numbers called, phone numbers received, and duration of call.

The affidavit should have the summary of the offense [9]. Suppose there was a murder and the phone was believed to have been used to take pictures of the body; that would be mentioned. Mr X was believed to have killed Mr Y, taken pictures with a cell phone, and then dumped the body somewhere. Then there needs to be a discussion of any privacy that may be lost and perhaps the local county prosecutor could give guidance on that. Records about calls may need to be subpoenaed from a phone company but pen and registers were not considered a privacy issue according to a Supreme court case "Smith v. Maryland, 442 U.S. 735, 99 S. Ct. 2577, 61 L. Ed. 2d 220 (1979)." Lastly, there is the search strategy and justification.

The search strategy might be something such as the following paragraph. I will knock on the door and ask the subject to give me his cell phone. At the surrender of his cell phone, I will put it in a Faraday bag to keep it from being tampered with and bring it back to the computer crime laboratory at the county prosecutor's office. At that time I will update the chain of custody and ask detective Jones to use XRY forensics, Susteen Secure View, Paraben's Device Seizure 4.0, and Mobil Edit to extract all the photographs, email (including JPG photos), and the call log. At that point the person would be arrested and criminal proceedings occur.

GETTING A SEARCH WARRANT: HISTORY OF THE FOURTH AMENDMENT SEARCH AND SEIZURE

There is much written about the history of the Fourth Amendment and after reading a brief history of its content, namely search and seizure, one can see why the history is so important. As one thinks about digital evidence being seized after a warrant is obtained, consider the long trail of political and legal history of the Fourth Amendment. It is important for the reader to think of how search and seizure is done in the twenty-first century in the United States and how that same dichotomy of issues evident in *Entick vs. Carrington* are still being played out over two and one half centuries later. Many judges will read historical commentaries of Fourth Amendment cases and writing of the founding fathers to help give them an insight into the spirit of the law and to help guide them as they make rulings.

It was said that a person had to first describe the goods that were stolen before a search could be made in someone's home. The search also had to be supervised with witnesses. This was the first recorded instance in history of people having some right to privacy and having their home protected from unreasonable search and seizure [11]. Many people today see ancient Rome as a model for authority with its law and senators debating laws and representing a diverse group of people spread out over a large area, much like the United States. Many of the buildings in Washington, D.C. are also modeled after ancient Rome with the large marble buildings and carved columns. Some of the literature of ancient Rome has also been read by legal scholars and politicians of England who often incorporated some elements into their system. England was also the mother country of the United States before the American Revolution. Most of the founding fathers of the United States were once citizens of the Crown of England. English Common

Law also found its way into the American system. If one goes to the courthouse in Morristown, New Jersey, one can still see old oil paintings of judges in white powdered wigs, a reminder of our English roots.

The famous saying, "A man's home is his castle," is often repeated in movies, cartoons, and in popular culture. The saying was used in a case in 1603 in England known as Semayne's Case [11]. The courts recognized that people had a right against unlawful entry by the King's men but there were also times where the King's men had a right to enter into the house by force if necessary to carry out some type of order of arrest [4]. Here is where we see a dichotomy of power discussed. The King's men cannot just break in and plunder homes but at the same time, there is a need at times to arrest people who harm other citizens and endanger public safety. The boundaries of this power and how it would be carried out are still discussed almost 400 years later by citizens, law enforcement, and governments in the United States. *Entick vs. Carrington* was a famous case in 1765 that took place in the British Court system during the reign of George III [12]. This case is really the father of the Fourth Amendment that is located in the United States' Bill of Rights, an addition to the United States Constitution. Entick was charged by the King to find documents of John Wilkes. Wilkes' writings attacked England's policies and the King personally. Carrington was an associate of Wilkes. Entick broke into Carrington's home and seized a variety of papers. The English Judge in Court said that Entick had no probable cause and there was no required list of what was taken. This case then set limits to search and seizure in that there needs to be probable cause, a warrant, and a list of things taken. This could be considered by many to be the case that inspired the chain of custody and the modern search warrant.

William Pitt was a person who lived during the time of the case of *Entick vs. Carrington* in England. William Pitt was a citizen of the Crown of England, understood the importance of the case, and then was a founding father of the United States. William Pitt wrote many papers such as the Pitt Papers that described the ideas for the new American Republic. In 1763, William Pitt said in Parliament of England, "The poorest man may in his cottage bid defiance to all the force of the crown. It may be roof may shake—the wind may blow through it—the storm may enter but the King of England can not enter all his force dares not cross the threshold of the ruined tenement" [11]. This was an idea that the Crown cannot enter a person's home. This idea seems extreme today in 2011 because some citizens may create or spread child pornography and need to be arrested so the government needs to enter someone's home, arrest them, and seize equipment. However, such an opinion helps guard against someone who misuses the law to tyrannize people.

William Pitt's speech in Parliament may seem extreme and a bit of an outburst but it is because it is difficult for people today to understand what life under the British was like. If one went to Martha's Vineyard in the 1990s, one may have gone to Captain Dagget's House. The foundation and some parts of the house were built in the 1660s. This piece of Massachusetts property was under British rule for about 100 years. It was a hotel for a while with people eating breakfast in the basement; one could occasionally see a book case move and reveal a secret stairway to a hidden room [13]. The hotel manager said this room existed because people wanted to have papers or items safe from British soldiers. British soldiers were also quartered in people's homes in the winter leaving people little privacy. The story was that colonists were tired of being subject to search and seizure and quartering British soldiers in their homes. It should also be noted that there was a lot of smuggling in the American colonies and avoidance of paying taxes to the Crown. The search and seizures were really in reaction to the amount of smuggling going on [14]. The British also used "writs of assistance" which are somewhat comparable to a search warrant today except that they seemed to give wide powers of search and seizure. Later, in language that would become the Fourth Amendment, the Massachusetts Constitution of 1780 declared in its Article I Section XIV that, "Every subject has a right to be free from all unreasonable searches and seizures" [11]. One can go visit the Independence Hall in Philadelphia and learn that after the American Revolution, there was much fighting about the United States Constitution and the delegates signed only on the provision that amendment or what we call today, "The Bill of Rights" could be added later. Elbridge Gerry from Massachusetts, Edmund Randolph and George Mason from Virginia were the three people who refused to sign the Constitution because it had no Bill of Rights. The American people owe them much for their liberty.

George Mason was one of the delegates at the Constitutional Convention in Philadelphia. George Mason and James Madison, who later became President of the United States, were the main writers of the Virginia Declaration of Rights [15]. Section 10 is especially of interest to anyone interested in search and seizure. "Section 10. That general warrants, whereby any officer or messenger may be commanded to search suspected places without evidence of a fact committed, or to seize any person or persons not named, or whose offense is not particularly described and supported by evidence, are grievous and oppressive and ought not to be granted" [15].

The Constitution and the Original Bill of Rights are on display in the area around Independence Hall. One can see where the Declaration of Independence was signed and the Constitution was signed. One can also go next door and to Congress Hall which is a building with the U.S. Senate upstairs and the United States Congress on the first floor. Philadelphia was a good central location for the United States and the fact that it was a port made it easy for delegates to ride a ship home. The Bill of Rights includes the first 10 amendments. The Fourth Amendment is interesting to anyone who visited the Captain Dagget House in Massachusetts and knows the story of the secret stairway because it says, "No Soldier shall, in time of peace be quartered in any house, without the consent of the Owner, nor in time of war, but in a manner to be prescribed by law" [16]. The Fourth Amendment says, "The right of the people to be secure in their persons, houses, papers, against unreasonable searches and seizures, shall not be violated, and not Warrants shall issue, but upon probable cause, supported by Oath or affirmation, and particularly describing the place to be searched, and the persons or things to be seized" [16].

If one is an educator and teaching a class on search and seizure, the Fourth Amendment, and the Constitution, a field trip to Independence Hall might be in order. There are free tours for the public at Independence Hall, The Constitutional Visitors Center, and Congress Hall. Each building has a national parks officer give a guided tour. One can ask questions, see the original documents such as the Bill of Rights, Constitution, Declaration of Independence, and the Articles of Confederation. There is also a gift shop associated with The Constitutional Center where one can purchase many books on American Law, the Constitution, and various works of the founding fathers of the United States. The visit is educational as well as fun and one can ask for suggested reading on the Fourth Amendment.

GETTING A SEARCH WARRANT: DEFENSE ASKS IF WARRANT IS TOO BROAD AND INVALID

If one takes a class with a law enforcement officer, they will often say to make the request for any possible micro SD cards or other pieces of external media that go with the phone because it could hold key evidence to prove the person guilty or innocent, exculpatory data. It was already discussed with the Fourth Amendment that items to be seized must be named. It would be an incomplete investigation if one only asked for the phone and received it, and then the investigator saw many digital media cards for the phone near the phone but could not take them. Some digital forensic investigators say that there exist templates with wide languages that allow law enforcement officers the opportunity to take all kinds of digital media and accessories that might be associated with the phone. However, one might ask, "What would Patrick Henry, James Madison, and George Mason think of that?" Is that in the spirit of the Bill of Rights? Could such wide language templates be questioned and evidence suppressed if one has a good lawyer who is an expert on the Fourth Amendment and digital evidence collection.

FOURTH AMENDMENT EXCEPTIONS

There are exceptions to the Fourth Amendment which are open fields, plain view, consent, motor vehicle exception, exigency, and border patrol exception. These exceptions have been defined by the court over time. In 1984, there was a case known as *Oliver vs. United States* and another case of *Hester vs. The United States* discussed on page 1245 [14]. The judges said that if something is in an open field, a lake, a wooded field, pasture, or a vacant field, it is not protected by the Fourth Amendment. If someone had a pile of stolen computers in the woods near their home, it is not protected by the Fourth Amendment. Plain view is another concept where the Fourth Amendment does not apply. In the late 1990s, many criminals were sitting by crowded toll booths with cell phone industry equipment and laptops. They were part of a phone cloning scheme. The person in the car would have the cell phone industry equipment in the car and collect mobile identification numbers (MIN) and electronic serial numbers (ESN) and put them in spreadsheets in laptops. This spreadsheet was sold to another person in the ring who put the ESNs and MINs on SIM cards and sold the cards to someone. Another person sold the cards to people who often used these cloned cards to call people in foreign countries. The bill would then come a month later to the person who had their ESN and MIN eavesdropped and stolen. There were stories that the police would walk up to the person, see them with cell phone industry equipment and laptops by congested places, and see spreadsheets of MINs and ESNs. Everything was in plain view. The people in the car said that they had nothing to do with the cell phone industry. Everything was in plain view and people were arrested.

Here is a case that has to do with cell phone cloning and warrants in Newark, New Jersey. There were elaborate investigations that had involved many law enforcement agencies in Passaic County and Essex County, New Jersey, working in cooperation with the cell phone company [17]. Three people were arrested and 20 cloned phones were seized. Since this was an investigation, there was probable cause, an affidavit was signed, a warrant was issued, almost two dozens cloned phones were seized, and three people were arrested [17]. Cell phone service stealing costs the phone company a lot of money [17].

Consent is another exception to the Fourth Amendment. A policeman sometimes stops a car for a violation and writes a ticket and sees a cell phone and may ask to see it. If the person says "sure" and then the policeman finds child pornography, for example, that is an example of *prima facie* evidence found in a warrantless search. Exigency is another time where a warrantless search may be employed. Suppose a person is unconscious and his wallet is stolen. The policeman sees that the person has a cell phone; it is reasonable to assume that the phone may give a clue to who he is. Therefore, the policeman goes through it and finds ICE, in case of emergency, it may have numbers of his wife and his parents. Then the policeman may call 911 to get help and one of the ICE numbers to get a family member to go to the hospital. The unconscious person may have allergies to medicine or need a consent for certain medical procedures. The circumstances were exigent and looking through the phone, perhaps, saved a life.

FOURTH AMENDMENT EXCEPTION: BORDER PATROL EXCEPTION

This is a very current topic that is often in the news. One may think that border patrol agents going through your cell phone or laptop is an invasion of privacy and something draconian since 9/11. An online paper posted at the University of Miami Law School says, "Searches at the United States border have not been subject to traditional Fourth Amendment protections since the Bill of Rights was created" [18]. If one reads the historical literature of the time, one learns that people were presumed suspicious when leaving or entering the country with goods. Smuggling was a problem for the British who often lost revenue to American colonists not purchasing tax stamps and illegally depriving the crown of revenue. In the twenty-first century, people are still buying carloads of cigarettes at Indian reservations and then trying to get them to Canada to sell. Containers have been examined at the border for contraband as people pass through the border. Many people and judges see the hard drive as a digital container which seems to make it fair game to inspection. Techdirt, a website that discusses digital privacy says, "For a while now, courts have said that you have no 4th Amendment rights at the border, and border patrol/customs officials have every right to search your laptop" [19].

Suppose a policeman has a gut feeling that someone named Fred is cloning phones but has no probable cause. Fred lives in the community and always takes his laptop with him. The policeman hears at the coffee shop that Fred and the guys are going on a fishing trip on Friday to Canada and coming back on Monday. The policeman calls a friend on border patrol and tells him about Fred, the laptop, and his suspicions. Fred returns to the United States on Monday and is stopped by Customs and Border Patrol. They examine his laptop because of the Fourth Amendment Exception and find evidence of phone cloning. Then Fred is arrested. This may be legal but would the founding fathers have thought about that with regard to the protections of the Bill of Rights and the freedom from unreasonable search and seizure? Many people express concern that rules that are meant for keeping us safe with regard to national security are misused for arresting people who cannot be trapped by conventional means. Some people express concern that this can be a path to tyranny. Some may say that is not the case and good law enforcement teamwork was done and no laws were broken. This is why there is the Supreme Court in the United States to look at constitutional issues and make a ruling. A convincing argument could possibly be made to suppress the evidence against Fred and a convincing argument could be made to imprison him.

CHAIN OF CUSTODY FOR PHONE

The chain of custody is important when taking any piece of evidence, but especially a phone, because it is small and can be lost. It can also be easily tampered by someone physically near it or by some wireless contact of some sort. That is why it is important to keep it in a Faraday bag and locked up in a safe place from tampering, and it should be in a temperature- and humidity-controlled environment. The chain of custody form is a tool that is used to keep people aware of securing the phone and free from the possibility of tampering. The form itself should have many things on it. The *Nebraska Lawyer,* October 2008, has an excellent article by Don Kohtz and Matt Churchill that gives an overview of what should be on the chain of custody form [20]. At FDU, I discuss the chain of evidence in the cell phone forensics class and tell people in my own words that the chain of custody is a document that lets people know where the phone was taken, where it was kept, how it was processed, and what other people received copies of the evidence and signed off for them. Anyone who looks at the form get a feel for the journey of that evidence from the time it was seized until the time it is displayed in court. Whoever looks at the chain of evidence should not doubt the integrity of the evidence since they can see how it was seized, secured, and processed. If other copies were made, the details of that should be there. There are plenty of samples online of a variety of chain of custody forms and one can easily find them by using a search engine such as Google.

SUSTEEN SECURE VIEW/PARABEN DEVICE SEIZURE, EXAMINE PHONE

Both cell phone examination products are very good and can do the basics such as collect the phone books with the names, addresses, phone numbers, email addresses, and addresses. Both might be able to collect SMS messages, MMS messages, text messages, pictures, call logs, and video on certain phones. The physical imaging might be better on Paraben's Device Seizure for some cell phones. This means that one can get the allocated space with files and unallocated space which is used for buffering files and existing files that were marked for deletion. There can certainly be a lot of evidence in the slack space too. If one only wanted to acquire all the existing phone books, email, messaging, pictures, video, and other existing files, then Susteen Secure View would be the way to go. The advantage of using Susteen Secure View would be its unique features such as being able to graph all the contacts who call the phone owner or who the phone owner calls. This graph of contact frequency is very useful. The other feature that is great is the graph that shows when the calling activity occurs. The other feature is the ability to extract the GPS data from a picture and link it from the phone.

The best situation would be to purchase Paraben's Device Seizure, Susteen Secure View, X-Ways Forensics, and Access Data's Forensic Tool Kit (FTK). Getting the physical image of the phone with Device Seizure would be great. Then, one could carve the file out from the space with FTK. If there were password-protected files, FTK would be useful in cracking them. FTK could also be useful in finding pictures files that were deliberately given the file extension of a word-processed document. X-Ways forensics is a great tool for collecting unallocated file space and slack space from small files within large clusters. It is also good for data validation to use more than one tool for the same task. Susteen Secure View is excellent for the graphs of contacts, the Google Earth maps linked to the location of photos, and for incorporating all the evidence of all the other tools into its report creation tool. This is important so that the jury just looks at one report and not five different reports. The consolidation of reports may prevent confusion and allow jurors and lawyers to focus on the content and not bureaucratic details.

USING EVIDENCE FROM A CELL PHONE AND GETTING SUSPECTS TO TALK

I want to discuss a fictional situation that is within the realm of possibilities that illustrates how technology, human behavior, some interesting concepts regarding cell phone forensics, interviewing, and interrogation could come together in a situation where a person is working as a mobile forensics investigator in a war zone. Imagine the story starts where a person in Iraq was brought in for questioning by an intelligence officer. A translator may also need to be present. The person said they knew nothing of a car bomb or person X. Then the person would talk about some other topics. In the mean time, a mobile forensic examiner connected to the suspect's phone by using the infrared connector in conjunction with an examination laptop and cell phone forensics tool such as Susteen Secure View. The contact book was seized by someone behind a two-way mirror. The intelligence officer was wearing an ear whig and the mobile forensic examiner told him that X was in the contact book. Then the intelligence officer says, "My piece of paper says that X calls you and you call X. Can you tell me about that?" The person starts to squirm and fidget in the seat. The mobile forensic examiner then might use the SVProbe feature of Susteen Secure View to graph all the contacts and display the frequency of those contacts [21]. The intelligence person then gets some information about the number of times that X and the suspect spoke. This is presented to the person and they get more nervous. Perhaps there is a hint about travel to another place for enhanced interrogation techniques. The person tells the truth about everything they know. A little information at a time can be fed to the suspect. As more is known, it is presented. The pressure is turned up and the person breaks and tells the truth. This is only a fictional story to show what is possible and of course this could only happen if seizing a contact book and call log in such a manner was legal in Iraq.

Here is another fictional situation that is within the realm of possibilities that illustrates how mobile forensics technology, the law, and investigation come together in our modern world in the United States. Imagine that there is a gang that drives trucks through fences and gates and sledge hammers holes in walls, and steals big screen televisions and laptops. This goes on for a while and the police cannot solve it. Then a cell phone is dropped at the crime scene and the police arrive afterwards. It is obvious to nearly anyone that the cell phone should have evidence in it relating to the crime. The policeman who finds it calls in the county prosecutor's office. An incident response team arrives shortly. One person carefully photographs the phone and fills out a chain of custody form. The phone is put in a Faraday bag and locked up so that it cannot be tampered with wirelessly or physically. The officer who finds it calls and wakes up the judge at night and says he needs a search warrant right away. He and the judge meet at a court. The officer fills out an affidavit that says that he found this phone and wants to use Paraben's Device Seizure, XRY Forensics, and Susteen Secure View to get the contact log and phone book. Then he wishes to bring in some of the contacts for questioning. The judge gives a search warrant.

The phone is searched and the last few contacts within the last 10 min of the robbery are asked to come in for questioning. Nobody knows anything. Each of the suspects is in a different room. Suspect 1 is told suspect 2 is making a deal to give evidence and testimony on suspects 1 and 3. Then suspect 1 says he wants to make a deal right away and takes a small amount of prison time in exchange for telling everything on suspects 2 and 3. Each person is made to think one of the others told on them in exchange for a plea bargain. The person who is the weakest link breaks and tells on the others.

POLICE HAVE A SUSPECT FOR A MURDER

A continuing education student once told a story in my class about a cell phone forensics investigation that he heard from someone else. The story was researched and nothing substantial could be found about the story. I enlisted the help of other researchers and a law enforcement officer and it seems probable that either the case was something that was not publicized online or that the person's friend must have told a story about mobile device forensics that he saw in a movie or on a television show such as *CSI*, *Criminal Mind*, or *Law and Order*. Though the story does not appear to be true, it does illustrate how cell phone forensics could hold the key information in a murder investigation that could help to locate a body and prosecute a suspect. What we do know for sure is that today, cell phones are considered to be the new "Evidentiary Gold Mine," as stated in an article by Don Kohtz and Matt Churchill [20].

The story starts with people noticing that a person has been missing from work and around the neighborhood for a few days. The police are called and investigation starts. The missing person's credit card has not been active prior to four or five days before being reported as missing. The bank ATM card was not used either. An investigation results in some reported alleged heavy gambling debts. The missing person's computer is searched in exigent circumstances because it is believed that conversations or activities could lead to where the missing person is being held. A tool such as Guidance Software's Encase recovers all the slack space and chat room conversations. The chat room uncovers some conversations between a person who appears to be a crime family enforcer with a screen name such as "Nickname" and the missing person.

Because of the Communication Assistance for Law Enforcement Agencies (CALEA) legislation that was passed, Internet Service Providers (ISP) must provide law enforcement with help in lawful investigations. The ISP was served a subpoena and provided results on the "Nickname." "Nickname" was asked to come in for questioning but nothing of substance was revealed during the questioning. The person's background had a long history of criminal activity, some of which was associated with people who allegedly ran illegal gambling establishments. The man said he often dialed some wrong numbers and accidentally called an illegal gambling establishment where bets were taken on a cordless phone in a backroom. The police said that raids only revealed buckets of water, perhaps where bets written on water-soluble rice paper were put in water to disintegrate. There was often a smell of burning paper at those locations. It was suspected that bets were written on magician's flash paper and then thrown up into the air thus bursting into flame if the police arrived.

SEIZE CELL PHONE

The story was that the police suspected "Nickname" and then filled out an affidavit saying that they believed that the person played a part in the person's disappearance and wished to search the cell phone with a tool such as Susteen Secure View, XRY Forensics, or Paraben's Device Seizure. The results then may show calls, or chat rooms conversations, giving a clue to the missing person's location. Since there was no body or ransom note, this was not a kidnapping or murder investigation. There was probably a broad template used so that all SD cards, wires, chargers, and manuals for the digital phone could be seized and examined.

COLLECT PICTURES WITH GPS

Tools such as Susteen Secure View have features such as Gallery where pictures can be seen as thumbnails. It is also possible to select pictures that have GPS data embedded in them and link them to Google Maps in the same way that GPS Visualizer works as mentioned in a previous chapter. The fictional story was that police did a knock and enter the home of "Nickname." The cell phone was surrendered to the police with the cables, docking station, manual, extra battery, and SD cards.

METADATA SHOWS LOCATION OF BODY AND GRAVE OF VICTIM

The story that the student told was that the police used one of the forensic tools for cell phones such as the ones discussed in the affidavit for the search warrant. Since the cell phone was a camera phone, many pictures were recovered on the camera phone. Some of the pictures appeared to show the victim dead on the carpet in his home. Other pictures showed the same body in a shallow grave in the woods before burial. The phone also had a call log with the victim and someone in a crime family associated with illegal gambling. The pictures, when right clicked, had GPS embedded data in them showing latitude and longitude points. A tool such as GPS Visualizer or Susteen Secure View revealed that the body of what appeared to be the victim had the GPS coordinates of the victim's house. The shallow grave had the GPS coordinates of a wooded area near the cell phone owner's home.

A county coroner and a group from the prosecutor's office went to the location of the shallow grave suggested by the GPS coordinates. They noticed some of the trees and landmarks in the photograph and then exhumed the body. The group suggested that the murder was a contract hit and that the person was killed for failing to pay excessive gambling debts. This sounded pretty good but there was still not enough evidence to conclusively link all the pieces. Another affidavit was filled out and another search warrant was obtained for "Nickname's" car.

This warrant was granted and the car trunk was searched. Some of the fibers found in the trunk matched the clothing of the victim thus tying all the loose ends together. The student ended the story in it being a slam dunk victory for the prosecution. If the story was not true, it is certainly in the realm of possibilities since all the technological, investigative, and legal elements are certainly items that can be found in journals, newspaper articles, and textbooks.

POST ANALYSIS

The term "knock and enter" is actually a legal term that one can read about in a case such as the United States of America, *Appellee, vs. Patrick Harm Keene*, Appellant, No. 89-5442. Police have to knock, announce their arrival, authority and purpose, and then enter when consent is given and they are carrying out the search warrant [22]. Such search warrants are said to be executed during the day.

Could the GPS data in the pictured have been spoofed? Consider the evidence of the pictures and the GPS data. If the GPS data shows the body at the house, that possibly could be spoofed with a program. That data must be scrutinized carefully. It was also discussed earlier in a previous chapter that GPS data may be inaccurate due to clouds, or a lack of clear view to the sky and satellites. However, if the GPS was used to locate the body, that is the proof of the matter.

Is the date and time of the pictures accurate? The dates and times of all the pictures should be checked about the system date and time in the phone to see if there is a differential. It could be off by a certain amount of time, for example, if the phone was very old and did not use the daylight savings time changed by President George Walker Bush. If the crime took place near a time zone change such as Hoover Dam where one side is 1 hour different than the other side, that should be considered. The cell phone should be checked to see if it auto corrects date and time. Some phones such as the Blackberry Curve will automatically correct the date and time to the locality it is in.

Could the pictures have been put on the phone? It might be a defense to try to refute that the pictures were taken by that phone. Each of the pictures has metadata in it. If one loads the pictures into the software program known as Picasa Two, for example, there is a properties option that will give the make and model of the camera as well as some specifics about the lens. On camera phones, there is usually a special ID associated with the user's phone account that is put in the phone. It was also noted in a previous chapter that the resolution of the picture should be supported by the camera that supposedly took the picture. There are also minor imperfections of each camera since various diodes burnout and those imperfections should be noted in the pictures in question. The phone may not have been set up for email so the pictures may not have been able to be placed there that way too.

Suppose the person admits the pictures were taken with his phone but it was not him. The person may have a transponder in the car that is used for paying tolls. If the person drove on a toll road during the murder, there should be transponder evidence as that car went along the road. There may also be security camera video of the road with him or her on the road at the time the transponder left data. The person may also have a GPS device in the car and the routes may show that the person drove to the house about the time of the murder. A period of time parked at the murder victim's house might be on that GPS device in the car about how long it took to load the body in the trunk. The times and dates on the trip to the shallow grave may also be supported by the time on the pictures. The person's cell phone may have GPS enabled and a navigation application may also contain the same data as a Tom Tom, Magellan, or Garmin device. There is an article on the Trucker Steve website that discusses a real case of how GPS is being used in a murder investigation [23].

The cell phone may have been left on because the person wanted to call the crime family or receive a call about the job being done and receive payment instructions. There should be a trail of towers that the phone pinged showing a time and date of the trip to the murder victim's home and to the shallow grave. Some towers only collect data on a person's phone and whereabouts when a call is made or a text message is sent or received [24]. Consider a real case of the recent cell phone tower evidence playing a key part in the location of Christopher Pullman who is on trial for the murder of Sylvestor Eddings and the manner in which towers show his location from a call [24]. This type of location evidence is frequently used in cases now. Since there is the CALEA legislation, cell phone providers must work with law enforcement investigations to provide this data from cell phone tower accesses. There is much data created by the cell phone, the toll-paying transponder, the roadside cameras, and any GPS navigation devices in the car. The amount of electronic data that is created from the devices people carry is amazing. Everything creates electronic footprints.

If "Nickname" connects his phone to his computer, it might be set up with automatic synchronization software that takes the latest files from his phone when he connects a cable from the computer to his phone. There may be pictures of the murder that are emailed to the crime boss. It would behoove the prosecution to seek a search warrant and collect any desktops or laptops that the suspect had at home. They may have emails with picture attachments from the phone as well as correspondence about the murder. There may even be Mapquest searches about possible burial locations or to the crime family's home. The desktop or laptop as well as any nearby media could contain a treasure trove of evidence relevant to the case.

When we analyze this case, we also have to consider any nearby homes and businesses that might have had security cameras. A digital camera and digital video recorder system might have caught the car stopping and then the person popping the trunk and taking the body out. Perhaps digging and dumping the body might be caught on film too. Some digital video recorders and high-resolution cameras can collect incredible detail and it may be possible to zoom in on a license plate if the angle of the camera was placed on a certain azimuth. It would behoove the detective to notice any local businesses or home security systems that might have cameras that caught the wooded area where the body was dumped. Many traffic lights also have security cameras. Some may have continuous feed while others may only take the picture of a car that goes through a red light. Looking for traffic light cameras near the crime scene might be worth it.

GOOGLE STREET VIEW

It might also be possible to use Google Street View to check the GPS location found in the pictures where the body was found. There might be images of the car stopped, the body dumped, or digging. This is not as far-fetched as one may think because a young man was able to get images involving a robbery of him on a bicycle 6 months later using Google Street View. The man gave this image to the detectives in his locality and they contacted Google. Google gave the detectives the original video and two men were then arrested for the crime [25]. It is possible to peruse the video for a particular location and collect evidence. Some privacy advocates are against this but it is an available tool to the public and no special warrants are needed. There is no expectation of privacy in public places and it is completely legal that such a video is taken and collected. Some literature has discussed that in the past we enjoyed more privacy because there was no technology to collect this data. It does not mean that people throughout history had a constitutional right to privacy but it was a perk or benefit of being part of a low-tech society. Google Street View collects a tremendous amount of video on our planet and this tool should probably be used in tandem with all GPS picture investigations because it produces some good results. Michael Miller wrote a book called *Using Google Maps and Google Earth*. There are some informative sections on Google Street View that any detective in the world who has Internet access might find very useful.

SkypeLogView

Before using Susteen Secure View, there may have been some Skype phone calls on the cell phone. Many people use Skype on cell phones, PDA cell phones, laptops, and desktop computers. It is wrong to assume that all calls are made only on traditional phones. A call could be made on a Skype application on a cell phone or even with an application such as Vonage mobile that can also be used on cell phones. This means that some calls made by a cell phone may be missed if one thinks of cell phone calls as going through the cell phone provider service. If one finds out that Skype is on the phone, then a program such as SkypeLogView should be run on the examination machine. Then all incoming and outgoing calls with Skype can be obtained. That tool will also find chatting sessions and file transfers. That might be important because someone could transfer pictures. In our discussion of "Nickname," perhaps phone calls were made with other tools such as Vonage or Skype.

SOME FINAL THOUGHTS

It is important to learn about the law and its history so that one can understand how it is interpreted by judges. It is interesting to see how *Entick vs. Carrington* and the speech of William Pitt in Parliament have been institutionally communicated to the Virginia State Constitution, The Bill of Rights, and ultimately the Fourth Amendment. As time goes on, more cases involving the Fourth Amendment and its exceptions have helped refine search and seizure in the United States.

Mobile device forensics is a crossroad of mobile devices, science, and the law. It is important to use accepted methodology and tools to collect evidence for use in court. The digital examiner must look for exculpatory data to find one innocent as well as data that is used to prosecute people. There are many tools to collect and examine digital evidence from mobile devices. It is good to get as many tools as possible because each has its special features or strong points. It is also good to have a tool such as Susteen Secure View that has a reporting feature that lets the examiner include all the various findings from different tools in one report. As one puts the report together, it is also good to think logically and include as many supporting pieces of evidence as possible. There are GPS navigation devices, cell phones, laptops, desktops, traffic light cameras, outdoor security cameras, toll transponders, Google Street View, and other sources of digital evidence that may be relevant to the case. It is important to get search warrants or subpoenas for as many sources of evidence as possible since justice is depending on it. It is also important to think, not make mistakes, and be careful so that evidence is not suppressed for someone's carelessness to rules.

REFERENCES

1. Brady, K. (June 23, 2011). Dropped cell phone leads to arrest of Valrico Walgreens robbery suspect. *Brandon Patch*. URL accessed January 1, 2011. Retrieved from http://brandon.patch.com/articles/dropped-cell-phone-leads-to-arrest-of-walgreens-robbery-suspect.
2. Sinins, R. (July 14, 2009). Communications data warrant for stored email permitted on lower standard than wiretap. *Advocate's Almanac A New Jersey Trial Lawyer's Blog*. URL accessed January 1, 2012. Retrieved from http://advocatesalmanac.blogspot.com/2009/07/communications-data-warrant-for-stored.html.
3. Wiretap definition (n.d.). *The Random House College Dictionary*, Revised edition, 1975. Random House, Inc., New York.
4. Government of India Ministry of Communications & Information Technology, Department of Telecommunications. Unlawfully attempting to learning the contents

of messages, Section 24 of Indian Telegraph Act of 1885. URL accessed December 29, 2011. Retrieved from http://www.dot.gov.in/Acts/telegraphact.htm.

5. Ribeiro, J. (December 30, 2010). India cracks down on unauthorized communication snooping. *PCWorld.* URL accessed December 29, 2011. Retrieved from http://www.pcworld.com/businesscenter/article/215180/india_cracks_down_on_unauthorized_communication_snooping.html.
6. Spy Factory, DVD, ISBN 978-1-59375-886-8.
7. *Ehow.* The history of the search warrant. URL accessed December 31, 2011. Retrieved from http://www.ehow.com/facts_6123799_history-search-warrant.html.
8. *LawBrain.* Search warrant. URL issued December 31, 2011. Retrieved from http://lawbrain.com/wiki/Search_Warrant.
9. US Department of Justice. Sample Language for Search Warrants and Accompanying Affidavits to Search and Seize Computers Provided by the U.S. Department of Justice. URL accessed January 1, 2011. Retrieved from http://www.forwardedge2.com/pdf/form-sw.pdf.
10. *LawBrain.* Pen register. URL accessed January 1, 2012. Retrieved from http://lawbrain.com/wiki/Pen_Register.
11. *NUSD.* A historical overview of the Fourth Amendment. URL accessed December 30, 2011. Retrieved from http://web1.nusd.k12.az.us/schools/nhs/gthomson.class/pol699.paper/pol699.hist.overview.html.
12. *Constitution.* Entick v. Carrington 19 Howell's State trials 1029 (1765). Retrieved from http://www.constitution.org/trials/entick/entick_v_carrington.htm.
13. Resortsandlodges.com. Daggett House (Edgartown, MA), Retrieved from http://www.resortsandlodges.com/lodging/usa/massachusetts/marthas-vineyard/daggett-house.html.
14. GPOaccess. *Fourth Amendment Search and Seizure.* Retrieved from http://www.gpoaccess.gov/constitution/pdf/con015.pdf.
15. *US Constitution Online. The Virginia Declaration of Rights.* Retrieved from http://www.usconstitution.net/vdeclar.html.
16. Levine, E., Cornwell, E. (1975). *An Introduction to American Government*, 3rd edition, p. 314, Macmillan Publishing, New York, ISBN 0-02-370300-8.
17. PRNewswire (May 18, 1995). Cellular companies team up with law enforcement to bust Newark cloning operation. *The Free Library by Farlex.* Retrieved from http://www.thefreelibrary.com/CELLULAR + PHONE + BANDITS + ARRESTED + IN + NEWARK-a016890835.
18. University of Miami Law Review, vol. 65, no. 3, p. 1003. URL accessed December 30, 2011. Retrieved from http://www.law.miami.edu/studentorg/miami_law_review/issue_archive/pdf/vol65no3/MIA301.pdf, page 1003.
19. *Tech Dirt.* Court says border patrol can take your laptop for off-site search if they have reasonable suspicion. URL accessed December 30, 2011. Reference from http://www.techdirt.com/articles/20100603/0036229666.shtml.
20. Kohtz, D., Churchill, M. (October 2008). Cell phone forensics: The New "Evidenciary" Gold Mine. The Nebraska Lawyer, pp. 11–14. *Continuum Worldwide.* URL accessed January 7, 2012. Retrieved from http://www.continuumww.com/Libraries/PDFs/cell_phone_forensics_the_new_evidentiary_gold_mine.sflb.ashx.
21. Doherty, E. (2011). Cell phone investigative tools for the counter terrorism professional. Retrieved from http://www.iacsp.com/digdev_sec.php?p=10.
22. *Resource.org.* 915 F.2d1164, 31 Fed. R. Evid. Serv. 64, United States of America, Appellee, v. Patrick Harm Keene, Appellant. No. 89-5442. United States Court of

Appeals, Eight Circuit. Submitted March 15, 1990. Decided September 25, 1990. URL accessed February 26, 2012. Retrieved from http://ftp.resource.org/courts.gov/c/F2/915/915.F2d.1164.89-5442.html.

23. Admin (January 20, 2011). GPS tracking aids murder investigation. *Trucker Steve*. URL accessed January 8, 2012. Retrieved from http://truckersteve.org/2011/01/gps-tracking-aids-murder-investigation/.
24. Wellner, B. (November 25, 2009). Cell phone activity evidence in murder trial. *Quad-city Times*. URL accessed January 8, 2012. Retrieved from http://qctimes.com/news/local/crime-and-courts/article_828ae766-d9f9-11de-823c-001cc4c03286.html.
25. Lur, X. (April 20, 2010). 5 examples of how Google Street View helped solve crimes. *Tech XAV*. URL accessed January 8, 2011. Retrieved from http://www.techxav.com/2010/04/20/5-examples-of-how-google-maps-helps-solve-crimes/.

CHAPTER 7

Recovering Existing or Deleted Data from USB Devices

USB FLASH DRIVES

Before one starts discussing USB drives and the misuse of USB devices, one must ask what is the meaning of the term USB? Imagine a forensic expert sitting on the stand next to the judge ready to discuss all of his or her investigation results and then the defense lawyers asks, "What does USB mean?" Can you give the court a definition of USB? It is possible that one can answer all types of questions but cannot answer the basic question. How will that be perceived by the judge and jury? What impact will this have on the case and will the court accept this person's findings? How can this person qualify as an expert and his or her work be believed if a basic question cannot be answered? If he or she answers with their own definition, the defense lawyer could read a definition of USB and discuss how the established definition is so different than the so-called expert's definition. How will that be interpreted?

The expert may say it is a universal serial bus. The defense may blurt out that I asked for a definition, not what an acronym is. If the expert is not ready, he may say it is a small storage device that plugs on the side of a computer. The defense lawyer may pull out an SD card and show the jury and then says, "This must be a USB device because it is a storage device and plugs in the side of the computer." The expert may say it is not. Then the defense lawyer says, "How can that be, this fits the definition you gave me?" Then the expert may start to sweat and squirm and has to come up with a new definition. How do you think this expert looks so far? This is a bit of a drama but it is to illustrate a possible scenario of how important it is to be prepared with the basics and to have definitions memorized and be ready to reply back with under duress.

It is important to start out with a definition of USB and it should be from a reliable source. There are many variations of definitions that one can find regarding USB but *The World Dictionary* gives a definition that is concise and simple. They say, "Universal Serial Bus: a standard for connection sockets on computers and other electronic equipment" [1]. Many of us are aware of USB 1.0 which started out in 1996 and then USB 1.1 was in the news in 1998 when 1.5 Mbps was announced. Then many people became aware of USB 2.0 in April 2000 when the new standard was announced with speeds of up to 480 Mbps. Then USB 3.0 was announced and supported data transfer speeds of 4.8 Gbps and USB 3.0 has been utilized by Microsoft as a standard for interfacing with the release of Windows 8 in 2011. USB 3.0 is known as super speed. The Linux operating system supports it too. USB 3.0 is also compatible with connectors for USB 1.0 and 2.0. This is commonly called backwards compatible. The USB 2.0 comes in three types which are USB 2.0 Standard A, USB Standard B, and Micro USB. The USB 2.0 Standard A has both power of 4.5–5.5 V and data. The USB Standard A is the rectangular end that fits in your computer and USB Standard B is the other squarer end that fits in your scanner or printer.

The howstuffworks website says that USB 2.0 cables can provide 500 milliamps at 5 V in one direction at a time. The breakthrough with USB 3.0 cables is that it provides bidirectional data simultaneously as well as increased power at 900 milliamps [2]. USB cables should not be longer than 5 feet or attenuation or degradation of the signal can occur. The USB cable has four wires in it. The red and brown wires are for the power. The yellow and blue wires are for the data. Many people and manufactures have adopted USB as a standard because if various hubs are used, one computer could connect as many as 127 devices. Other previous connector standards such as parallel might allow one to use an A,B,C,D,E manual switch box with four devices. These devices also required external power supplies and their own outlet connections, and consumed more electricity. USB connectors and cables should not be confused with Firewire which looks similar but is different and governed by IEEE standard 1394.

If one is going to become an expert on USB flash drive forensics, it is best to start by going to the USB standards group and read what the standards are. It is generally described by computer technicians as NAND gate-type storage technology that is nonvolatile. There are various types of gates and circuits that hardware makers can choose from. Logically, a NAND is a combination of the NOT, an inverter, and an AND gate. The other gates that hardware designers could have chosen from are the OR and NOR gates. Designers prefer the NAND gate because of its speed. The NAND becomes a logical choice for use in the manufacture of flash drives. When it comes to NAND technology and nonvolatility, it means that if there is no power, the flash drive may still retain data. That is why one will often see advertising claims from China saying that data may be retrievable up to 10 years. Then it is advisable to go to the IEEE website and see if there is a standard for USB flash drives. Some similar technologies such as "Firewire" have standards such as IEE 1394. Then it is good to look at consumer ads and see the terms that exist such as "Hot Swappable." This term means that a USB flash drive can be pulled out or pushed in a USB port at anytime during its use [2]. An external hub is a device with a male USB connector that connects to a USB port in a computer. This external hub has a number of female USB connectors that in turn allow a number of male USB connectors from devices to be plugged in. One can visualize it as a USB power strip that allows us to plug many USB devices in it. If you do not know the lingo or standards, it is easy to appear less believable in court if one is asked questions.

The wires that are known as number one and four are red and black. They carry the power to the USB device as stated. The white wire is known as D– and green wire is known as D+. They carry the data to and from the flash drive to the computer and vice versa. Beyond Logic has a section that simplifies signaling http://www.beyondlogic.org/usbnutshell/usb2.shtml. The signal known as one in binary can be explained by changing the voltage signaling across two wires. The green wire also known as D+, 2.8 V is pulled across a wire with the use of a signal and a resistor of 15,000 ohm and goes to a common ground. There are color coded bands on the resistor so that is how one can figure out the value. The white wire known as D–, has 0.3 V that is pulled up to 3.6 V. Another set of signaling values explains a binary value of zero. It is amazing that every picture is nothing more than a set of ones or zeroes that describe each pixel's location, color, and brightness. These ones and zeroes are arranged in formats that are represented by agreed-upon file extensions such as JPG.

USB FLASH DRIVE: LITERATURE

Jan Axelson has a book called *USB Complete, Developers Guide, Fourth Edition* which is excellent because it describes the hardware in depth as well as how to create programs that utilize the port. Jan previously had an excellent book on the DB 25 parallel port that was very complete and used by me when I programmed a robotic arm through a Visual Basic Six program I created with my class. There is another very good book that describes the hardware and software principles associated with universal serial bus technology called *Design by Example* by John Hyde. Anyone who is going to testify about USB technology would be advised to take a look at both of these books.

USB FILE SYSTEMS: OPERATING SYSTEM

One will often find that the latest file systems of the USB flash drives are FAT 32. Some of the older flash drives may use FAT 12 or FAT 16. That means that they use a file allocation table to organize the file names. There are two copies of the FAT table in case one is corrupted. Each entry of a file that is associated with the file allocation table has a starting cluster associated with it. Each of the clusters is chained together. If a file is deleted, the first character of the filename is replaced with an E5 hex character, sometimes known a sigma (σ), and the cluster associated with that file is now available to the operating system to use. It is important to learn about the algorithm used for file storage with your particular device. Many of the USB flash drives are just plug and play on the new systems, but sometimes on older systems such as Windows 98 before special edition, a driver is needed.

In many versions of Microsoft Windows such as 95, 98, Millennium, XP, and Windows 7, one will occasionally find that the USB flash drive is not recognized by the examination machine. The examiner can quickly determine if the drive is working or not by trying the flash drive on another computer. However, the other computer should have a USB hardware write blocker on it so that the flash drive is not modified and that no malware could be transferred to or from the flash drive. The USB hardware write blocker is a device that connects between the USB port and the flash drive. If the drive can be accessed on the computer with the write blocker, it shows that the USB flash drive is operational and that there is a problem with the USB port on the examination machine. There are three remedies to fix the problem. The first is to try using another USB port. The second option is to connect a USB hub, then a write blocker, and then the USB flash drive. Lastly, one can go into Microsoft Window's Control Panel's Device Manager and then remove the USB driver from the USB controller and shut the machine down. Then one goes back to the device manager after restarting it, and goes back to the universal serial bus controller and right clicks. Then one should choose the option known as "scan for hardware changes." This should fix the problem and allow the USB port to work with the write blocker and flash drive. Problems can occur when software gets corrupted.

There is at least one antiforensic website that has a statement that says, "a community dedicated to the research and sharing of methods, tools, and information that can be used to frustrate computer forensic investigations" [3]. These websites are interesting because a new digital evidence investigator can learn where to look for evidence by reading the antiforensic tips. Suppose a new investigator wonders, "Does a USB drive leave any trace in the operating system of a computer that it was once connected too?" There is not much useful information on this topic posted on the Internet. However, if one reads the antiforensics websites, then one can see what files the antiforensic writers suggest deleting. A countermeasure would be to try and go to those places to examine existing evidence. The antiforensic websites also enlighten the reader to interesting and useful items that might not be evident. Here is one example. The system restore function might have many copies of previously configured registries. That means that it is possible to go back a month, 6 months, or more and possibly find a previous point when a USB drive was connected.

The Setupapi.log file has a USBSTOR registry key that may show a particular brand and USB flash key was installed [4]. The antiforensics website is also so kind as to show a large printscreen of the Setupapi.log file with the USBSTOR registry key circled so we cannot miss it. Many of us may not know that CCleaner backs up the registry, another bonus of where to seek information. In Windows 7, two files that contain information are C:\WINDOWS\INF\setupapi.dev.log and C:\WINDOWS\INF\setupapi.app.log.

The antiforensics websites are useful to see what the other side is doing to remove information and then one can research that topic and see if they forgot something. If they did, you as an investigator know one more location to check if they got rid of all the other information. This will also make you a better investigator. The technique is much like being a double agent where one looks at the other side to spy on their techniques so that you can revise your spying techniques to countermeasure their techniques. Every time a file is created, there is a creation date and time. There is also a last accessed date, a last modified date, and a file size measured in bytes. These are examples of metadata. Metadata is created by the operating system and is important evidence in an investigation. If one takes a picture with a pen cam, the metadata concerning the creation time of the picture is in the file. There is also software that could be used to remove metadata. Some websites give advice on this topic. There is software that can be used to look on someone's flash drive and/or desktop that can be used to remove metadata. In Windows 7, one can easily right click on a picture or document, click on properties, and remove one item of metadata, many items, or more. This has a legitimate use which is to ensure privacy. Perhaps, someone wants to be a whistle-blower and report some crime but he or she wishes to be anonymous; removing metadata is essential in such cases. If one has a system such as Windows 98 or Millennium, the process is difficult. That is why there are programs such as Batch Purifier Lite that can be used to remove metadata from files. There are some people who could use this software as antiforensics software to cover up tracks and remove metadata to frustrate investigators.

USB HARDWARE THAT COULD BE MISTAKEN FOR A FLASH DRIVE: KEYGRABBER NANO USB

There is a device worth mentioning because it looks so similar to a USB flash drive but it is not one. The KeyGrabber Nano USB is a wireless keylogger. It works similar to the old PS/2 port Keycatcher that was connected between a keyboard and a desktop computer. Keycatchers were used on laptops that used an external keyboard. Everything was caught on the Keycatcher and could be viewed later. The KeyGrabber Nano USB connects between the laptop or desktop and the keyboard. It looks as if it is a small adapter for the keyboard. This device has wireless connectivity and can send email reports via WiFi about the data it collected. The legitimate uses of this device are by corporate investigators who may use this to monitor the activities of an employee suspected of running their own business while on the company payroll. It could be used to monitor someone who allegedly surfs porn at work or talks inappropriately to kids on chat rooms at work. It is also a useful surveillance device. USB keyloggers could also be used by parents to monitor their children's computer or to supervise someone who uses the computer that they have a custodial relationship over, such as having power of attorney over a relative with a mental impairment.

The device can also be misused by people in a corporation for stalking. Please consider this following example. Perhaps, there is a very attractive secretary that some employee is interested in. A cyberstalker may first check out the person's profile in social media such as LinkedIn, Facebook, TopFace, Xanga, and then do a Google search on the person. This in tandem with an online background search from a company such as Intellus is comparable to performing a recon mission in the military. Then the cyberstalker may install and use the wireless keylogger as some type of surveillance tool to follow the secretary's activities. In the twenty-first century, secretaries can be men or women and the same is true for cyberstalkers. There is a report from the Attorney General to Vice President Al Gore in 1999 that one of the growing challenges for industry and law enforcement is same-sex cyberstalking and women stalking men [5]. Cyberstalking in the office can progress to a criminal offense covered by 47 U.S.C. 223 and carry a penalty of 2 years in prison. If a corporation finds a wireless keylogger, it is time to call the local law enforcement. They will probably proceed to call the local cybercrime division of the County Prosecutor's Office.

WIDE RANGE OF USB FLASH DRIVE PRODUCTS

The USB flash drive is said to be popular because it contains no moving parts, is somewhat durable, and allows people to take a lot of digital multimedia such as presentations, documents, videos, and pictures with them. There are said to be five types of these drives. They are "generic, high performance, ultra durable, secure, and novelty" [6]. The secure drive such as the Defender F200 will be discussed later. The generic USB flash drive is one of the many types of flash drives that we may be familiar with. They are the inexpensive mass-produced drives that are given out at trade shows and university open houses. An example of one of the ultradurable category of USB flash drives is the Patriot Memory Pef32gusb Xt Boost Water-Resistant USB 32 GB Drive. Then there is the novelty drive which will be the focus of much of this chapter because it is a disguised device with other practical uses and could be used for purposes that violate corporate policies in a corporation. The novelty device if used incorrectly, could allow one to be the focus of a criminal investigation. This chapter will discuss the various types of USB flash drives, how they may be misused, and how to investigate them with various tools.

USB DEVICES THAT CAN BE MISTAKEN FOR FLASH DRIVES

Some devices that plug in the USB port are not novelty flash drives but gadgets that come apart and blow away dust on the keyboard and screen. It takes some studying to learn the difference between what is a novelty USB flash drive and what is only a gadget with no storage capability that only utilizes the power aspect of the USB port. Some devices appear to be round USB flash drives but are actually only small coffee cup warmers with no storage capability. It is important for anyone who does investigation or corporate security to understand the products that are on the market and get an idea what people might bring to the office. If one was trying to forensically image the round coffee cup warmer thinking it was an old larger flash drive, it could be an embarrassment in court or if the case was picked up by the television or newspaper. It becomes important to look at mail order catalogues, engage people in polite conversation about things on their desk, and Google search items such as "USB Gadgets." An investigator needs to get an idea of what is on the market and to know what the capabilities of the devices are. I would also suggest joining a group such as IACIS or ASIS International or the HTCIA and talking to members of these organizations about their investigations, latest trends, and any odd things that one sees. People are often quick to share a story about something strange in investigation as long as it is not going to be in the media. Someone told me of a device that they saw on an auction site online that plugged in the USB port and looked like a drive but it was actually an audio recorder. This could have been connected to a computer and left in a meeting room and used to illegally wiretap a meeting, perhaps, even be used for insider trading. If there was an illegal wiretap in a New Jersey office, it would violate N.J.S.A. 2A: 156A New Jersey Wiretap Statute and would require law enforcement to be called. It would also pay to go through Amazon and eBay and search USB to see what one finds. It pays to stay informed.

EXAMPLE OF AN ULTRADURABLE USB FLASH DRIVE

The Patriot Memory Pef32gusb Xt Boost Water-Resistant USB 32 GB Drive is an example of an ultradurable USB flash drive. This drive is considered ultradurable because it is water resistant and shock resistant. There exist some blogs and articles online that discuss how various water-resistant USB flash drives have been left in the pockets of pants and have been washed with a load of laundry in the washing machine on more than one occasion. Amazingly such flash drives often still function. When a USB flash drive is described as shock-resistant housing, some digital examiner bloggers say that in laymen's terms it means that it can fall from heights of 6 feet or less to the floor on more than one occasion and the device is still usable.

USB FLASH DRIVES: SECURE

The USB flash drive is a wonderful invention that allows people to carry large amounts of information in a small space. This device may enable people to easily transport documents, pictures, PowerPoint presentations, multimedia files, and share them with others at a conference, symposium, sales presentation, or collaborative multiagency project. The problem is that people can use these small devices to covertly remove a large amount of data from an organization without detection. USB devices can also be encrypted with open source products such as TrueCrypt. Encryption has also hindered law enforcement because they often cannot devote resources running for 24 × 7 for close to a year to decrypt a drive or file. This chapter will discuss some of the new techniques and tools for recovering deleted files on USB drives and breaking encryption for lawful investigations.

THUMB DRIVES CLASSIC STYLE

The USB flash drive goes by many names. The 64 MB flash drives were once in a rounded shape at the top. The shape resembled a nineteenth-century tombstone but was only about the size of a human thumb. This resemblance to a thumb inspired people to name them thumb drives. Other people called them jump drives or disks on keys. It is possible to know the period of time that individuals first learned of these devices by the name they call it. That is linguistic history and something that profilers might pick up on. Computer science students at Fairleigh Dickinson University had USB thumbs drives in 2001 with a capacity of only 8 MB. Some people called them thumb drives but those who had them mounted on key chains referred to them as disks on keys. Regardless of what students called them, they seemed to find them convenient because they held the equivalent of 5 high density or 10 double density 3.5-inch floppy disks. They still held many times less than the 3.5-inch zip disk which came in 100 or 250 MB disks. Students said that they preferred the thumb drives because few labs at school were equipped with zip drives. I had purchased a 64 MB thumb drive which was made by the Singaporean Trek Company. Thumb drives looked like a small rectangular plastic case with a USB male connection. The drive had what appeared to be a microprocessor of some sort on a printed circuit board. They used USB 1.0 which is considered slow in 2011.

The USB then progressed to USB 2.0 and flash memory chips were used. Later releases of such technology allowed for the sale of flash drives to capacities such as 1, 2, 4, 8, 16, 32, 64, 128, and 256 GB. As of January 2012, 512 GB, 1 terabyte (TB) and 2 TB USB flash drives are in the planning stage. It seems that each release of technology doubled or increased by a power of two, a noticeable property of binary number systems. Incidentally, the numbering system and organization of the flash drive is with binary addressing. The later USB flash drives used a FAT 32 file system and many computer scientists assert that the first sector includes a master boot record and partition table since there can be more than one partition on a USB drive. In an investigation, it is important to check the partition tables. Pedophiles and spies can use the no man's land between partitions to hide data.

POKER CHIP USB DRIVE: NOVELTY

The poker chip USB drive is an ingenious concept because if it is lost, someone will probably think it is just a poker chip and either dispose of it or put it in a box at home with some poker chips (see Figure 7.1). The disguised USB drive is sold with various storage capacities of 1, 2, 4, 8, 16, or 32 GB. They can be purchased new and online from eBay for amounts varying from approximately eight to sixteen U.S. dollars from a supplier in the Kowloon section of Hong Kong [7]. The transfer speed from the computer to the USB drive is 8 MB/s and the reading from the drive to the computer is quite fast at speeds of 16 MB/s since it is a USB 2.0 interface. They are plug and play with Microsoft

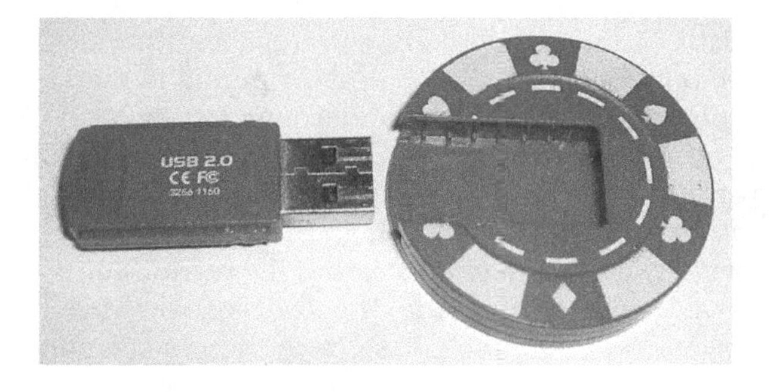

FIGURE 7.1 USB poker flash drive.

Windows 98, ME, 2000, XP, and Vista. The drives can also be used with Linux 2.4, Mac OS9.0, and above [7]. If the data was accidentally or purposely deleted, there are a few tools that can be used to recover the files. One program that has worked very well is called RecoverMyFiles by GetData. If one needed to image the entire USB flash drive, X-Ways Forensics works very well. It also has a hex dump reader and allows one to examine the metadata within a file. X-Ways forensic tools also allows one to collect the slack space within a file. Suppose a file is 1000 bytes and your cluster size is two sectors, then the cluster is 1024 bytes. The slack size is 24 bytes and is located between the end of the file and the end of the cluster. It can contain items from RAM which might include usernames, passwords, or encryption strings.

Because of the application of Moore's law, the drive capacity doubles every 18 months. The company that sells the drive also says it is possible that the drive may be used for periods as long as ten years, though there is no battery in it. This may be true because of the NAND gate architecture. These flash drives work with the voltage supplied by the USB port which is 4.5–5.5 V. They are said to be able to work in environments from –40°C to 70°C and may be stored in temperatures of –50°C to 80°C according to some posting online by the company. The USB poker drive uses a FAT 32 file system which allows the large capacity. Some people have used an open source tool called TrueCrypt to encrypt the drive in case it is lost.

Poker chip drives have the potential to be misused by someone who might want to do corporate espionage. Perhaps, a person visits a salesperson at a company that he or she wishes to attack. This visitor may use a cover story such as being a buyer and appears to wish to purchase one large item to get access in the office. The visitor may try a diversionary tactic such as spilling a glass of water on a computer or desk. A natural polite reaction may be for the employee to leave the office to seek a paper towel for drying the spill. The visitor, who is actually a corporate spy, has a period of time from when the employee went to get the towel until the password protected screen saver locks the machine. During this time, the spy may insert a USB flash drive into the office computer. If the computer's operating system has the autoplay feature enabled, a keylogger, rootkit, and malware may be installed. Some keylogger spyware periodically sends every key stroke to the attacker's organization in an email as long as there is an Internet connection. If a rootkit is successfully installed, it may allow hackers a backdoor access from any remote location as long as an Internet connection exists. If an investigation happens far after remote access of the computer or file theft has occurred, the harm is done and it may be too late to reveal how this simple attack on intellectual property was carried out. The information security department of the company would be better off disabling the autoplay feature as well as the USB ports. If the USB ports are needed to transfer files, it would make more sense to require the computer user to make a telephone call to the system administrator (sys admin). The sys admin may unlock the ports from a remote console or a short period of time so the files could be transferred. File transfer is important for telecommuters who do work from home. There is a very good book that discusses polices and technology for employees who work from home or telecommute. The book is titled *Home Workplace, A Handbook for Employees and Managers* by Brendan B. Read.

There is a movie called the *Recruit* starring Al Pacino and Colin Farrell. It is a fictional story but was made with the help of a CIA consultant. The story is about the activities of some people who join the CIA. One person in the movie demonstrates an intellectual property theft as part of a security exercise by copying all kinds of documents from a computer on to a USB flash drive. Then the character smuggles the flash drive out of the CIA office in the false bottom of a thermos. The thermos is x-rayed but the drive is not caught. The audience learns that prosecuting the person does not help the loss of very important information to any enemy. The harm is already done. Proactive security is needed to stop the problem of information loss via USB ports. This type of war games scenario can be used to test the security at an organization.

LAW ENFORCEMENT'S STRUGGLE WITH ENCRYPTED DEVICES

There is an article that reported that the INC and FBI could not decrypt a hard drive that was encrypted with TrueCrypt after a year of trying [8]. The drive was seized from a banker in Brazil named Daniel Dantas. FBI director Louis Freeh said that robust encryption products are a serious threat to public safety because they do not allow for timely access and decryption [9]. However, many computer scientists are trying to address the problem of robust encryption with tools such as rainbow tables, dictionary attacks, and running attack software on what appears to be many computers running in parallel. This is done by using the processors on graphics cards in parallel. The Russians broke some encryption on WiFi by using the power of a graphics card, namely, "The GeForce 8800 GTX delivers something like 300 gigaflops" [10].

I once used Access Data's Forensic Toolkit (FTK) to collect the unused space and slack space on a small drive and build a dictionary. I had used the dictionary with the FTK software and broke a file that was encrypted for an academic experiment. Quite often the key to encrypted files are located somewhere within the hard disk itself unless a lot of activity occurs on the drive and the unallocated space and slack space is rewritten. Encryption may not be unbreakable but the idea is that it protects data for a period of time that makes it unusable in an investigation and consumes too many resources that must be devoted to the decryption breaking task.

WRISTWATCH USB DRIVE: NOVELTY

The wristwatch USB flash drive is a useful device because it includes a wristwatch and the USB flash drive. Some people who use this device indicate that they prefer this type of device because the portable USB flash drive is housed within the timepiece. It also becomes difficult to lose the USB flash drive because the watch is always on one's wrist. There have been reports about many people losing USB flash drives. The USB watch seems to prevent the problem of loss due to accidental dropping or stepping on. There is nothing wrong with a USB watch and one could make an argument that it could help special needs populations with physical or memory impairments to safely carry data without loss. In fact it could be a good device for forgetful people or those with a disability who tend to drop things and cannot easily pick them up. USB flash drive watches are available on the eBay auction site. Here is an example of one [11]. The USB watch costs $16.50 and approximately $7.00 for shipping from Shanghai to the United States. This watch has a 4 GB capacity and includes a USB cable to and from the computer for USB 2.0 ports. It is said to be able to hold data for periods of up to 10 years and may be written to and read approximately one hundred million times [11]. The device also supports password protection and it may be possible to make it bootable. The bios of a computer, laptop, notebook, or server may be configured to first search the USB devices for an operating system to bootup with. Flash drive USB devices support file transfer with Windows 7, Vista, XP, and 2000. This device can be used to carry a presentation to a board meeting or securely hold medical documents for a special needs person who may otherwise drop it. However, like any device, there exists a potential for misuse which will be addressed next.

I, from Fairleigh Dickinson University, was the main author of an article called "Watching the USB Watch" that discussed the security risks associated with USB watches in corporations [12]. People can walk in an office, distract someone at the computer, and then pull out a cable from the wrist watch. This cable may be inserted in the USB port of the workstation and documents may be stolen [12]. This unauthorized action may be caught on bubble cameras which are often installed high up on walls in corporations. The unauthorized copying of files should be recorded in the operating system log files. Other log files should hold the time and date of when the device was plugged in and then removed from the USB port. A tool such as EventLog Analyzer by Manage Engine should be a useful tool to consolidate the numerous types of log files that exist and to simplify the analysis of security breaches. The DVR associated with a bubble camera should have video stored for periods of at least 30 days. This archived video may link the person who copied the files with the log files on the operating system. This evidence combined with the USB flash drive device information located in the registry should give all the evidence needed for dismissal of the employee and prosecution in the legal system.

USB WATCH: INVESTIGATION

Suppose someone who is an employee in another department comes into the research and development area of the corporation and copied documents at the company. Perhaps someone saw the activity and questioned the action and called the authorized requestor (AR) or security department. The person who copied the documents knew that they were going to get caught. Perhaps they got to a computer, connected a cable, deleted documents, and then gave the watch a whack. Then the incident response team (IRT) at the corporation arrived. They may call the AR on an encrypted cell phone to a landline that utilizes a communication security product such as phonecrypt. It is possible that an employee may have an accomplice with a wiretap so communication on an encrypted line is advised. The incident response team may ask for identification if the person is not wearing a photo badge showing that he or she is an employee of the same company. The IRT may call the AR who creates a conference call with the general counsel and the HR department. Once it is determined that all policies are signed, the IRT may ask the suspect to hand over the watch.

The wristwatch should first be seized and secured and then a chain of custody form should be filled out. The seized wristwatch may be put in a secure container and taken to the digital forensics lab in the corporation or contracted out to an investigation firm. As custody changes hands and evidence is processed, the chain of custody must be updated. The investigation machine would first be disconnected from the Internet and all forms of Bluetooth, wireless, and infrared would be verified to be turned off. The latest version of antispyware and antivirus would be run on the investigation machine. A Tableau USB write blocker might be used or the registry changed to make the USB port read-only. Then a tool such as Avanquest Perfect Image may be used to image the drive within the watch. BadCopy Pro may be used to recover the files even if there were bad sectors from the physical impact with the watch by the employee. The reader may see an example of a USB watch in Figure 7.2.

FIGURE 7.2 USB watch.

USB WATCH WITH DIGITAL VIDEO RECORDER: NOVELTY

The USB watch was very popular but manufactures responded to consumer demand with a new larger-capacity USB flash drive as well as a tiny camera that recorded video from the watch face. This device has many legitimate uses. Suppose one was going to their child's school to view their son or daughter in a school play. They may want to film them but not distract the performers or the audience. The USB watch with a digital video recorder and camera may be the perfect solution. These devices are available from China through the eBay auction site. The specifications for one example will now be surveyed [13]. The item has a 4 GB storage capacity and is rated as daily waterproof in case of bad weather. It has a small light for operation in low illumination. The camera takes pictures in a JPG format with resolution at 3264 × 2448 pixels [13]. The video coding is M-JPEG and the video is saved in an AVI format for easy viewing on Microsoft Windows computers. The video frame rate is 30 frames per second and includes a 960P HD Mini DV. The watch comes with a USB cable and supports TF and Micro SD cards for external memory. The device runs on a lithium-on polymer battery and includes a CD. The price is $21.99 and $5.99 for shipping to the USA. The device works on Windows Me, 2000, XP, and Vista, as well as MacOs and Linux.

Though there is nothing wrong with having such a device and its uses could be honorable as described above, there is a potential for misuse by criminals in a corporate setting. It is possible that a person could walk in a corporation with the camera function enabled and then proceed to record video of the inside layout of the building. The length of hallways and the size of rooms could be ascertained fairly accurately by counting paces. If one measures one pace, then it is easy to calculate distance. If a person has a stride of 1 m and one walked 100 paces down a hallway, then that hallway is 100 m. If one was caught with such a watch doing surveillance, there are a number of tools that could be used to recover the video. The list would include Guidance Software's Encase, RecoverMyFiles, Access Data's Forensic Toolkit, Paraben's Device Seizure, X-Ways Forensics, JufSoft BadCopy Pro, DataLifter 2.0, and Avanquest's Perfect Image for capturing the drive contents.

LEGO BRICK USB DRIVE: NOVELTY

There is nothing wrong with a lego brick with a USB drive in it. In fact it could be put on a keychain and serve as a reminder that one has important documents and a child in day care, for example. However, the device has a potential for misuse and illegal activity by people with bad intentions. The lego brick USB drive looks like a children's toy and could be incorporated into a child's lego building. Many people bring their children to work with them when a baby sitter is not available. Some companies also offer day care facilities to employees who have both a mother and father working. A parent who wishes to perform corporate espionage could easily incorporate the USB lego brick within a child's building. Who has ever seen a security guard ever check the child's lego? The brick can be pulled apart and then the USB connector is exposed. The drive can have a capacity of 2–8 GB. The drive is manufactured for or by Samsung. The lego bricks can come in a variety of colors such as blue, yellow, green, orange, and red. The drive is considered compliant with USB 2.0 and the speed for reading is anywhere from 12–22 Mbps. The speed to write to the device from a computer is much slower at 7–12 Mbps. The USB drive is considered plug and play and no drivers need to be manually installed if one uses it on a system that is Microsoft Windows ME, 2000, Vista, or CE3 for small portable devices. The device is reported to be able to retain data for 10 years be encrypted more than 1,000,000 times.

Corporate security should find a socially acceptable way to check children's lego buildings and blocks for USB drives hidden in with thousands of other blocks. This may be difficult because there may be objections by parents. Children may not understand why their toys are being searched. There is a poster that was made by the Office of National Counter Intelligence that says, "The research project of a lifetime just ran out my ... usb port!" [14] There is a course called Corporate Espionage 201 offered by SANS that discusses all these USB toys, pens, and devices as a means to steal data [15]. The security professionals could disable the USB ports with Microsoft Window's device manager or configure the port in the registry to be read-only and not allow data on them. Digital forensic examiners need to regularly check an event viewer to learn if the logs of computers have unauthorized USB activity from unrecognized flash drives. An examination of the registry and log could show if specific flash drive devices were put in USB ports on that computer at a particular time. Even a simple application that makes a long 4 s beep when someone writes data to something connected to a USB port would greatly increase security.

PEN: DIGITAL VIDEO CAMERA AND USB DRIVE—NOVELTY

There are four types of pens with USB drives. The classic one that is widely available is the pen with the 64 MB USB flash drive. These were sold on eBay as well as from a variety of promotional products organizations that would customize the pen with the name of your business. The second type of USB flash drive pen included an audio recorder and a larger capacity USB flash drive. A third iteration of the device was sold with a larger capacity flash drive and a video camera. A fourth iteration of the device on the market now includes a laser for pointing out things while giving presentations in PowerPoint. On December 21, 2011, a person could go on the eBay auction site and purchase a 1 GB flash drive pen with laser pointer for $6.85 and $1.99 shipping. The item was from Santa Clara, California.

All the USB flash drive pen devices with whatever extras they have, such as audio recorders, lasers, and cameras, are all useful devices that can be a great convenience in the office. The laser pointer is good so that a professor or salesperson can point out a useful fact or drawing on the screen with a red dot. An audio recorder could be good for someone who is blind and needs to take notes on a lecture. The video camera might be good for someone in private security or law enforcement who wishes to film an area or collect video of a crime in progress. The same devices could also be used improperly by people in ways that could violate company policies or the law. The laser component could be misused by someone to point at airplanes. Pointing lasers is not only a misuse but a criminal act that can get someone many years in prison since the laser travels far and can damage the pilot's eyes in some cases. The Federal Aviation Administration had reports in June of 2011 that approximately 2800 pilots reported getting targeted with lasers [16]. Many USB flash drives with pens are called pen cams.

The following misuses of small cameras and recording devices may or may not be from pen cameras but do illustrate possible abuses of such devices in corporate environments or public places and illustrate potential high-profile criminal investigations that may result. The digital video camera component of the pen might be used to take inappropriate pictures of passers by who walk over an air vent grate in the street or in an office. This is known as a skirt cam and there was in investigation of a skirt cam in New York City [17]. Small cameras with any type of recording device have a potential for inappropriate uses and thus cause criminal investigations. In a case of 417 F. 3d 879—*Wright vs. Rolette County*, there was some discussion of a potty cam which has to do with a hidden camera in the bathroom.

The pen cam can also be used by people who want to collect evidence and report elder abuse with a grammy cam or with nannies abusing children by use of a nanny cam. Some websites have reported that the pen cam may be left in a room near grandma to catch video of nannies who abuse the elderly. There may be an investigation after the video is recovered and the results are discussed with grandma and the police. Governor Richardson of New Mexico signed a law that allows people to install a hidden camera in grandma's room at the nursing home because of all the reports of elder abuse [18]. A United States Senate Special Committee on Aging estimates that as many as five million elderly people are victims of some type of abuse every year [18]. Other states are expected to follow. If a person was going to set up a grammy cam, it is advisable that the family first discuss it with a private investigator and grandma, otherwise it could be considered an invasion of privacy and grandma may object because the video could expose her to getting changed, washed, or other personal activities.

CORPORATE INVESTIGATION: UNDER THE TABLE CAM

On December 21, 2011, if you Googled "Under the Table Cam," you would get over fifty-four million results on the Bing website for this topic. It is a type of voyeurism that can be done with webcams, pen cams, or other types of miniature portable video cameras that can be put under the table. The motivation seems to be to catch some video of a woman in a short skirt who is sitting at the table. If this was done at a corporation meeting, a corporate investigation might occur if the woman with the short skirt had a miniature camera detector and it went off. Then she could politely excuse herself from the table, call the security department, and they could contact the AR. The AR could dispatch an incident response team with camera investigators to catch the culprit.

CORPORATE INVESTIGATION: CHANGING ROOM CAMS AND INVESTIGATIONS

A middle-aged man was caught filming a woman and her 5 year old son in a changing room of a leisure center in England with a device that was described by the victim as a miniature camcorder. The article described the man as a middle-aged pervert and he was caught with adult items [19]. Many of the investigations of mobile device abuses and forensics may expose the investigator to the sex offender community and practices that are not discussed in classes. Students who are very religious may complete a degree in digital forensics and not fully know what they may be exposed to later. The topics on the last two pages may be uncomfortable to discuss in a classroom and perhaps cannot be discussed in mixed company. However, it could be discussed first with a committee and perhaps there should be a pamphlet that can be handed out, and privately read that discusses the deviant behavior that one will experience in one's career. There could be options for either counseling to learn to handle it or seek another line of work. I took a class on mobile forensics at a police station near Boston in 2005. The instructor, who was a policeman, talked about some of the strange practices and exemplars of evidence that mobile devices examiners might find in an investigation. All the other students who were already employed as policemen said they have come in contact with all kinds of sex offenders at some point in their careers. In a school, such topics are often avoided and teachers discuss the safe comfortable topics such as the technical aspects of recovering evidence. There is a whole psychological and lifestyle component that may shock some students when they get to the real world of investigation.

Suppose a person is caught filming with their pen cam in a changing room. Perhaps they did this activity at a variety of places. Some films on the drive may be genuine films of victims who were spied upon but some films may be part of a personal collection of downloaded adult films. There are places online where amateur adult film makers post films of people supposedly caught in changing rooms. Some type of expert on the subject may need to be brought in on the investigation to sort out the adult material from the real evidence.

PEN WITH HIDDEN CAMERA DETECTOR: COUNTERMEASURE TO NOVELTY USB FLASH DRIVES

There are countermeasure devices that one can purchase to find out if someone has an active camera near them. I met someone who worked in intelligence that had a pen that gave off a beep if someone had an active camera near him. The camera detection pen was actually quite common in online security magazines and websites for a number of years. There are many online auction sites and private security places that sell these devices. An example of one can be seen on the ecrater website [20]. If a woman feels that there may be a pervert among her coworkers, she may wish to purchase a pen camera detector so that she can detect anyone filming her with an under the table cam and then alert the AR in the corporation. If a high-level manager was caught using an under-the-table cam, it could bring all kinds of unwanted publicity and perhaps a high-profile lawsuit that would be expensive to the company in both legal fees and lost business. The better option might be to have the camera detectors installed in lunch rooms, board rooms, and meeting rooms and let everyone know they are there. Then the camera detector could be a deterrent to deviant behavior.

PEN CAM ABUSE AND PUNISHMENTS

In Milwaukee, a man was reported to have filmed his roommate without her consent with a pen camera. The video contains footage of "her using the toilet and dressing after a shower. She turned it in to police and on Thursday, prosecutors charged her roommate with two counts of capturing images of nudity without the subject's consent, a felony punishable by up to 18 months in prison" [21]. The interesting part of the case is that the woman found the pen cam and plugged it into a computer. Then she not only saw the video of her on the toilet and getting dressed but also saw footage of it being set up by the roommate and taking the pen from the shower.

RECOVERING THE FILES FROM THE USB FLASH DRIVE DEVICES

The flash drive pen devices seem to have FAT 16 or FAT 32 file systems. Therefore, the techniques used to recover files on any FAT 16 or FAT 32 file system device such as a digital camera, USB flash drive, PDA, or SD card will all work. A FAT 16 file system, for example, uses 16 bits for addressing and there is a limit of the partition having 4 GB if a cluster size of 64 KB is used. FAT 32 uses 32 bit and 28 of those bits are used for cluster numbers. Clusters are made up of even multiples of 512 byte sectors. It is also possible to have the new large USB external hard drives use FAT 32 to address up to 2 terabytes but then the sectors are no longer 512 bytes but now use 4096 bytes. This can waste space with small files because even a 200 byte file needs a 4096 byte sector. That can leave lots of slack space holding potential evidence.

SURVEY OF TOOLS TO RECOVER DELETED FILES

Some people will ask what is the difference between computer forensic software and data recovery software? The difference is that computer forensic software is software that is used to collect digital evidence in an investigation to help prove someone innocent, exculpatory data, or data that can be used to prosecute someone. The best computer forensic software has been rigorously tested, accepted by courts for use in computer investigations, and does not alter the original evidence. That means that the products are accepted by one's peers in the investigation field and pass the Frye test. Many products such as Paraben's Device Seizure and some older products such as cell seizure and PDA Seizure are some forensic software products that include write blockers and do not change the original evidence. Digital forensic software products support the preservation, collection, analysis, and reporting process of computer forensics.

Many people will argue that some software is not digital forensic software and is only data recovery software. However, many small corporations and municipal police departments cannot afford the Guidance Software's Encase, the full version of Access Data's FTK, and other such premier digital forensic products. Some places that are under financial constraints will change a key in the registry to make the USB port read-only or they may have purchased a used USB write blocker. Once the software or hardware write blockers are used, data recovery tools such as Avanquest Perfect Image, Avanquest Data Recovery Professional, JufSoft BadCopyPro, and other less expensive data recovery tools can be used. One could also make the argument that these data recovery tools are forensic because data recovery is an element of computer forensics. This part of the chapter will discuss some of the available products that can be used to support an investigation of a USB flash drive.

AVANQUEST PERFECT IMAGE

There are a number of tools that one may use to copy all the allocated files to a flash drive. One inexpensive tool that is easy to use is called Perfect Image and it is made by Avanquest. It is 160 MB and works on Windows XP, Windows XP-64, Windows Vista, and Windows Vista-64. Though it was not supported on Windows 7, I did get it to run on Windows 7. Avanquest also has a very nice support feature which is an online chat about specific products. The software is for backup but Avanquest reports that it can be used to clone a drive or a partition. The two features that are interesting to the digital forensic investigator are Hdd raw processing which also clones unused sectors, and partition raw processing which copies the entire partition including unused sectors. The software is under $50 and seems ideal for digital investigators with limited budgets. The software is very easy to use and has lots of easy-to-read documentation and gives a lot of information about the flash drive. There is also an image integrity checker which checks the quality of the obtained image. This seems a necessity if the investigation proceeds to court and the quality of the image is questioned. One can also pay an extra $9.95 and have a DVD of the software sent to one's office. This can be a good backup. Many investigators have an album of CDs so they can easily restore a particular software product if a hard disk gets corrupted and a program can no longer run.

AVANQUEST: DATA RECOVERY PROFESSIONAL

This is good because the second most important task after collecting the raw image of a disk or partition is to recover or carve out the files that were deleted. This software is also under $50 and also seems to be a great bargain for a digital investigator with a limited budget. There is a chat feature for Avanquest to get questions answered. The product has good online documentation for the features and that is included in the software. There is also an option for extended download which many people may opt for. Suppose one loses the CD, the program gets corrupted, and the office server has difficulties, one could still download the software for an extended period of time and put it on a new examination computer. This software can help the investigator recover over 300 different file types on a variety of file systems such as FAT 16, FAT 32, VFAT, NTFS, and NTFS5. Digital evidence examiners would most likely appreciate the software because it can be used with a CD and DVD and a variety of memory cards such as XD, SD, microSD, USB drives, flash memory cards, MP3 Players, iPods, hard drives, and digital cameras. There is also a disk cloning option.

TRACKER BY PHASEWARE: INCIDENT MANAGEMENT SOFTWARE

There is now a software called PhaseWare that can be used to track incidents and investigations in corporations, law offices, intelligence agencies, and in private investigation. If an incident is reported, the initial call starts a ticket number or case number. Ticket numbers would be if it was something such as a digital camera, PDA, or netbook hardware failure, then that could be a help desk item. Ticket numbers are for equipment failures and not policy violations. A failure of a mobile device requires a different type of incident response. Perhaps the person making the call could drop off the item or mail it to a repair facility. That type of incident would be tracked differently. That incident lifecycle would be an item reported as broken, item sent to repair facility, loaner mobile device given, original device repaired, original device sent by registered mail to the user, repaired device accepted, loaner device sent back, and loaner device received.

However, if the call included an allegation such as employee X was emailing changing room pictures or making biased remarks, this would start an email investigation. The tracker program seems to be a combination of record management and project management and empowers ARs and managers to track the entire investigation lifecycle. This can be good for legal compliance management too since many industries where the incident takes places are governed differently. The software also provides the AR an easy way to show management how the incident is being prioritized in case other investigations are concurrently being conducted. Reports and analytics can provide the general counsel with the information to show the investigation is being conducted so it follows regulatory requirements. This could be used for audits and in universities where such data would have to be placed in reports for prospective students and parents as required by the Clery Act [22].

PPM 2000'S INCIDENT REPORTING AND INVESTIGATION MANAGEMENT SOFTWARE

Here is another program that could be used by the AR handling the incident response team to track the investigation. The investigation lifecycle could be tracked with this software and reports could be issued to managers, security professionals, and committees who oversee investigations within the organization. This program also provides a lot of capability to create metrics which could be used for compliance regulatory audits and for forecasting trends and hiring new incident response personnel. Forecasting trends could also let management compare their crime to other corporation's crime to see if it is just business as usual or new alarming events are happening. Trend reports could also guide the chief information officer in what type of policy violations are occurring on mobile devices and desktop computers and then countermeasures can be developed to inhibit those trends. Software investigation tools for those types of events could also be purchased.

JufSoft BadCopyPro

This software is reasonably priced at approximately $40 and seems to be a good option for a digital investigator with financial constraints. Version 4.10 has an easy-to-use interface and has options to allow the investigator to recover files from a CD/DVD, or a floppy disk, a flash drive, zip disk, or other media. The software has some documentation and help features with it. This is important. The software can run on an examination machine that has one of many operating systems such as Microsoft Windows 9x/ME/2000/NT/XP/2003/Vista/7. The hardware requirements are not demanding either because only 32 MB of RAM is needed and 1 MB of free hard disk space for installation of the product. The software gives two options for USB flash drive recovery. The first option is to recover lost files and the second is to recover corrupted files. The software ran efficiently in a test session performed by me, and I was able to recover hundreds of files in a short period of time. The interface was simple to operate and this would probably be a good software tool to run under the duress of a corporate investigation.

UNDELETE IN DOS 6.22

If one has a computer with a version of DOS 6.22 on it, there is a command that is very useful for recovering files that were deleted and not written over yet. If one installed Windows 95 or higher versions, this command does not work. This command is called "Undelete" and can recover files by seeking files in the file allocation table and then replacing the sigma with a valid character that allows for the file entry to reappear. Undelete/List will list all the files that are recoverable. Suppose my USB flash drive was drive E. The undelete command to recover all files is entered at the command line as follows: C:\> undelete E:*.*/ALL.

FORENSIC REPLICATOR V4.3

This is an important tool that I believe is very useful for forensically acquiring a USB flash drive. Many people in the legal community and digital forensics will say that getting the forensic image of a hard drive, phone, or flash drive is the most important part of the case. Paraben's software tool known as Forensic Replicator version 4.3 could also be used in conjunction with a hardware device known as Paraben's Lockdown used for forensically imaging any type of media in a drive that was connected to the Lockdown device. Some USB devices that could be connected to the Lockdown include floppy disks, CDs, and netbook hard drives. Batch assistant mode is used to help automate and speed the process of floppy disk copying when a large group such as a shopping bag of disks must be copied. If the digital media to be imaged is too large, the Forensic Replicator tool by Paraben can be configured to segment the captured media into certain size files which can be reconstituted in an environment such as Encase. The segmented files can also be compressed and used with a variety of tools on the market such as Access Data's FTK.

The program can also be used with the forensic imaging of the PDA/organizers that use EXT2 and EXT3 partitions. Suppose the digital examiner has a Tableau USB write blocker connected between the USB flash drive and the laptop; then he or she can use Forensic Replicator to image that flash drive without changing it. If a WiebeTech (such as a Forensic UltraDoc v5) was connected to the forensic workstation, an SD card in a universal card reader could also be connected to the WiebeTech. Then Forensic Replicator could be used to image the SD card. Forensic replicator also allows for viewing various directory structures in a tree format much like the DOS Tree command. One can also preview FAT files that are considered active. The program also has some support for NTFS systems.

One of the other good things about Forensic Replicator version 4.3 is that when it is used without WiebeTech devices and Paraben's Lockdown; it can be used to wipe digital media for preparation for receiving a forensic image. The wiping of forensic media is done to the Department of Defense standards. Then the other interesting feature is that it can be used in a drive to drive image option. Suppose there was a Paraben Lockdown device with the Suspect's USB flash drive in it. That could be connected to the forensic workstation. Then a clean USB flash drive of the same size could be put in the other USB port. Forensic Replicator could be used to forensically wipe the clean drive and then a drive to drive image option could be done to copy the suspect's drive to the clean drive. These drive wiping and drive to drive copying tasks of Paraben's Forensic Replicator is exactly what the LogicCube MD5 device does too.

The other important feature of this tool is that the image of a flash drive or any other media can be a raw image and then encrypted at 128 bit encryption. This is important for sensitive data that might get stolen and is too valuable to be released to thieves. There is also a capability to change the configuration to allow the digital examiner to compress images on the fly. Suppose the USB flash drive had a partition on it that was considered above top secret and another partition that was unclassified. Forensic Replicator can create an image of an entire physical drive or a partition This is important for examinations where the material may have a higher clearance than the investigator has. Forensic Replicator comes with 1 year of free support. That means software updates, technical help, and newsletters will arrive for a period of 1 year without a bill.

BitFlare

The initial download and usage of the BitFlare program is free. If evidence is found, then it is suggested that one purchase another module called the eDiscovery Pack (EDP). The EDP can help the investigator to recover data to a USB drive as long as the drive is formatted with FAT, FAT 32, ext2, ext3, or the NTFS file systems. It is also recommended that USB 2.0 is used since USB 1.0 is so slow. Later, the investigator should take the data that was collected to the office to organize, collate, analyze, and report the evidence for use in an investigation. BitFlare was created and supported by SunBlock Systems in McLean, Virginia. BitFlare is part of an ISO image file that is downloaded and then burned on a CD with a tool such as Nero or Roxio. An investigator can bootup with BitFlare and then safely look at deleted emails and some that have been partially overwritten [23]. It is a good previewing tool for viewing existing, hidden, and partially written over files too [23]. Bitflare has an online tutorial video to help one with the process. Bitflare has an automated chain of custody feature which is important to investigators for data integrity. The website says that a comprehensive EDP could cost as much as $900. However, if one figures the cost of hiring a certified computer examiner to do the task, purchasing the EDP might be a better option.

BitFlare TRAINING AND CERTIFICATION

The tool appears simple to operate, but the company has training options for those who want advanced training. It is possible that the company can send a trainer to a class on their site for a fee. This is good because it is expensive to fly a few employees to a remote site, pay for classes, lodging, and meals. BitFlare has a certification for their product and certified computer examiners working for them. It is good to get certified to make one more credible for performing computer forensic investigations and testifying in court.

CD BOOTING, GENERAL THOUGHTS

A computer can be booted to the CD if the computer supports the El Torrito bootable CD extension of ISO 9660 CD-ROM. It is important that the bios of the computer that will be investigated support the booting up of a CD because old systems from the early 1990s do not. The BIOS bootup order may be configured to check the CD before the hard drive. If so, then one puts the CD with BitFlare on it in the CD tray, and the bios looks for the operating system first on the CD. The next step is that the computer will boot up to BitFlare. It is good that investigators bootup with BitFlare so that no temporary files are created and so that no evidence is changed on the hard drive.

INVESTIGATION OPTIONS

It is also important that a professional certified computer investigator be used for the examination of the forensic image. If the organization doing the seizure with BitFlare does not have a properly trained and certified examiner, then arrangements can be made with SunBlock to contract the services of an investigator who can examine the USB drive once it is received.

ByteBack 4.2.1

The ByteBack program is a forensic tool that has been around for many years. Early versions ran in DOS. Version 4.2.1 is reported to have the disk cloning feature as it did before. Many digital examiners appreciate a tool that does the most important forensic task which is to clone or forensically image a disk. The latest version has support for ATA, SATA, and UDMA. The UDMA is ultra-DMA which supports burst speeds of transfer up to 33 Mbps. Serial ATA or SATA supports speeds of 150 Mbps. The ATA is a standard for use with IDE hard drives. Some websites that offer the latest version of ByteBack say that it can be used to perform a low-level format or wiping of forensic media for use in preparation of an image.

There are also many logging features which is good. One could use the tool to prepare forensically wipe some media such as USB flash drive. Then one could have a USB hub with one flash drive ready to receive the image and another connected with a USB enclosure with a suspect's IDE hard drive connected. ByteBack could then be run to clone the disk and the disk compare feature could be used to make sure it was all done. In the mean time, the logging feature could document that all activities were successfully done. The forensic mode which has to do with a write block could have been used on the suspect's drive so that the evidence was preserved. One website says that under certain configurations of ByteBack with certain types of hardware, support for 2 terabyte drives is possible [24].

There is also a disk editor that may be useful for examining the partition table and boot sector. In the old days of computing, people could hide small amounts of data there, such as a Swiss bank account number. ByteBack also has functions for master boot record repair and partition table repair. This is important because some people may try to delete a partition in order to hide data from an investigator. ByteBack also has features for hiding and unhiding partions. This is known as BPTM or Basic Partition Table Management. There is also advanced boot sector repair for various types of file systems such as FAT, NTFS, and FAT 32. Previous versions of ByteBack had string search capability to find certain words or terms in files. The search could be extended to unallocated space too. The latest version is one of those tools that should be considered for use in the incident response toolkit. Investigators say that wiping media, cloning a drive, and then repairing the image so that it can be booted and examined are invaluable capabilities. This may be considered to be a type of Swiss Army Knife of computer forensics and netbook forensics and could make a great addition for the toolbox. ByteBack is listed on the tools website http://www.e-evidence.info/other.html. The Computer Forensics Jump Start text by Solomon, Barrett, and Broom discuss this tool too. There seems to be many people using this tool so one should be able to post questions on the CCE list server and find a community of users that one can ask questions too. It is very important to be able to ask other professionals about new ways to use tools or if tools are valid for use on new types of hardware just released. It is also good to ask people if they know of any cases where ByteBack was used so they know if it passed the Frye test.

RECOVER MY FILES

The GetData corporation's Recover My Files product is now at version 4.9.4.1343 as of December 22, 2011. One of the great features of the product is the free trial preview. Digital forensic investigators may download the product and run the trial version so that one can preview thumbnails of the pictures that one may recover with the software. If the investigator is satisfied with the results, then he or she can purchase the one time license, copy in the license key, and perform the final step of recovering the files. The files can be recovered to a variety of locations and options, including burning them to a CD/DVD. This option is good for both corporate investigations and law enforcement investigations because the defense and prosecutor can both get the same results burned on CD or DVD. That is part of the ediscovery process. The software is reasonably priced at approximately $70. This version of the software can recover over three hundred different file types from hard drives, camera cards, USB flash drives, Zip disks, floppy disks, or other digital media. The software runs on Windows 98/ME/2000/2003/XP/Vista/Windows 7 platforms. This recovery tool is flexible because it works with a variety of classic and modern file systems such the FAT 12 found on the original floppy disks as well as FAT 16, FAT 32, NTFS, and NTFS5 file systems. The software is menu driven and very self-explanatory.

RECUVA VERSION 1.42.544

Digital Forensic Investigators with a very limited budget can download Recuva for free. There are inexpensive versions that provide support that might be worth it for a digital investigator who wishes to ask technical questions and reduce his or her time spent on learning tools. Recuva allows one to recover multimedia files and a number of other files from external media and USB flash drives. It supports a number of file systems such as NTFS, NTFS + EFS, NTSF5, and then the FAT 12, FAT 16, FAT 32, and exFAT systems. The interface is easy to use and the tool can be put on a thumb drive and be used as part of someone's incident response toolkit. Recuva is made by Piriform and the free versions can be downloaded from the FileHippo website.

STELLAR PHOENIX 4.2

This software is interesting because it is low cost, has an easy-to-use interface, and can recover encrypted files from the USB flash drive. One reviewer reported that the program occasionally crashes and the "file recovery is not perfect" [25]. However, the program can help one recover a flash drive where the files were deleted by formatting or lost to a power outage. The program has very good documentation and is often useful in recovering damaged or corrupted files.

DIGITAL RESCUE PREMIUM VERSION 3.1

This program is easy to use and has a menu-driven interface with plenty of documentation. There is much support for the program and one can use it to recover a variety of encrypted and nonencrypted files from USB flash drives. There is also a tool within the program to recover email which many programs do not do. The price for the program was inexpensive and could be put on a USB flash drive as part of one's incident response kit.

SNAGIT 10.0

This is a program that allows for the screen capture of pictures and sessions to create videos. Many investigators will use Snagit to show the Windows desktop of an investigator who opens up a window with FTK running and then investigates a CD, USB flash drive, or other piece of media. The process can be used in court to educate the jury to the investigation process. Many people who use Snagit may also use other screen capture programs such as Camtasia Studio 7.0 to create video documentaries of investigative work. There are excellent short training films on YouTube that show one how to use video documentation tools such as Camtasia and Snagit. There are also videos on how to use various versions of these programs which is important in case the examination has an older version of the software and the lab image of the investigation workstation may not be changed.

VIRTUALIZATION SOFTWARE: A BREAKTHROUGH FOR DIGITAL FORENSIC LABS

The idea of virtualization has been discussed by computer scientists since the 1950s and was actually in use in the 1970s with virtual memory. Each user of an IBM 360 and 370 seemed to have their own memory and environment. Virtualization has been utilized in some form by computer hardware and software manufacturers since the 1970s. By 2007, many virtualization products were available to the digital investigator as well as the computer scientist hobbyist. Due to the continued actualization of Moore's law, laptops with large-capacity hard drives and CPU processors with speeds in the gigahertz has allowed virtualization software to be run efficiently on the average consumer's laptops. The benefits of virtualization were now democratized and available to anyone with a modern laptop and not just with a large server or mainframe computer. One could download and install virtual software, add an image of an operating system, and now open a window with Windows 98, another with Windows XP, and another window with Windows Millennium. One computer could have various windows opened and all containing different applications and operating systems. It was no longer necessary to have a laboratory of old computers with different operating systems and applications. This saved room, electricity, and maintenance costs! One no longer needed to maintain a museum of working machines and parts that were expensive to energize. This was a breakthrough.

Another benefit of virtualization was that digital forensic investigators using products such as Citrix XenServer 6 might now have the ability to emulate the large labs with many types of operating systems and workstations. Each virtual environment runs in its own window on a laptop or workstation and is also run on a computer with processors that process code at speeds which are hundreds or almost a thousand times faster than the original. Consider the original IBM PC. It had a processor that ran at a speed of 4.77 MHz and laptops can sometimes run at speeds of almost 4 GHz. VirtualBox, Microsoft Virtual PC, VMware, and QEMU are four of the Virtual Machine environments one can research, download, and use for single nonuse examination machines. These environments also allow one to share peripherals among various operating systems and allow for investigation of USB drives and media in a safe environment.

CITRIX XENSERVER 6

Citrix has a trusted brand name that lawyers, private investigators, and digital forensic investigators are already familiar with, from their GoToMeeting program. New computer forensic products, computer forensic training sessions, and meetings are often conducted among small groups of geographically diverse people with GoToMeeting. XenServer 6 could be used installed on a server and networked with workstations with virtual machines deployed on them. This type of large-scale setup might be good for a large-scale forensics lab that needs to have a storage area network and some type of networked virtual workstations working simultaneously on cases.

VIRTUALBOX

This is a product that is made for large industrial uses and for the home user. VirtualBox could be used to demonstrate forensic principles on a large scale in computer labs in large schools with site licenses of Microsoft Windows 7. There is a free version that allows a limited group of VirtualBoxes working at the same time with certain versions of Windows 7. VirtualBox is an Oracle tool that works with many old and new operating systems. The old systems are DOS/Windows 3.x, Windows 2000, and the new ones are Windows 7, Linux 2.4, and Windows XP. VirtualBox is very good for computer forensic examiners because you can set up VirtualBox on your Windows 7 machine and then create an instance of it for a certain operating system such as Windows XP. Then when you set it up, you could give it an image (ISO file) of a seized Windows XP operating system and run it so that the investigator may safely examine the seized system. VirtualBox allows the examiner to have a window with another operating system and session of someone else's seized computer to examine. It is an amazing product and could be considered a green product because it allows one examination machine with numerous virtual boxes of various operating systems instead of maintaining and energizing a multitude of systems with various operating systems to examine software and data from suspects.

A person can also mount floppy drives, USB ports, networks, and other choices within a VirtualBox. There are many videos online in YouTube demonstrating the installation of the VirtualBox and using a Windows environment. One can even access the Internet if it is installed properly and the option for network access was selected during installation. VirtualBox can be used with an operating system that is activated in a window to examine malware, external media, and learn exactly what the suspect saw on their computer. A program such as Snagit can be used to document the process of using a VirtualBox to examine a suspect's hard drive and then a jury could watch a resulting video to understand the investigation process as well as the results.

VMWare

I had performed an experiment in the Cybercrime Training Lab at Fairleigh Dickinson University in 2008. A LogicCube imaging device was used to create a forensic image of a 2 GB hard drive with Windows 98 installed on it. The resulting image was one dd file of 2 GB. Then I took the dd image and followed the directions from the VMware documentation and created a vcf file. Then I ran the VMware software and used the vcf file to create a virtual session that looked just like the Windows 98 environment that I seized. Other computer scientists looked at it and said it would be a great environment to investigate malware that was quarantined on the USB flash drive without harming a school network.

QEMU: VIRTUALIZATION SOFTWARE AND AN EXPERIMENT FOR THE DEFENSE

QEMU is an open source software tool that became available in 2011 and allows one to create windows where one can install an image of an operating system into a virtual environment so that one may investigate the effects of malware, the contents of USB devices, and run older versions of applications. The virtual software allows one to easily evaluate the capabilities and vulnerabilities of older operating systems and applications. For example, it is easy to run a virtual session with Windows 98 and a firewall. Then an additional videoconferencing program called Netmeeting can be selected and run. This could be part of an experiment to support the defense of someone who said that he was not negligent because his high-security machine was hacked. This experiment could be used to demonstrate vulnerability in Netmeeting that kept 64,000 ports open on the firewall while he did a videoconference with a colleague.

Since QEMU is open source software, this means that one can get the source code for the software and learn how it works. This is important because one can then be able to answer the questions of a defense lawyer or prosecution about some of the technical details if they ask. It also means that the digital investigator may make changes to the software so that possible objections to the use of virtualization are addressed. In a paper by Bem and Huebner, they assert on page one that some may question virtualization as an investigative environment because "analysis by VMware is likely to taint the evidence" [26]. However, if one knows how virtualization works, one may argue that an instance of it is altered temporarily as one investigates applications and files and that such concerns against virtualization are merely sophistry.

MICROSOFT VIRTUAL PC

The Microsoft Corporation saw that others were offering virtual software that allowed for the emulation of various operating systems on x86 processers and came out with a virtual environment in 2007. Since Windows 7 and Windows XP are Microsoft products, many people would trust running these operating systems in a Microsoft virtual machine environment. Since all the products would be made by the same manufacturer, any bugs or issues could be addressed quickly and easily since the one corporation has all the source code and can both create and test changes to the VM environment. Microsoft Virtual PC also supports many types of USB flash drives, cameras, printers, and other pieces of hardware. There was some concern expressed in the Bem and Huebner paper that virtualized hardware may not be the same as those on the suspect's computer that is being investigated [26]. The wide range of supported USB devices might appease some of the objections of those who feel that perhaps a generic virtualized hardware may not be the same as the suspect's hardware.

IrFan VIEW

This is a program that one can download for free and is considered a picture manipulation program. It can be used to organize and view massive amounts of recovered images as well as thumbnails. Many law enforcement investigators and corporate investigators say that they will often recover thousands of pictures from unallocated space but it is too time consuming to view them in a slide show which often takes two seconds to load per image. IrFan View is reported to be a good tool. Exemplars of evidence can then be moved to a DVD or CD for a case. The software also allows viewing a picture and rotating the views. There are also many videos on YouTube that demonstrate the capability of the tool and give an overview of how to use it.

GUIDANCE SOFTWARE ENCASE

This is often said to be the premier law enforcement digital forensics tool. There is a short video on YouTube called "Computer Forensics: Recovering Deleted Files with Encase" [27]. This video uses an example of deleting a file on a USB flash drive and then recovering it. The interface in Guidance Software's Encase program seems user friendly and shows the file that was deleted as well as the contents below in a window with a physical storage dump of the contents. Encase has many advanced features that are great for working with small or large amounts of data. Results can be exported to a variety of media and a large forensic image of a drive can be easily managed by creating a series of sequential files. One of the numerous features that are useful to the investigator is one that lets him or her know how long it will take to copy a file. If a file takes a long time, the investigator can sit there and make a call or check notes so that multitasking can be done. Multitasking is common for both corporate and law enforcement investigators. Most digital evidence investigators say that nobody pays for you to just sit and watch a machine process data. You are expected to follow up leads, read files, and multitask during long periods of time when images are being copied. A person cannot leave the evidence unattended because that would allow the possibility of tampering with evidence but doing other things while supervising a process is expected.

ACCESS DATA FTK AND FTK IMAGER

FTK Imager is a tool that is free. The interface is intuitive and menu driven. FTK Imager allows one to image a device and collect all the allocated and unallocated space of a digital device such as a USB flash drive. Once the image is obtained, one can use Access Data's FTK to carve out all the images, documents, and evidence. One can then use FTK with the captured image to collect the free space, build a dictionary, and then use it to attack password-protected files. There are also many instructional videos on

YouTube that teach one how to use FTK Imager and FTK. The software also has a hash database which lets one quickly determine if there are hacker tools and child pornography images. The menu-driven software is user friendly but one may wish to seek the ACE certification to be proven proficient with the tool. The full stand-alone version of FTK version 3.0 is $3835 as of December 22, 2011. The full version needs a license dongle to run.

CORPORATE INVESTIGATIONS OF HOSTILE ENVIRONMENTS (USB FLASH DRIVE)

A corporate computer forensics investigator told me that occasionally there are hostile environment investigations where women report that men have pornographic screen savers on their computer. The screen saver may even be linked to a file on a USB flash drive so that the file or files are not detected on the desktop. An incident response team can come by and put in a flash drive with Paraben's Porn Detector Stick. This program uses AI techniques and skin filters to seek and find pornographic images. Paraben's Porn Detection Stick is available online installed on a USB flash drive for $99 [28]. This tool is easy to use. Put the porn stick in the USB port, double click on the program, and click start. Then the results appear shortly. The device is reported to have a 1% false positive rate and can find deleted as well as existing images. The device was reported to have examined 70,000 existing and deleted images in 90 min [29].

AUTOPSY BROWSER

There is a software suite of security and forensic tools incorporated within Backtrack 4. There are many videos on the subject within the popular website called YouTube. One such video is titled "Backtrack 4 R2 Digital Forensics Autopsy—Case Management." This video discusses a digital evidence tool called "Autopsy." There are many tools on the Backtrack 4 CD and it takes some navigation to find Autopsy. One has to first start Backtrack, and then choose the K menu, Digital Forensics, Forensic Analysis, and then Autopsy. When the forensic browser called Autopsy is started, one must first use a wizard to help create a report later about the investigation. The name of the examiner and other pertinent information will be asked. Then one must link the USB drive image to the browser with the "add evidence" button. There are many tools that can be used to image the USB drive such as FTK Imager, Avanquest Perfect Imager, IX Imager, and others. Then one selects evidence and adds it to the case. The file name is shown and there are many tabs that can be selected such as content and metadata. Metadata can include GPS data suggesting the location of where a picture was taken as well as creation, modification, and last accessed dates for files. The results of the report can be burned to a CD and truly help organize the ediscovery process for everyone.

USB BIOMETRICS FLASH DRIVE AND CORPORATE INVESTIGATION

There is a new type of flash drive that is on the market known as the "SanDisk 1 GB Cruzer Profile Biometric USB Flash Drive." This device has a USB connection at one end, a finger print reader in the middle, and the flash drive at the other end. It is used as a secure storage device to store data as well as passwords. There are other companies that put the fingerprint reader on the top of the drive so that it is more robust and does not break. The Defender F200 has a built-in encryption program and uses 256 bit AES encryption and includes virus scanning software [30]. The Defender F200 can be used with Windows XP, Vista, and Windows 7. The device has security features to limit the access such as requiring a minimum length for the password. There is also a threshold feature that can require two factor authentications such as a thumbprint and password. There is also an option known as data destruction that may be configured to destroy the data if a certain number of incorrect passwords have been exceeded. There can be an administrator account as well as a certain user profile account which means that the device could be used by more than one person and the administrator might have more access then someone else.

If the focus of an investigation is someone who is using one of these devices, then there is a strong possibility that AES 256 bit encryption is being used with two factor authentication. The data destruction feature may also be on and could activate after 10 tries. A similar feature exists in the Blackberry devices. If the investigation is important enough such as with industrial espionage, then one might want to consider getting the State Police and a Regional Computer Forensics Lab RCFL) involved early in the process before evidence is lost. However, there should be some evidence that it is a high-profile investigation and not just allegations to take such a measure.

VALIDATION PLAN AND TESTING YOUR TOOLS: TABLEAU IMAGER V 1.1

One of the questions that often arise at academic conferences and in classrooms where digital forensics is taught is "How accurate are these imaging tools?" Is there a testing plan? Is what is on the drive actually what is imaged? Are there many errors? Is there benign filler placed in the image if there is a bad sector on the physical storage being copied and that location cannot be obtained? What about the various operating systems; how does that affect the media being imaged? The National Institute of Technology and Standards (NIST) has a document that shows a test plan for a tool among USB flash drives and many other types of digital media. The imaging tool known as Tableau Imager version 1.1 was tested against the Digital Data Acquisition Tool Assertions and Test Plan Version 1.0 [31]. The methodology of the testing plan can be adapted to other tools for those who need to validate their tools.

IXIMAGER: DIGITAL ACQUISITION TOOL

The NIST did a series of tests to check how accurate the digital data acquisition tools were for USB flash drives and many other pieces of digital media. The test results for the tool IXimager version 2.0 can be downloaded from http://www.cftt.nist.gov/disk_imaging.htm and one can read how accurate this tool was against the testing plan. One can compare the results of the report against the Tableau Imager V 1.1. It saves corporate investigators tremendous amounts of time in validating their tools if a reliable source such as the NIST has already done the work.

DataLifter 2

There is a software tool called DataLifter 2 that sells for approximately $199 and allows one to carve many types of pictures, spreadsheets, and documents from free space. First, one must collect all the unallocated space and slack space with X-Ways Forensics, IXimager, FTK Imager, Avanquest Perfect Imager, or a number of other imaging programs. Then one takes that ISO or file(s) and uses DataLifter 2 to carve out a particular file type within that collected space. One could easily collect the raw disk space with Avanquest Perfect Imager, for example, and save it as a file. Then the second part of the investigation could be to look for JPG or JPEG files with DataLifter 2 to investigate if illegal pornography or photographs of trade secrets were present.

HARDWARE WRITE BLOCKERS FOR USB FLASH DRIVES: WiebeTech

There is a company called WiebeTech that makes a variety of forensic write blocker products that have been used in court and are accepted by both the computer forensics and legal communities. One might say that those products have passed the Frye test. The one product that is very useful for cell phone, USB flash drive, and digital camera investigations is the USB write blocker. The write blocker connects to the USB port and blocks any attempts of writing to that device. The price is sometimes advertised on websites for $151. This seems reasonable since it can be used for digital cameras, PDAs, cell phones, smartphones, flash drives, and hard drive enclosures. The WiebeTech was also tested with many computer forensic tools and a full report is on the NIST website. There are many computer forensic tool test results in the report http://www.cftt.nist.gov/hardware_write_block.htm.

WiebeTech also makes the ComboBlock. This product prevents writing to devices connected by the following ports: mini USB, USB Standard A, USB Standard B, and IDE/SATA hard drives. The exact model is called the FCDKV4 Versatile Bundle Writeblock FW800. This sells for around $275 but it includes the various types of USB ports and the common hard drive interfaces in one piece of equipment. Some digital forensic examiners who work for government agencies or private industry say that they prefer multifunction devices because it reduces the need for more paperwork and purchase orders. Many people in a variety of agencies and private companies say that it is better to purchase all the items one can while the funds are available because they may not be later. Some investigators say that a budget that is not spent in a year will shrink the following year since obviously that much money was not needed.

HARDWARE WRITE BLOCKERS FOR USB FLASH DRIVES: TABLEAU T8 USB FORENSIC BRIDGE

This write blocker for the USB port was also tested by the computer forensic tool testing group and a full report is available online at http://www.cftt.nist.gov/hardware_write_block.htm. This device is very good to have because it is always in write blocker mode. Just like the WiebeTech, it connects between the device being imaged and the USB port. The device is special in that it also blocks firewire ports. Those are the ports that look very similar to USB ports but have a different-shaped connector and support IEEE standard 1394. This device is just like the WiebeTech in that it is widely used and accepted as a forensic tool for acquiring images in a forensically sound methodology from devices connected to USB ports.

SOURCE SAFE DATA GUARD: USB FLASH DRIVES AND EXTERNAL MEDIA CARD WRITE BLOCKER

This tool is made by a group known as Salvation Data. The write blocking device tool is quite interesting because it can be used with digital cameras, cell phones, PDAs, hard drive enclosures, and a universal card reader. The universal card readers are frequently used with SD, Micro SD, XD cards, MMC cards, and a variety of others. The device could be used for a variety of mobile device forensics and therefore it might be easier for an investigator to justify the expense. It seems that many of the hardware devices offered for sale in 2011 have multipurpose uses. The device also gets its power from the USB Standard A port wires and does not need an external power source. There are also LED light signals to keep the user informed of processes. The LED or light-emitting diodes use very low power and last a very long time. Transfer rates are at 1.8 GB/min which means that the popular 2 GB USB flash drives can be imaged within 3 min or less.

SOFTWARE: WRITE BLOCKER

There is a software blocker program that is free and can be downloaded from the Mid Michigan Forensics Group. The program was first posted in 2004 but then updated and reposted in October 2011. This M2CFG Writeblock Utility is for Windows XP and one must first install all the CDs for Service Pack Two or install the download for Service Pack Three. The reason for the service pack requirements is that Service Pack One does not have a special feature to write protect all the USB ports. The program download is only 208 KB and the forensics group asks that one validates the program by testing the writeblock ability and not to take anyone's word. They also provide the MD5 hash for the program which is MD5 = e6a350f914a3140c6ce12e4b672b8f9d. Many people hash the program so that one can recheck it and show the defense team it was not compromised while downloading it from the source.

SOFTWARE: THUMBSCREW USB WRITE BLOCK

This type of program can be found on the web and it offers no guarantee of its forensic write blocking ability. The program is called "The poor man's write blocker." This type of write blocker could be downloaded and tested and perhaps used in an emergency situation. A device that provides some protection is certainly better than imaging a drive with no protection. If one also did 10 attempted writes to the USB drive when Threwscrew was enabled, this could provide some type of baseline if the results were questioned.

AUTHORIZED REQUESTOR REVIEWING THE USB INVESTIGATION

In football, a coach may send players out in the field with a plan and they execute it. Then, at half time, they may look at how they did, discuss new plays, and then go on with the rest of the game. Periodic review is necessary in team sports and it is necessary in corporate investigations of USB flash drives with other media. This would be a good moment for the investigator to look at the NIST website about the Information Technology Lab's Computer Forensic Tool Testing Program. Methodologies, tools, and recent concerns could be reviewed and discussed. Perhaps there could be a slight modification in the investigation plan.

The incident response team may be investigating a USB flash drive for company secrets but they may also wish to check some other media with email files on them for communications with a competitor at the same time. The AR may be speaking to the digital forensics investigator in the corporation and reviewing some of the process with him or her and then discussing how they may proceed. Let us suppose that there may be only one computer. The investigator and AR may first check the chain of custody form to make sure that everything looks fine, and then update it as they look at the USB flash drive.

Before looking at the USB flash drive, they may use a Tableau USB write blocker, change the registry to make the USB port read-only, or use a software write blocker. The NIST has a computer forensic toolkit testing area of a website that has reports of reviewed write blockers [32]. After selecting the best one available and implementing it, they may decide to prepare the forensic media. The same site has reports on forensic media preparation using WiebeTech equipment. The LogicCube may be used to wipe a small new hard drive for accepting the image from the USB flash drive. The forensic workstation may be checked again that it is not connected to any networks and that all connectivity is blocked, no infrared, no Bluetooth, no wireless, and so on. Then the antivirus and antispyware may be run again on the prepared drive and the workstation. The write blocker may be tested with another test USB flash drive. Then, FTK Imager might be run and the image of the USB flash drive may be put on a hard drive that was wiped with the LogicCube earlier. This wiped drive may be connected by a drive enclosure to the examination machine. Perhaps the FTK imager was chosen after all the reports for imaging tools on the NIST website was checked [32].

Then the AR and the investigator want to check another drive image and recover email. They may add another hardware write blocker to the investigation machine and connect the other hard drive. They may then use a program such as VirtualBox and have a window open with a virtual machine running. It is separate from FTK imager and the other processes going on. Then they may run some antivirus in VirtualBox to make sure there is no malware. They then run Digital Rescue Premium (DSR) version 3.1 to recover the email. On the regular desktop, they may now run FTK and review the image of the USB flash drive that is on the other hard drive. They now have a window open with VirtualBox and DSR running, and outside the Window the FTK is running and they are reviewing the drive. Perhaps the AR and investigator find many instances of the same picture with different names and file sizes. They notice all the same looking pictures have different MD5 hashes. It may be a case of steganography. The next part of this chapter will look at steganography, how it can be done, and then discuss how to detect it.

STEGANOGRAPHY: INTRODUCTION

Steganography is commonly known as the hiding of data in a picture. In the days of ancient Greece, a person's head would be shaved and a message would be written on the head. Then the hair would grow in to cover it. The person would be sent to a far-off place to deliver the message. This was reported to be one of the first uses of steganography [33]. Then there are other cases where a layer of wax would be scraped off a picture and a message would be underneath. Then in the modern world where mail is censored and phone calls are monitored, messages can be hidden in pictures. People may have gone back to another country to live after receiving promises of a better life. However, if life is really bad they may have a previously agreed signal such as leaning against the refrigerator if life was not as it was promised.

MODERN STEGANOGRAPHY AND COMPUTERS

The modern forms of steganography involve digital photographs and a computer. One such program that allows people to hide documents is called JP Hide and Seek. It can be commonly found for download on many places within the Internet. The program is simple to use. One runs JP Hide and seek. It asks for the document that you wish to hide. Then it asks for the picture which is known as a container, where you wish to hide the document. Then the program asks for a pass phrase for security. The document is then distributed across the least significant bits of the photograph known as a container. It is often said that the picture should be at least eight times as large as the document that one wishes to hide. If a person wishes to hide a 125 KB document, for example, the picture or container should be at least 1 MB or larger. A byte looks like 10001000 with eight bits being either a 1 or a 0. Some digital photographs use anywhere from 1 to 3 bytes to hold information about the pixel and its color. If the least significant bit is used for hiding a document, it will affect the colors of the picture but it will not be enough to be detected easily by the human eye. The interesting thing is that the file size with the document embedded in the picture is often smaller than the same picture without the document. The image had less bytes after a document was embedded in it with the JP Hide and Seek Steganography tool. The readme.txt file was 972 bytes. The same picture with the embedded file was 1.08 MB. Viewing the pictures side by side on a large monitor, no difference was visible as can be seen in Figure 7.3.

STEGANOGRAPHY: EXAMPLES OF MOTIVATION FOR USAGE

It is important for a mobile forensics examiner to understand the motivations and circumstances when a person might use steganography. Then one can consider if such an information hiding and communication technique may be something to be considered in the present investigation. Suppose a person such as Aldrich Ames or Robert Hanssen were spying today in 2011. Let us also suppose the KGB existed in 2011 and was actively paying spies. Then it would be possible for a person such as Ames and Hanssen to put text documents within harmless pictures and email them to their spy handlers in the KGB. People email pictures all the time as attachments from work so if someone uses a

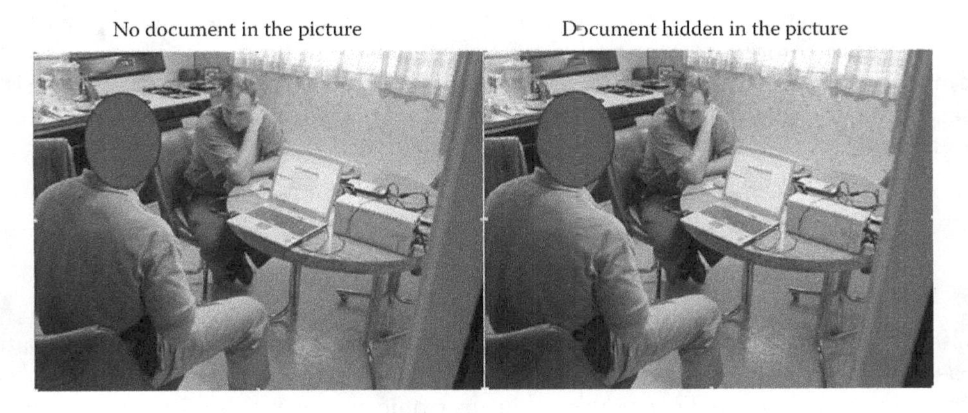

FIGURE 7.3 Embedded document in a picture on the right and the original picture on the left.

program to embed a document in a harmless picture, who would know? The practice of steganography is to pass documents from one party to another via pictures. The pictures do not have to be emailed to the other party. They could be posted online among a group of seemingly harmless pictures such as a company picnic.

Other uses of steganography could be for better purposes. Perhaps a person lives in a country with a poor human rights record and wishes to pass a report to the United Nations about conditions in that country. The person in that country who was considered a dissident could use a steganography program to hide a report within a seemingly innocuous picture such as saluting the flag or celebrating a national holiday. Then the picture could be posted on a website or emailed.

The reasons for hiding a document are numerous. It could also be for smuggling out a report of noncompliance to Department of Defense standards for a classified project. The person might be a whistle-blower. Basically, the motivation could be anything from corporate espionage, to human rights activist, to being a whistle-blower. Steganography itself is neither good nor bad. The tools are neither good nor bad. It is how the tools and methodology are used and for what motivation that matters.

THEORY TO FIND PICTURES THAT MAY HAVE DOCUMENTS HIDDEN WITHIN THEM

This is an important consideration since we now know that it is possible to use steganography for corporate espionage or espionage. The first line of defense might be a visible inspection of a person's computer. This could happen as printer drivers are updated and new versions of software are installed on company computers. The first red flag might occur when a technician sees multiple copies of what appear to be identical pictures on a person's computer. The second red flag would be if the identical pictures have different file sizes and the same creation date but all different modification dates. There are many programs available including freeware such as WinMD5 by Edwin Olson that allow one to create MD5 hashes of many files. If one has identical pictures with all different file sizes and MD5 hashes, it might be time to look further into the possibility of steganography and corporate espionage.

The second way would be more high tech and somewhat automate the process. Consider that in a normal picture without hidden documents in them, the least significant bit should randomly be assigned a 1 or 0. In a picture with documents hidden within it, there should be more information hidden in that least significant bit. Therefore, a program that examines the least significant bit in a picture and does a statistical analysis on it should suggest it might have a document in it. It is helpful to read some academic literature from Gary Kessler to learn more about the theory of steganography [34].

FBI ARRESTS ELEVEN PEOPLE SUSPECTED OF USING STEGANOGRAPHY WITH DIGITAL PICTURES

In 2010, the FBI arrested a group of 11 people who were suspected of passing information to a Russian intelligence agency with the use of steganography tools. The article said "Deep-cover Russian intelligence agents hid electronic messages behind computer images" [35]. This is a real case that discusses how documents were allegedly placed in pictures and put on websites for Russian intelligence types to download. The pictures were in plain view but the documents were not visible since they were embedded in the pictures. Ms Higgins who wrote the article also stated that there are approximately 800 easy-to-use tools which can be downloaded and used to hide documents within pictures.

LOCATING STEGANOGRAPHIC TOOLS ON A SUSPECT'S STORAGE AREA

Access Data makes a product called Forensic Toolkit (FTK). This product is interesting because it has a library of digital signatures or MD5 hashes of hacker tools which can include steganography tools. A CCE can use FTK Imager to make an exact clone or image of the digital camera's SD card, internal memory, XD card, or the suspect's laptop. One of the numerous steganography tools may be found on any place where digital files may be stored. Encase by Guidance software can do the same thing. The problem is that some tools on the web offer the source code for programmers to compile. If one alters the program ever so slightly, the resulting digital hash will not be the same. This means that the library of hash functions will not find it. If a CCE opens a picture in MS Paint and puts the smallest dot somewhere on the picture and saves it, everyone will get a completely different MD5 hash for that picture when WinMD5 is reused to create the hash.

WetStone Stego Suite has a set of tools that allow the CCE to find steganography tools and possible pictures that are used to hold documents. Stego Analyst is one tool within this group that analyzes pictures using statistics and analysis on the least significant bit. If the distribution of 1s or 0s are far from the mean of normal pictures, this suggests a hidden document. Stego Watch and Stego Hunter can work together to find steganography tools or their artifacts. Stego Break can help find the pass phrase used as the cipher to create the cipher text that is distributed across the least significant bits.

STEGANOGRAPHY: PICTURES WITH EMBEDDED DOCUMENTS

Steganography can be considered to be the hiding of digital computer-generated word-processed documents in digital photographs. The hiding of documents within a picture is easy to do with a tool such as Pict Encrypt. This tool is only for the Mac computer. The program is less than 1 MB. It should be noted that there are so many tools to perform steganography and very few tools and techniques to identify it and even less to recover it. If anyone has a math, computer science, and cryptography background and needs research ideas, this is an area in great need. There may also be chances for people with small businesses to partner with schools and apply for grants to create tools to meet this need. Steganography is a problem because people in a criminal or terrorist organization can covertly communicate command and control decisions as well as send plans, intellectual property, and other items. These items could be emailed or put on a flash drive in a false-bottomed thermos cup and then sent to someone. The threat of steganography may not be fully realized because there is no accurate statistics of how widespread it is.

STEGANOGRAPHY: MISUSE OF A PHOTO EDITOR AND MAGNIFIER

There are simple methods of hiding data in a picture that are quite effective. One could misuse Faststone Image Viewer and violate the user agreement by magnifying the picture and then adding extremely small text in an area that most people would miss. One could email the picture to someone who then magnifies the picture to a prearranged location and then zooms in on the message. One could also misuse the product to access and change metadata to put a message in there. People could also have a prearranged signal that is not text to mean something too. One could cut and paste a butterfly on the picture and that could be a signal to go through with some event tomorrow such as a drug deal or robbery. There are other signals that can be put in the picture without editing. Since the mail was often censored in the old Soviet Union, I was told that relatives would signal that life was really bad with a prearranged signal such as a picture with one leaning against the refrigerator. People can have a sheet with lists of verbal signals and predefined meanings. Words are not always needed to transfer the message. The picture file could be on the flash drive.

STEGANOGRAPHY: MISUSE OF MICROSOFT PowerPoint

One of the simple things that one can do is have a PowerPoint slide show and put a message on a slide. Then one covers the message with a picture. One can also have part of a slide with a light blue font and text in blue font so that nobody sees it. One can also include a slide with a JPG picture in it that has a document embedded in it with a tool such as JP Hide and Seek. One could also have a sound file with text in it. One could also have a picture that has the small print on the picture that is only viewable with a magnifier. There are so many ways to hide messages in pictures that one would have to know how sophisticated the message hider is. The person who is hiding the messages may also put three messages that he or she knows the investigator will find and then two really hidden ones. Sometimes the same technique is done with hidden cameras where people will have two semihidden ones so that the person who might look for them feels they were all found and then feels confident and stops looking for more.

There are so many tools and nesting techniques to hide information within pictures and put more information under those pictures. Unless one has a tremendous amount of time and staff, it is nearly impossible to stop a determined person who wishes to pass a message.

CRYPTOGRAPHY: MISUSE OF MICROSOFT WORD

A person can easily have a document such as a novel or anything downloaded from the Internet that was written that might contain a hundred pages. Certain pages and certain paragraphs could maintain the message. Perhaps, pages 50–55 contain the message in the second paragraph. These paragraphs may not have writing hidden but rather the message is hidden in plain sight. One could also use a substitution code with A replacing R, B with X, T with 4, U with @, until the whole alphabet is decoded. The code could be complicated with foreign alphabets. This means that if one goes through the whole document and even finds where the correct material is hidden, it still has to be decoded. The other type of steganography that could be done is to put a message on page 50 and then perform a substitution of the entire hundred page document. The file could be on the flash drive. This type of steganography would probably best be recognized by a manual process of having a document reader go through all the evidence. Once such documents are located, perhaps contracting the work to a data carver specialist or cryptographer would be the best option.

STEGANOGRAPHY IN WAV SOUND FILES ON A USB FLASH DRIVE

I witnessed a demonstration of a written message being place in a JPG file with a program such as MS Paint or any image editor. The file was called Message.jpg. One could have just as easily saved the written message as a bmp file. The result is that a written message is saved as a JPG picture file. Then one opens up a program called Coagulla 1.6 Color Note Organ and then one opens up the Message.jpg file. Then one clicks on an icon that says "Render Selection without Blue/Noise." Then save this as a wav file. The result is Message.wav which is a sound file but there is a picture of written message embedded in it which was rendered without blue noise. Then one opens Sonic Visualizer, a program copyrighted to Chris Cannam and Queen Mary, University of London. Then one opens Message.wav and chooses Layer, Add Spectrogram, and all channels. A sound that reminds one of buzzing bees plays but during that message, the visual message could be seen. It did work. The LifeHacker website did describe the process accurately, quite amazing [36].

DATA CARVING AND TOOLS TO DETECT STEGANOGRAPHY

Data Carving is the skill of detecting that a document is hidden in a picture and then removing the document from the picture. How can we detect if there is a possibility of steganography? One can use a program that does statistical analysis on the least significant bit of each byte in the picture. If the distribution of ones and zeroes is far from fifty/fifty percent distribution on the least significant bit, than it may indicate that the picture may have a document in it. Suppose I want to check a picture with a gif extension for a possible hidden document using the least significant bit method. One may use a program to check the least significant bit of a statistically significant sample of gif pictures and find out the ratio of ones and zeroes in the least significant bit. Suppose it is 49–51% ones to zeroes. My hypothesis is that if the number of ones and zeroes has a distribution that is significantly greater than 51%, the picture in question has a document in it. One can test that type of hypotheses in an experiment. There is a tool called Outguess that can be downloaded, compiled, and run which may aid in this process.

WHETSTONE TECHNOLOGIES STEGO SUITE

There is a program known as Stego Suite that is sold by Whetstone Technologies. Stego Hunter is one of the programs within the suite that is used for locating steganography from an image. The program checks for clues left by any of 500 steganography programs. The program is able to scan JPG, png, GIF, and audio files for hidden payloads such as documents. Stego Watch can be used to scan the system and flag suspicious files. Stego Analyst is a tool that one can utilize to examine the visual cues of files and check the least significant bit distributions to locate suspicious files. Stego Break is a program that is designed to get the password or pass phrase used to encrypt the document in a picture. Then it is up to the investigator to find the program that the person used with the pass phrase to hide the document in the picture. Perhaps the registry in the image of the seized computer shows what steganography program was used.

The program does not need many resources to run. There is a minimum of Microsoft Windows 98 and at least 100 MB of free disk space to install the program. One needs a

Pentium III 1 GHz processor or faster. The cost is not bad since the product is well supported and is very unique. Some bloggers estimate that the suite may cost about $950.00.

The FBI helped bringing 11 people to justice recently for spying. A group used a steganography program that was allegedly developed in Moscow and linked Russians and the Russian Foreign Intelligence Agency. There was a 27-character password and two hot keys to bring the program from memory to the active screen to communicate secret messages [37]. This goes to show that steganography is really used and is not just an academic exercise. Music as well as pictures can be used to hide documents that are passed between people. There are websites that describes how to hide documents and messages in digital music. The techniques are not difficult to do.

OTHER IMPORTANT TOPICS RELATED TO USB DRIVES

The following sections have to do with some important topics related to USB flash drives but do not fit neatly anywhere else. Therefore, they are in this section known as other important topics. It is important to know that some drives have programs which act as investigation tools on them and are basically read-only. It is also important to know what to do when a flash drive gets wet so that one does not burn it out or short out the USB port on an examination machine. Lastly, it is also important to know some of the important changes in cyberlaw that are on the horizon as well as some of the new issues facing investigators. This last part of the chapter will face these miscellaneous topics or as some would say, other important topics.

BLACKJACK USB FLASH DRIVES WITH SEARCH CAPABILITY

A lot of time can be lost during a lawful investigation looking for certain pieces of evidence that are relevant to a case. Many people read and write tens of thousands of emails over the course of a few years as well as create hundreds of word-processed documents. People may have also copied hundreds of documents to a USB flash drive over a period of years. The Blackjack USB flash drives could be used to reduce wasted time and help investigators find documents, emails, and spreadsheets that are related to a certain investigation [38]. With the help and instruction of the vendor, one might be able to configure one of these Blackjack USB drives to search a computer and any connected devices such as USB flash drive to find related materials with keyword searches.

USB FLASH DRIVES THAT GET WET

The USB flash drive can get wet and if one plugs it in, the computer and flash drive can be shorted out. The best thing to do is to dry the USB flash drive by putting it in a plastic bag in a sunny place with those silica gel bags that come packaged with food to absorb moisture. A former private detective said that a bowl of uncooked rice also works well for drying flash drives. The important thing is not to use a hair dryer because the high-intensity heat can damage parts. It is often said that if the electronics are wet, one can put some isopropyl alcohol on it which will help it dry up faster and remove any impurities left by the water. The USB flash drives often get wet and lost; in 2010, it was claimed that over 17,000 USB flash drives went through the wash and were left at launderettes in the United Kingdom. The drying techniques previously mentioned are tested quite often.

CORPORATE INVESTIGATION OF LOST DRIVES

Mat Bettinson said that a security firm named Sophos examined 57 lost USB flash drives that were purchased at an auction. The drives were lost on an Australian train network. Two thirds of the drives had malware according to an investigation [39]. There were no credit cards, top secret information, but there was a lot of personal information on the drives. The lesson is that people need to encrypt their drives so they do not fall into the hands of people who may wish to exploit the information. This is also a lesson to run antivirus and antispyware software on these drives to reduce the possibility of tampering by outside forces.

INVESTIGATION OF USB FLASH DRIVES WILL BE MORE DIFFICULT IN THE FUTURE

Ewa Huebner and Derek Bem wrote an informative paper about the analysis of a USB flash drive in a virtual environment and some of their future concerns for investigations of these devices. One of the valid concerns seems to be the lack of evidence that will be left on a desktop computer if one will attach and utilize a flash drive with the U3 standard [26]. The U3 standard will allow the running of what appears to be a full virtual environment using applications and data from the files that are contained on the USB. This is very convenient but it allows one to plug in their own USB at someone else's computer, do a series of activities, transfer files, do email, and then remove the drive without leaving any personal data. There may have to be a law in the future that every computer connected to the Internet is watched by a camera and digital video recorder. This may seem extreme but go to any university and try to use a computer. You must put in a valid username, password, and the session activity is logged. There is also a bubble camera connect to a DVR someplace. If an investigation occurs, a person, the activity, and the video can be assembled and should be able to be successfully prosecuted.

CHANGING LANDSCAPE OF CYBERLAW

Suppose a person has an online company that digitally signed an online contract to provide customized USB flash drives to the customer. In our theoretical example, the USB flash drives are supposed to say "Billy and Betty's 25th Anniversary Party," but never arrived. Then there is an investigation of the incident The online flash drive provider may try to argue that he did not understand the contract, that the e-signature was not valid, or that an online contract is not enforceable like a paper contract [40]. The world of e-commerce creates a new environment that the law is catching up to, but the rest of the online community needs to educate itself on the topic. People have been working with paper documents, contracts, and physical documents since the time of the ancient Romans. However, this cyber environment is a relatively new paradigm that needs a new common rule set. Many of the assumptions of the physical world do not always translate as we expect to in the cyberworld. The book *An Introduction to U.S. Telecommunications Law* by Charles H. Kennedy is very enlightening and discusses some of the new and existing laws and some of the issues they address, and do not address.

Another book that addresses this changing environment is *Cyberlaw Text and Cases* by Gerald Ferrera, Margo Reder, Stephen Lichtenstein, Robert Bird, and Jonathan Darrow. The book takes people through the business cycle and discusses ethics as well as the law. Before there can be an investigation of a breach of the law, one has to see what the laws are. One has to understand some of the implications of having an email account in one state, a website in another country such as China, and perhaps a payment mechanism such as PayPal in another state. One should not only read books on the subject but also join a group such as the American Society of Digital Forensics and E-Discovery where one can read about new laws and how it impacts both the cyber environment and the traditional business environment.

REFERENCES

1. USB. (n.d.). *Collins English Dictionary—Complete & Unabridged*, 10th edition. Retrieved December 23, 2011, from Dictionary.com website: http://dictionary.reference.com/browse/usb.
2. Brain, M. (n.d.). How USB ports work. *Howstuffworks*. URL accessed December 23, 2011. Retrieved from http://computer.howstuffworks.com/usb4.htm.
3. *Anti-forensics*. URL accessed December 24, 2011. Retrieved from http://www.anti-forensics.com/about-anti-forensics.
4. Max (April 21, 2009). Delete USB device history from the windows registry usbstor key and the setupapi.log. *Anti-forensics*. URL accessed December 24, 2011. Retrieved fromhttp://www.anti-forensics.com/delete-usb-device-history-from-the-windows-registry-usbstor-key-and-the-setupapilog.
5. Department of Justice (1999). 1999 Report on cyberstalking: A new challenge for law enforcement and industry. Accessed December 26, 2011. Retrieved from http://ncsi-net.ncsi.iisc.ernet.in/cyberspace/law/responsibility/cybercrime/www.usdoj.gov/criminal/cybercrime/cybersta.htm.
6. *Everything USB*. USB flash drive thumb drive FAQ. URL accessed December 23, 2011. Retrieved from http://www.everythingusb.com/flash-drives.html#1.
7. *eBay.* URL accessed December 20, 2011. Retrieved from http://www.ebay.com/itm/Poker-Chips-Design-5-USB-Flash-Memory-Drive-Stick-Thumb-Pen-FD1617-/290618422597?pt=LH_DefaultDomain_0&var=&hash=item8960574ddb#ht_1838wt_981.
8. *Webcite* (June 25, 2010). Not even FBI was able to decrypt files of Daniel Dantas. URL accessed December 20, 2011. Retrieved from http://g1.globo.com/English/noticia/2010/06/not-even-fbi-can-de-crypt-files-daniel-dantas.html.
9. Statement of Louis Freeh, Director of the FBI, July 25, 1996, before the Committee on Commerce, Science and Transportation, United States Senate, Regarding the Impact of Encryption on Law Enforcement and Public Safety. *EPIC*. Retrieved from http://epic.org/crypto/export_controls/freeh.html.
10. AssertTrue (October 11, 2008). Russians use graphics card to break wifi encryption. URL accessed December 20, 2011. Retrieved from http://asserttrue.blogspot.com/2008/10/russians-use-graphics-card-to-break.html.

11. *eBay.* URL accessed December 21, 2011. Retrieved from http://www.ebay.com/itm/4 GB-USB-Flash-Memory-Thumb-Drive-Bracelet-Watch-w-Crystal-Set-w-Fine-Gift-Box-/110768503432?pt=LH_DefaultDomain_0&hash=item19ca513a88.
12. Doherty, E., Devine, J. (2008). Watching the USB Watch. *Security Magazine*, pp. 70–71.
13. *eBay.* URL accessed December 21, 2011. Retrieved from http://www.ebay.com/itm/Wrist-Watch-4GB-USB-Mini-DV-DVR-Hidden-Spy-Cam-Camera-Camcorder-/130688317033?pt=LH_DefaultDomain_0&hash=item1e6da17669#ht_4051wt_1035.
14. *NCIX*. URL accessed December 20, 2011. Retrieved from http://www.ncix.gov/images/publications/posters/EconEspionage-USB_Port1.jpg.
15. Robinson, S. W. (n.d.). Corporate espionage 201. *Reading Room SANS*. URL Accessed December 20, 2011. Retrieved from http://www.sans.org/reading_room/whitepapers/engineering/corporate-espionage-201_512.
16. Ahlers, M. M. (June 1, 2001). FAA to crack down on people who target planes with lasers. *CNN*. URL accessed December 21, 2011. Retrieved from http://articles.cnn.com/2011-06-01/us/laser.flight.safety_1_laser-beams-faa-administrator-randy-babbitt-federal-aviation-administration-officials?_s=PM:US.
17. *NY1 News* (May 18, 2005). Skirt cam found under subway grate on upper east side. URL accessed December 21, 2011. Retrieved from http://www.ny1.com/content/top_stories/50957/-skirt-cam—Found-under-subway-grate-on-upper-east-side.
18. *Online lawyer source* (January 27, 2005). Nursing home and elder abuse. URL accessed December 21, 2011. Retrieved from http://www.onlinelawyersource.com/news/elder-abuse.html.
19. Daily mail reporter (n.d.). Middle-aged pervert who filmed women in leisure centre changing room blames Viagra, wife's menopause for 'change in behaviour'. *Mail online*. URL accessed December 21, 2011. Retrieved from http://www.dailymail.co.uk/news/article-1345861/Pervert-Neil-Smith-filmed-women-changing-rooms-blames-viagra-wifes-menopause.html.
20. *Ecrater.* URL accessed December 21, 2011. Retrieved from http://www.ecrater.com/p/11663300/pen-hidden-camera-detector.
21. Vielmetti, B. (September 8, 2011). Complaint: Men taped roommate nude (not so subtle) with pen cam. *JS online*. URL accessed December 22, 2011. Retrieved from http://www.jsonline.com/blogs/news/129469443.html.
22. Complying with the Jeanne Clery. In *Security on campus, Inc.* URL accessed January 20, 2012. Retrieved from http://www.securityoncampus.org/index.php?Itemid=60&id=271&option=com_content&view=article.
23. What is bitflare. *Bitflare*. URL accessed February 9, 2012. Retrieved from http://www.bitflare.com/whatisbitflare.php.
24. *Fyxm*. URL accessed February 12, 2012. Retrieved from http://download.fyxm.net/ByteBack-38605.html.
25. Stellar Phoenix 4.2. *Toptenreviews*. URL accessed December 23, 2011. Retrieved from http://www.laptopmag.com/best-data-recovery-software/stellar-phoenix-review.aspx.
26. Bem, D., Huebner (2007). Analysis of USB flash grives in a virtual environment. *Small Scale Digital Device Forensics Journal*, 1(1), 1–6.
27. *YouTube*. Computer forensics: recovering deleted files with Encase. URL accessed December 22, 2011. Retrieved from http://www.youtube.com/watch?v=33HS50gQOEQ.

28. Porn Detection Stick: as seen on TV. *Proof pronto.com*. URL accessed December 22, 2011. Retrieved from http://www.proofpronto.com/porn-detection-stick-by-paraben.html.
29. *Gizmodo*. Retrieved from http://gizmodo.com/5484601/the-porn-detection-stick-is-like-the-hot-tub-time-machine-for-smut.
30. *Imation*. URL accessed February 25, 2012. Retrieved from http://www.imation.com/Global/en-US/Mobile%20Security/Products/Secure%20Mobile%20Data/%28rev2.10.2011%29Defender%20F200Bio.pdf.
31. NIST. Test Results for Digital Data Acquisition Tool: Tableau Imager (TIM) Version 1.1 Retrieved December 23, 2011 from http://www.nij.gov/pubs-sum/233984.htm.
32. Computer forensics tool testing program NIST. URL accessed December 24, 2011. Retrieved from http://www.cftt.nist.gov/.
33. Doherty, E., Liebesfeld, J., Purdy, G., Liebesfeld, T. (2008). *Computing and Investigations for Everyone*, p. 231, Authorhouse, Indiana, USA, ISBN 978-1-4343-7231-4.
34. Kessler, G. C. (September 2001). Steganography: Hiding data within data. URL accessed December 24, 2011. Retrieved from http://www.garykessler.net/library/steganography.html.
35. Higgins, K. J. (June 29, 2010). Busted alleged Russian spies used steganography to conceal communications 'deep-cover' Russian intelligence agents hid electronic messages behind computer images. In *darkReading*. URL accessed February 25, 2012. Retrieved from http://www.darkreading.com/security/article/225701866/index.html.
36. How to hide secret message and codes in audio files. *Lifehacker*. URL accessed December 25, 2011. Retrieved from http://lifehacker.com/5807289/how-to-hide-secret-messages-and-codes-in-audio-files.
37. Admin (September 7, 2010). Thwarted Russian spy ring communicated using steganography. *Vincents forensic technology*. URL accessed December 24, 2011. Retrieved from http://www.vft.com.au/computer-forensics/thwarted-russian-spy-ring-communicated-using-steganography/.
38. Jill, B. (September 28, 2010). Blackjack USB flash drives ready for government cyber security. *Allusb*. URL accessed December 24, 2011. Retrieved from http://blog.allusb.com/2010/09/blackjack-usb-flash-drives-ready-for-government-cyber-security/.
39. Betinson, M. (December 8, 2011). Two thirds of lost USB sticks are infected. *PCR*. URL accessed December 24, 2011. Retrieved from http://www.pcr-online.biz/news/read/two-thirds-of-lost-usb-sticks-are-infected/027623.
40. Kennedy, C. (2001). *An Introduction to U.S. Telecommunications Law*. p 171 Artech House, Boston, MA, ISBN 0-89006-380-X.

CHAPTER 8

Places to Work at Investigating Mobile Devices

INTRODUCTION

There are many places to work as an investigator of mobile devices. There are contractor positions with the Department of Defense and private investigation agencies; one can own a consulting business; and there are corporate as well as IT positions. Do not forget being a researcher, academic, and developer of mobile device training courses. More opportunities will emerge as time goes by and more investigations may need to be done on mobile devices. Approximately three out of five people on planet Earth own a cell phone. Some of these people also have PDAs, secondary cell phones, digital organizers, netbooks, palmtops, and many other devices. Many of these devices will have data that is related to activities associated with crime or infractions of policies in a corporation. The number of mobile devices that will need investigation or checking and the circumstances for these checks will become too numerous to discuss as these devices become more and more part of our activities of daily living.

Whatever type of career that one selects that is associated with mobile device forensics, that person will need a variety of certifications such as the CISSP, ACE, CCE, and Encase certified, just to name a few. These certifications are to demonstrate a proficiency in all aspects of preserving, collecting, analyzing, and reporting on digital evidence with tools and methodologies accepted by one's peers and the legal system. Degrees and certificates will be needed to show benchmarks in the journey or acquired knowledge. Many people without degrees and certifications will sometimes have the same knowledge and ability to do mobile device forensics for their employer or customers. However, the work may not survive the scrutiny of the courts or academia if the person does not have the correct certifications. It is also important to have the American Society of Crime Laboratories Director's Certification (ASCLD) for the digital forensics lab or else the work of high-profile cases may be challenged.

As a person chooses his or her career path, he or she should think about the type of work environment and subject matter one wishes to work with. If one likes working on cases that involve matters of national security, than a DoD lab or the lab of a federal agency is the place to work. One will have the best of tools, access to great training, and work with great minds on famous cases. This may allow one to transition to some type of high-level consulting group later such as the Carlisle Group. This is good for an ambitious hard-working intelligent person. However, high-security labs may require one to take polygraphs on an annual basis and have to fill out forms that seem intrusive on one's privacy. Vacations to certain parts of the world may not be possible for security reasons and associations with certain people may have to be self-restricted to avoid finger pointing as a possible security leak.

Working at a digital forensics lab for a loss-prevention lab is still important, but will have less pressure and security than that of a lab doing military and national security cases. Many of the cases will involve small groups of thieves stealing products to finance drugs, drug habits, and perhaps some associations may be to organized crime or drug gangs. The environment will be strict but the tools and budget may not be as great as the federal labs. However, with that being said, there are some exceptions; the digital forensics lab for Target is known to assist many law enforcement agencies at times with pictures such as license plate magnification and clarity [1].

PRIVATE INVESTIGATORS

Many private investigators work from home or on the road. They may have a small office to collect mail or perhaps have someone sitting there to answer the phone, coordinate a calendar, and arrange meetings with clients and potential clients. Some of my students are private investigators and enjoy the freedom that they have to work whatever hours they want, and take off when they need to. Private investigators assist clients in finding lost relatives, locating people to give them inheritance, or to see that they are notified of financial obligations. They also assist companies and the public by finding out about false workmen compensation claims which can prevent those resources from being available to those who really need and deserve compensation. Private investigators also do the traditional type of fidelity investigations that we may see on television. They may also investigate personal computing devices that were surrendered to security professionals during an investigation of suspected mass shoplifting known as flash robbing. Private investigators may also be employed in the casino industry to check the mobile devices of employees who may be suspected of working in teams with patrons to have an unfair advantage against the casino and steal money. Private investigation could be an exciting career for someone who has an inquisitive mind, enjoys mobile device forensics, and has the research and writing skills of a reference librarian.

When we think of private investigation, we may think of a person hired by a spouse to confirm that a husband or wife is not faithful and is having an affair. The first reported case of a man spying on his wife was in 1884. Mr Cane hired nine detectives to perform surveillance on his wife who occasionally visited brothels [2]. The results of the activities resulted in a case of *Cane vs. Cane* in a chancery court [2]. Private investigation still includes that sideline but is much more than that. A person may get strange calls on a cell phone in the middle of the night. Perhaps there are recordings with voices and gunfire. The person may take the phone to the police department and speak to a detective, or he or she may wish to keep it quiet and hire a private investigator. The private investigator can use a service such as Intellius to do reverse lookups and find out who is calling. The private investigator may also find out what relationship the caller has with the person who retrieves the phone. The previously discussed situation happened to somebody I knew. The caller never used the person's name. Though it could not be proven, the intelligent guess was that the person I knew received a cell phone number that may have previously belonged to a gang member. Some of the calls were described by the private detective to be butt calls. This is a street term for a situation where the phone is unlocked and calls either a recently called number or someone randomly from the phonebook. The *Urban*

Dictionary says, "A Butt Call is an unintended cell phone call made by sitting on or otherwise putting pressure on the touch keys" [3].

PRIVATE INVESTIGATORS: SKIP/TRACE

One of the areas of private investigation that people specialize in is called skip tracing. It is here that people look for others who might have been roommates or life partners who took off without leaving a note, paying their share of the bills, or saying if or when if they would ever be back. With skip tracing, one may start with the examination of commonly owned cell phones and computers. Perhaps there are numbers of future landlords, new employers, or someone new they met on the Internet. A phone number can be reverse lookupped to see where the owner of that account lives. A phone number to a real estate broker in Huntsville, Alabama, for example, could indicate relocation to an apartment in that area. There is often some form of communication in the electronic devices around the home. Those devices could be PDA, cell phone, netbook, laptop, IP phone, or some piece of evidence that was not printed because the printer ran out of paper. If the person has a thermal fax, then all the documents that were sent out or received can be seen on a large roll if one removes the ribbon and puts it up to a light source.

Skip trace specialists may also help locate dead beat dads who owe child support [4]. They may find tenants who did not pay the last month's rent. There is a whole chapter in Steven Brown's book on private investigation that concerns finding runaway teenagers. Private investigators may be working in tandem with police on that one. Since the mobile devices are owned by the parents and the monthly bill is paid by them too, looking through any device left in the living room or family room by the teenager should be no problem. Skip trace specialists can often perform a great service to the person they are seeking when money is distributed to long-lost relatives. I knew a man who hired a private investigator to find his mother-in-law's ex-husband to leave him a large sum of money. The man was living on an Indian reservation with no phone or Internet service. Some families get separated by war and relocate. Skip trace specialists can reunite families. This an area of private investigation that requires one to know how to use online search engines, reverse lookups, and various investigation tools such as Lexus Nexus. A working knowledge of how to investigate mobile forensic devices is good too.

Many private investigators also help with insurance fraud investigations. There are many times where people claim they hurt themselves at work with a back injury. Then the person with the back injury may be photographed in his yard hoisting up an engine block that he will put in his antique muscle car restoration. A company cell phone, PDA, netbook, or desktop computer may be brought to the private investigator for examination. The PI may use tools such as X-Ways Forensics to image the drive and look for key words with regard to insurance payoffs. There may be chat room conversations remnants about how sad it is, that the company fell for the scam, and other damaging evidence that can be used by the prosecution.

Private investigators may also be asked to investigate employees in a casino. Many high-trust industries may have an agreement between employees and employers that personal mobile devices are subject to inspection. As long as polices are signed and people agree to this, surrendering a device for examination is no problem because everyone knows and agrees that there is no expectation of privacy. There are times where employees who are suspected of cheating and working in teams with patrons must submit their mobile device for inspection.

Some private investigators may be called to investigate a commonly owned cell phone by a husband and wife. Perhaps Tiger Wood's wife could have taken Tiger's cell phone to a private investigator to find out how many mistresses he had. Brett Favre's wife could have called a private investigator, commonly called a PI, to find out if the rumors were true about her husband sexting and sending pictures of his "private part" to a former cheerleader named Jenn Sterger. PIs do not just look into cell phones for investigation; they look into other mobile devices such as PDAs or netbooks. Some wives may also hire mobile forensic investigators to see if their husbands are having a virtual affair in an online environment such as Second Life. "The virtual world provides a place for individuals to create an avatar and engage in most everyday activities, including attending concerts, conducting meetings, meeting new friends, and apparently having virtual extra-marital affairs" [5]. One such virtual affair led a woman named Ann Taylor to divorce her real-life husband David Pollard [5].

Private investigators often do background searches on people for new jobs in the health care industry, daycare work, investigative positions for insurance companies, and for admission to colleges and universities. It is unfortunate that teenagers use mobile devices to post pictures of themselves to social networking sites where they are drunk, passed out at a party, or with others while engaging in some type of deviant behavior. PIs will often find these pictures and put them in a report for the customer requesting it. Such reports will often result in a denial to a prestigious university or position in a company. However, sometimes there are fake social networking pages made and it is so prevalent that Christiano Renaldo wrote a page on the subject [6]. Private investigators may be hired to prove that a social networking page was faked and clear a person's name. It is very easy for a mobile device to get lost and fall into the hands of a prankster. An investigation might show that all the pictures were posted after the device was reported to the phone company as stolen. Various log files might show that the IP address of where the pictures were posted were in Ohio and the subject lives in New Jersey and has not been to Ohio. The metadata in the pictures may also show that the pictures are not genuine and were created with a photo editor. PIs can do great service in helping clear a person's name.

A classic timeless book to help you get started is named *PI, A Self Study Guide on Becoming a Private Detective* [7]. The book discusses office equipment and supplies as well as the education requirements to get in the field. It also discusses the licensing requirements for each state since they vary considerably. The book discusses some of the lesser-discussed important considerations such as marketing your business, dressing the part, and what the pitfalls are to avoid. It is important to think about digital forensics as a scientific investigation to answer a hypothesis such as "Did this phone send a SMS harassing message to my client?" A classic book that an investigator gave me is called *The Art of Scientific Investigation*. This book helps one think logically, form hypothesis, test them, and decide what is true [8]. *Business Background Investigations, Tools and Techniques for Solution Driven due Diligence* by Cynthia Hetherington is a good book for learning about background investigations.

LAWYER

Working for the legal community can be an exciting way to make a living as one may work on famous cases in the news. One may also discover the potential dangers in social media such as posting pictures with GPS data embedded in it. This could potentially be used by stalkers. As a person in the legal field, you may discuss it with your employer and then seek to notify the legislature to pass a bill requiring cell phone camera resellers to notify customers about such a hazard. There are many areas of mobile forensics that one may be employed within the legal community. A person could be a document reader who is employed to read email, documents, and spreadsheets. A person could be a defense lawyer specializing in digital evidence and seizure.

Working for a law firm is one option for those seeking employment in the field of investigation because they can work under the lawyer's license to help him or her investigate the case. I have been approached by many lawyers at seminars, symposiums, and conferences and asked if I wish any part-time work. Lawyers who deal with divorce often have customers who are involved cases where husband and wives are sending each other nastygrams via SMS messaging. Digital cameras commonly owned by both parties may show activities with a love interest of one of the partners who is in divorce proceedings. Digital devices may show activities that reveal one partner as unfaithful or communicating abusively.

The divorce rate in the United States is very high. Each of those people getting divorced will need a lawyer. Each one of these couples probably has digital devices that need to be investigated which holds pictures, SMS messages, or email to support some assertion that they made. That is a lot of potential work. A teenager may run away because of abuse, and then may seek emancipation from the parents. Perhaps the grandparents wish to adopt the child. Digital devices often hold evidence of abuse and may play a part in the emancipation and adoption hearings.

People may like working for lawyers because they know the law and will probably do a good job of protecting both their clients and those who work for them. It is easy to do something that puts oneself in harm's way when doing investigations. Please consider the "silver platter doctrine" again that we discussed in a previous chapter. Many of us are good citizens who want to help the police bring criminals to justice. If the police ask us to keep investigating the matter in our corporation after the silver platter doctrine was invoked, and then give them a call if we find something, that may be interpreted by some as violating someone's Fourth Amendment rights. That may put us at risk for a big lawsuit. If we work for a lawyer, he is going to want to avoid getting sued and supervise his employees and warn them. There is a certain amount of protection, advisement, and direction that one gets when working as an investigator for a lawyer. However, it is possible that one works for an unethical lawyer, so one should always get his or her directives in writing so that the employee is not the "fall guy" in an investigation. If one always keeps written documents concerning directives, then the employee can produce a document that the lawyer/employee ordered that task to be done by the employee.

Many people also like working for a lawyer because they get exposed to a wide variety of opportunities to earn money. I knew a private investigator who worked for a lawyer and would serve subpoenas before taking his wife to dinner. The serving of subpoenas was another area of the law that he learned about and it allowed him to purchase many dinners with his wife. Some people who work for lawyers may also make some money as document readers going through volumes of material in emails, computer, cell phones, and other electronic devices.

LAWYER: DEFENSE LAWYER SPECIALIZING IN DIGITAL EVIDENCE

A person might want to go into a county college computer technology program and then get an associates degree in such a technical field to get the understanding of how the technology works. The same person might enjoy getting one's hands on the technology to get practice installing networks and fixing computers. He or she could take classes on programming, file systems, and computer hardware. Then, one could choose a four year college and take a liberal arts degree to learn about English, philosophy, sociology, and logic. One could learn how to write, develop ideas, and present them. One could also take classes about history and then learn to present an article, debate, as well as write reports on nearly any academic subject. The youth might then join a debate club and learn to argue a set of historical points as with the Lincoln–Douglass debates. Then one might take the LSAT and apply to law school. It has been said that law school in New York City can be done in a few years of night school. After taking the bar exam, one may then progress into a firm of lawyers who defend people who were wrongly accused of cybercrimes. Some lawyers charge two hundred dollars per hour. It could be an exciting career as well as a rewarding one.

Perhaps one may specialize in mobile forensics and challenge law enforcement's handling of the evidence. The defense lawyer could discuss the Fourth Amendment, the chain of custody, and concepts of privacy discussed in the Virginia State Constitution. It might be exciting for some to blend the topics of digital evidence with legal concepts discussed by James Madison, George Mason, and Elbridge Gerry. This could be a very good career path for young people who want to blend liberal arts and technology. Since the law and technology are both ever changing, it would be a dynamic career, and never stagnating. After a long career, one might go into teaching at a law school or university and embark on another career. The defense lawyer who specializes in computer forensics or mobile device forensics could also later become active in politics. Once a person learns about politics, he or she may be recognized for his or her great understanding about the law, technology, and become appointed to a policy-making role in an organization such as the Department of Homeland Security position. The possibilities are endless.

SURVEY OF TOOLS COMMONLY USED BY LAW OFFICES AND THEIR CCE CONTRACTORS

Some law offices perform email investigations and use a variety of tools. They may be defending someone who was accused of sending a sexual harassment email or defending someone accused of selling trades secrets to another country. Here is a survey of tools that may be used to recover and examine email that was on an image that was collected by the computer forensic professional or mobile device forensic examiner. The following tools might be used with some netbooks and laptops.

ADVANCED OUTLOOK REPAIR

There are many times when digital evidence examiners must recover email from Microsoft Outlook and read through emails for a corporate investigation. Microsoft Outlook keeps its email in a file with an extension of pst. The pst file often gets inaccessible if it gets too large. Microsoft Outlook 2003 has a limit of 20 GB and if the file exceeds that, you cannot open it [9]. The incident response team would first image the person's hard drive and take the forensic image back to the office for examination. They might have to repair the email program's mailbox file in a virtual machine before they could examine the email content. That can be challenging and exciting. Perhaps the incident response team wanted to check the emails for a policy violation. This would be OK since there is no expectation of privacy in a corporation. Everyone usually signs policies affirming this. The incident response team might just have to export some of the emails and folders to another pst file first. Once the Outlook 2007 or Outlook 2003 pst file is less than 20 GB, it can be opened up and the emails can be examined. Some law firms, private investigators, and corporate security departments use document readers to go through all the email and attachments. There is often too much to read and highly paid staff cannot be used for reading through as much as 50 GB of email on programs such Outlook 2010. That is why document readers are paid low and often used to read lots of items on the suspect's computer. In Outlook 2010, an examiner can export folders and email to another pst file quite easily.

However, sometimes the pst file gets damaged and cannot be read. Then it is time to use a tool such as Advanced Outlook Repair which claims to have the best recovery rate of 95.70% [10]. Large files can easily get corrupted as clusters and sectors get marked as bad. The program uses an advanced algorithm to scan the file and recover as many emails and attachments as possible. Data Numen Incorporated makes this program and one can purchase it online. In the early 1990s, one used to have to call up software vendors, find the price, cut a purchase order, and one might have the software a month later on CDs or floppy disks. The package of software came by mail or United Parcel Service. Investigations often took a long time. Now, with high-speed broadband Internet, it is just a matter of clicking on the "Buy Now" button, putting in a name, address, phone number, and credit card number, and then downloading the tool. Investigations in a corporation or law enforcement are much faster than the early 1990s because there are many tools available almost immediately by broadband Internet download from vendor websites.

PST RecoveryPro

Many examiners use PST RecoveryPro for recovering email from corrupted pst files. Not every tool can successfully solve the problems because some problems such as a read/write head hitting the platter and causing a hard drive crash, or a wiping of a disk are beyond the scope of any tool. PST Recovery can often help recover files when the examiner tries to look at the email and receives errors such as 0×80040116, 0×80040119, and 0×80040600 or if one get a nebulous error such as "An unknown error occurred." It is also good to use this program when one reaches the file limit of 20 GB for Outlook 2003 or Outlook 2007 and the file is no longer accessible. It may also help when the Outlook. pst folder just seems inaccessible. The program is also good for recovering corrupt files. It is a good program to have in one's toolbox for email investigations. Many certified computer examiners (CCE) do not rely on one tool but use many tools. Some tools recover more data than others and using many tools is a good part of a tool verification plan. What would one say in court if the defense lawyer asked, "Did you use more than one tool and look for exculpatory evidence to find my client innocent?"

OUTLOOK PST RECOVERY

There is another mailbox recovery tool for Microsoft Outlook 2000, 2002, 2003, 2007, and 2010. It is called Outlook PST Recovery and can be useful when a pst file gets damaged or corrupted. It also a useful tool for occasions when the file is too large and cannot be opened. The PST file contains many things that are useful to the investigator besides email. There are tasks, notes, folders, contacts, and journals. This software tool has a free trial version that is in a downloadable file of only 4.5 MB and can be run to recover the pst file. It will show what is recoverable and to complete the process one only needs to purchase the product registration. It is suggested to have at least 50 MB of free space for the program to use. It also works on a variety of platforms such as Windows 7, NT 4, Vista, Windows XP, and Windows 2000. PC Help Soft is the company that makes and sells the tool. The website shows 10 industry awards from *PC Magazine* and other prestigious places. This seems like it would be a good tool to run on the examination machine in its own virtual machine to recover PST files.

RECOVERY TOOLBOX FOR OUTLOOK

This is another tool that is good to have in the computer forensics toolbox of email investigations. The tool was available for download and can be used for Microsoft Outlook pst files. The pst file was created by the Microsoft Corporation and many computer technicians say it is an acronym for a personal information store file. It is advisable to prepare a few basics such as the definition of a pst file and when it was first on the market in case one has to testify in court. This toolbox is very interesting to me because it can take the ost files, offline storage files, from Microsoft Exchange and convert them to pst files to view. It would be good to discuss the motivation for this file conversion with the general counsel in the corporation so there is no spoliation of evidence. Perhaps the corporate investigator only has tools for the pst files and it would otherwise be impossible to examine ost files. This should be discussed before conversion and examination. One of the good things about the Outlook recovery tool is that it can recover Outlook Messages from Outlook 97, Outlook 98, and Outlook 2003. This is important because many people have older laptops in the home or corporation that have email programs from the late twentieth century and they could contain information about developments between people that later turned into something that needs investigation. Recovered emails from old corporate or defense contractor laptops could also show relationships between people in different parts of the company who are supposed to be separately working on parts of a top secret project. The recovery of email is important and it is important to have tools that can recover both old and new email. The Outlook recovery tool can also recover pst files from Outlook XP, Outlook 2000, and Outlook XP SP2. The tools run on a variety of platforms and sound like a good addition to the computer forensic toolkit.

ScanPST

This is a tool that many people first try when email cannot be accessed. Scanpst.exe is an executable file that is the Inbox Repair Tool for getting back folders and email from offline folders .OST files and .pst personal files. When Windows is installed, some files are marked as system and others are marked as hidden. ScanPST is a hidden file that is installed with Windows and could be the first line of defense to recover corrupted files and folders. Many people like it because it is free. However, often more powerful tools need to be used. A chart from DataNumen Incorporated gives ScanPST a low recovery rate of 1.26% as opposed to Advanced Outlook Repair having 95.7%.

PST WALKER

The PST Walker software is another important tool that should be in the toolbox of computer forensic professionals who have to recover Microsoft Outlook and Microsoft Exchange emails and attachments. The tool can also be used for repairing, viewing, or searching ost and pst files. A search tool is very important since so many people have tens of thousands of emails. It could cost a fortune for a document reader to go through everything. A good search tool is very important. Then the documents that have certain key words can be viewed, printed, or exported to another file for review. This seems like a good tool to run in a virtual machine with the suspect's hard drive image, and this way nothing is permanently changed. Another good thing about PST Walker is that is has a newsletter and support options. Being able to see what others are doing or to be able to ask a question is very valuable. The newsletter has simple straightforward talk that could be very useful to a digital evidence examiner. One of the key features of PST Walker is that it can open password-protected files without a password [11]. It can also work with Outlook 97–2007 formats and all types of encryption [11]. The program is often good at finding permanently deleted items and it can process files that went over the 2 GB limit. It can also find the orphaned item. Orphaned items mean that the item is not associated with any folder of Microsoft Outlook [4]. PST Walker also helps with one other item that is useful to a digital investigation, namely the simple mail transfer protocol (SMTP) header information. The SMTP header often contains the routing path of the email, including an IP address, source email address, message number, and the time and date stamps. Sometimes some of the SMTP header information is obfuscated. People can use a remailer such as Hushmail which hides much of that information.

KERNAL FOR OUTLOOK

This tool has a free demo version. This is great because one can try it out before buying. This tool helps recover email that was from corrupted pst files. It is also good for password-protected pst files. The program works with a diverse group of Microsoft Outlook programs which is important. It works with Microsoft Outlook 97, 2000, 2002 (XP), 2003, 2007, and 2010. The free demo version uses an algorithm that is considered fast. It also shows the number of items and what will be recovered in the full paid version of the software tool. It is possible to include drafts, contacts, task, notes, and calendars. This could be another good tool to put in the computer forensic toolbox.

INTELLA: EMAIL INVESTIGATION TOOL

The Intella email investigation tool is quite a time-saving tool for computer forensic examiners because it speeds up the e-discovery process. Intella can use its analytical engine to examine SMTP headers and contact lists so that relationships among emailed people can be graphed. Reviewers may also read certain emails and annotate them for the manager. Consider this fictional example of an intellectual property theft case where it is important to show relationships between a trusted employee with access to trade secrets, and employees of a rival research and development firm. Perhaps the tool could be used to generate a graph that illustrates the roles of people that follow a pattern typical of those engaged in intellectual property theft and resale. There may be emails with the trusted employee and other people who are known for illegal gambling operations. There may be emails with mortgage companies and credit card companies asking for money. Then the emails can be read with the viewer and details of resulting correspondence can be obtained to show the roles of each person. The graph could indicate a person who has a need for lots of money quickly and engaged in illegal operations of selling trade secrets to competitors to obtain money to avid threats from people who will collect at any price.

There is version of the product called Intella Team which appears ideal for a large law firm. Intella Team would be good to use on a case such as our fictional example. It would work well with the case on a storage area network and a group of people sharing the case, reviewing it, and making comments. Intella Team is slightly less than $7000.00 but greatly saves a lot of time and allows the graphing of relationships and lets investigators quickly see who the main players are in an investigation. The Team version of Intella also allows collation of data that exceeds 250 GB and can be used on a local area network (LAN) for group reviewing and annotating a case. Intella Team has two large-scale pieces [12]. The first piece is the main one called Intella Team Manager. This piece prepares and indexes the evidence and allows its sharing among team members. This piece also allows for the approval of work performed by the reviewing team members after the manager reviews their work. Intella Team Manager also allows for the combining or perhaps merging of team members' work.

The other component of Intella Team is Intella Team Reviewer. There are three reviewers that work under the authority of the Team Manager. Each reviewer can do work by themselves that was approved by the Team Manager. That work may consist of searching case data, filtering it, bookmarking it, tagging it, and commenting on the case [12]. The software is said to be extremely easy to use and lets people focus on the work and not learning how to use the tool. This seems to be a great tool to organize data visually with the main players labeled. The graph can also show links to each player and then display annotations concerning actions and relationships. It would also seem to be a good tool to make simple charts explaining a case to a jury with exemplars of emails, attachments, and time stamps. The report could be focused on in detail to give the jury key points. It could be considered an email analysis tool, data collection tool, report writing tool, and evidence presentation for court tool.

There is also a small one-person version of the tool called Intella 10 and Intella Viewer. These two programs work together to allow one person the ability to organize, analyze, and prepare graphs on email pst files of 10 GB or less. This might be a great email investigation tool for older laptop investigations that have Outlook 97, 98, and 2000. There are other versions of Intella that are based on the amount of email that may be processed together in one session. The cofounder and partner of Intella, Peter Mercer, showed this product to many lawyers, private investigators, law enforcement types, and academics at the 2009 High Tech Crimes Investigative Association Conference in Hong Kong.

OmniX EMAIL ANALYTICS

Email analytics is an analytical tool that allows the digital examiner to identify people of interest from a group of email traffic. Email analytics can also be used to see their activities by a timeline. The tool is made for lawyers, corporate investigators, and law enforcement personnel who may have to create a report on a certain person's activities in a certain time period or show certain patterns of behavior. The reports can be used as an important communication tool for hearings, inquiries, depositions, and investigations [13]. The person of interest can also be illustrated as the center of a wheel graph with spokes going out to other parties. It is possible to get a quick picture of how often certain people communicate with the suspect and possibly identify patterns of behavior. Patterns of behavior can be things such as seeking advice or approval from a spouse, boss, parent, or some authority figure.

In a corporate setting, it would be easy to create reports that illustrate regular patterns of employees engaging in personal email and avoiding work. This email policy violation time might be before the boss gets in each morning. The tool could also be used as an intelligence tool. Perhaps there may be more email traffic among employees if a strike was about to take place. Excessive traffic could be a sign of planning. There might also be red flags indicating corporate espionage if excessive documents were emailed to competitors and then deleted by the suspect. Such behavior suggests trying to cover an audit trail. The tool is similar to Intella by Vound in that it helps organize voluminous email and create customized reports to support an investigation.

AMICUS ATTORNEY

This is a project management program for corporate investigation firms or lawyers who must manage an investigation and bill people. The program is also good to show clients, managers, and others exactly where the investigation is at and what the present cost is. Emails, calls, and preparation time can be tracked for both billing and to assess manpower. The program could also be used to manage intelligence projects. This program has a trial version which can be downloaded and tried by potential customers. The calendaring feature is very good. In the past, many people would draw Gant charts manually to project timeframes for phases of projects to be done and then make estimates of investigation completion dates. This is a good tool to use to apply the principles of project management on an investigation. Such topics are explored further in the book *Intelligence Led Policing* by Jerry Ratcliffe. Project management is an area of mobile device forensics that needs more attention. It is important to be able to forecast the use of resources, as well as track and control costs, throughout in an investigation. It is important to be able to break an investigation into components that may be assigned to various people for completion, and then create a timetable in case something has to go to court.

MICROSOFT OFFICE PROJECT 2010

Many people might disagree that Microsoft Project 2010 is a forensic tool, but it is a tool to manage the project of a forensic investigation. An investigation of a policy infraction such as email nastygrams and harassment at work may start out as a series of phases. The first would be to collect evidence. One could use the software to create a Gant chart and assign people such as document readers, interviewers, computer forensic technicians, and lawyers to each phase. Phase one, the collection of evidence might take 1 week. Phase one might consist of the authorized requestor (AR) initially dispatching an incident response team to the suspect's computer, imaging the suspect's hard drive, and filling out a chain of custody in case it becomes a criminal case. Phase could also include human resource (HR) personnel interviewing the victim, the suspect, managers, and other supporting actors. Phase two could be called "Analysis." It could be graphed and assigned a time span. During this phase, a tool such as Intella Team could be used. The AR could use Intella Team to assign various parts of the investigation to document readers, the general counsel, and HR. Each of the members of the team could use Intella Team Reviewer to annotate exemplars of harassing emails located from the pst file or ost files seized from the image. The third phase of the project might be writing a report and getting it ready for review. Phase four might be punitive action or turning it over to law enforcement for a criminal matter with the silver platter doctrine. In the end, the amount of manpower and costs can be tracked and controlled. The project that is not managed with project management techniques can quickly get out of control with costs, human assets, and other finite resources.

Aid4Mail eDISCOVERY SOFTWARE

Aid4Mail eDiscovery was created by Fookes and is used by law enforcement, corporate investigators, private investigators, intelligence agencies, and lawyers [14]. This program is useful because it allows for the accurate conversion of email from a proprietary format to a standard format such as a pst file. The converting of file formats allows for more flexibility with tools so that a law firm can use their own tool such as Intella Team. Outlook Express version 4 files would be in .idx and .mbx file extensions and could be changed to pst files. Many of these files and programs could be sitting on laptops. Outlook Express versions 5 and 6 would have .dbx, .eml, and .nws file extensions. These files could also be converted to .pst files. Onyx Email Analytics or Intella Team could be used to analyze the resulting pst files.

There was an era when people heard about all these viruses directed at Microsoft Outlook or Outlook Express. The resulting behavior was that many people switched over to other systems such as SeaMonkey, Eudora, Apple Mail, Opera, and many others. Aid4Mail can help change the format of dozens of proprietary formats to a variety of others formats that investigators can import into their investigation tool. The other challenge is that some people with netbooks or laptops may not always download all their webmail onto their laptop or netbook. It may be necessary for investigators to seek permission to obtain the email from the webmail client. The other problem is that many people have a plethora of email accounts. Some may be personal while some may be for work. There may be also a mix of both personal and work-related email further complicating investigations. There is a book called *Technostress* that suggests that people should keep business and work separate. Even those who work at home should have a separate office at home for work and create borders so that their whole life does not become work related and impossible to separate from their personal life. What people should or should not do is an academic issue. The fact is that people use netbooks, laptops, and mobile devices for both business and personal use. Further complexity is added by the fact that people purchase these devices and use them in both personal and work workspace thus blurring borders. It should become apparent that every corporate, law enforcement, military, or intelligence agency needs a good lawyer on the staff.

OUTSOURCING THE EMAIL INVESTIGATION

There are times in a corporation where the AR may realize that they do not have the tools, the properly trained investigators to use those tools, or the necessary legal expertise. It may be time to outsource the email investigation to a qualified investigation agency that is insured. The corporation might also want an outside vendor to conduct the investigation in order to appear impartial, especially if the investigation is politically charged and high profile. The corporation also transfers the responsibility and liabilities associated with the investigation to the outside agency. This would please the insurance company of the corporation where the object of the investigation is being conducted. There are many service companies that conduct computer forensic and email investigations for a fee. Some are international and employ people who read many languages. Vogon Forensic Investigation Services can conduct a variety of investigative services regarding corporate policy violations. It can sometimes be cheaper to let someone else do the investigation rather than conduct it improperly and risk bad publicity, decreased consumer confidence, and possible litigation from the subject of the investigation.

ONTRACK FIRST VIEW

This software is made to work in conjunction with Kroll's Ontrack Power Controls. Together they are useful for recovering a corrupted file containing email and analyzing it. There are many tools that can do key word searches but what makes Ontrack First View special is its "conceptual searching." One can type in a concept as well as key words and it will use a type of analytic algorithm to produce other terms or concepts that might be of interest and then apply them to the large volume of email being searched. That might produce more results than programs that just doing straight key word searches. The results might also be more meaningful and allow the investigator to find email that might otherwise be missed. Many companies experience a plethora of email being received daily. Employees of large companies also send out a plethora of email daily. The intelligent search engines are a must in many situations where there is voluminous email and only a few document readers and investigators with limited time to look at evidence. The software also creates good visuals and reports for use with a jury. The visuals could also be used to demonstrate relationships between certain parties. The graphs could also be useful in simplifying the results of reports to demonstrate concepts to jurors with limited education. Besides having a good analytic engine for finding relevant email, there is an advanced visualization feature that gives what is commonly known as a "10,000 foot view" or high-level view revealing connections that were not necessarily apparent using other analysis tools.

EMAIL DETECTIVE: FORENSIC SOFTWARE TOOL

The Email Detective Forensic Software Tool is made and sold by Hot Pepper Technology. This tool is very useful because it allows the investigator to capture the AOL email and its embedded components. Many times examiners can capture the email but lose all the little pictures and embedded objects that go with the mail. That can be a large loss in an investigation because much of the important content is lost and some of these objects help show the context of the email. It is certainly possible that pictures that might even include child pornography that could be emailed from some email system to someone in AOL and become important illegal evidence in a case. There could be some embedded objects and embedded links that show one was rebounded to an adult site where driveby downloads occurred. Adult as well as child pornography could have downloaded on someone's computer due to the drive download thus showing exculpatory data. Thus, programs such as email detective could become very important in helping capture all the email, pictures, and embedded objects that sometimes get lost with certain tools. Perhaps it could help prove someone innocent.

FINAL eMAIL

The URL for Final eMail is http://finaldata2.com/products/products_finalemail.php. This tool is another helpful item that one might find in a law office where computer forensic professionals, lawyers, and eDiscovery personnel work together.

YOUR OWN BUSINESS

It is very important to ask yourself a few questions and see if you are willing to do what it takes to rise in your field. These are not items that take great intellect but take great courage and will dramatically affect one's life. Am I willing to get in shape, purchase new clothes, shine my shoes, and always look like I am ready to go to court or a chief of staff meeting? Am I willing to move throughout the United States to any location that advances my career significantly to the next level? If I have a business, will I wear the best clothes, have an expensive orderly office in an expensive neighborhood, and project a professional image that will get me the best customers? If I work for people who limit my career because of personal factors, or because of a lack of resources, am I willing to move and change jobs? One will meet people who fall short of their potential because they were comfortable in a community and perhaps did not want to take a risk, move their family, or take a more challenging job. You have to decide who you want to be, nobody can do it for you.

This may be an option for many people because it offers the opportunity for flexibility, independence, travel, and making considerable money. One may choose to resell many of the mobile device forensic tools such as Mobil Edit, XRY Forensics, Device Seizure, Susteen Secure View, Cellebrite, Cell Dek, and a variety of other products. Reselling is good because people often like a local person they can go to with problems, concerns, and to compare one product with another. The reseller may not make much profit from the reselling a product but can profit greatly from running training classes and offering consulting services with those tools.

Small business owners may often partner with a university or government agency for small business innovate research (SBIR) grants. Perhaps a local municipal police department has a problem investigating mobile phones because the criminals shove a screwdriver in the mini USB connector. As a small business owner with a computer science background, you may have an idea of how some type of wireless connectivity could be added to the mobile forensic investigation tools. Then you could seek a grant to direct the companies to make the changes and then test it with a university and the police. This is exciting work and could be profitable.

As a business owner, you may decide to sell your services to the Saudi Royal family and help them develop mobile device forensics training with their police. You may partner between an American university and a Saudi university to help them develop a mobile device forensics lab and training program for their country. This might not only help another country investigate cell phones but also help advance the technical part of their justice system and promote more security. It might also be very profitable. When you have your own business, you set the limits of what type of company you wish to have. You may want to be international or just local. One of the big considerations before going international is if the technology is exportable. Device Seizure, for example, can be sold to Saudi Arabia. The United States Commerce Department sometimes has prohibitions about certain computer processors being exported from the United States. They should be contacted.

If you travel overseas you should check in with the American embassy in other countries and give them a copy of your itinerary and contact numbers for friends, business contacts, and relatives in both countries. It is also good to check the State Department website for warnings or concerns for the country that you are going to visit. There is an Overseas Advisory Council (OSAC) that gives good security advice. In some countries, one may wish to avoid low floors, rooms with windows that overlook a street, and one may wish to be on an upper floor near a stairway in the core of the building.

When you have a business and go international to do computer forensic or mobile device forensic jobs, you should have the original receipts as well as copies for all equipment going overseas. If you do not, you will have problems at customs and border patrol in the country you are going to, and then in the United States when you return. You may be expected to pay duty fees and some equipment may be questioned as spy equipment. The sending of equipment to other countries and then returning is known as a boomerang shipment. One should get what is known as an ATA CARNET [15]. This is a set of forms that streamline a shipment of equipment that will be used in one or more countries outside the United States and will return back in 12 months or less. The CARNET is a set of forms and accompanying paperwork that are accepted in 71 countries [15].

If you have your own business, it is important to get a retainer of at least 30% of the expected cost so that you may get paid. I was talking to a group of private investigators who did technical security countermeasures (TSCM), private investigation, and security assessments. The question posed to the group was "What is the most difficult thing about private investigation?" The immediate answer from the group was "collecting the money that we are owed from the client." One would think that operating technical equipment to collect data, or perhaps some technical aspect of the work was difficult. One man said that his best clients pay the bill in 90 days while many just do not pay. Then one has to go to one's lawyer and get them to write a letter to those who owe. Sometimes, one gets a collection agency involved or just writes it off as a business loss.

There is one other consideration that is often understood in the mobile forensics world but often not discussed. If one has a business that works with lawyers who defend sex offenders, your company's working relationship with people in law enforcement may deteriorate. When many law enforcement personnel who are in computer forensics retire and start their own business, their colleagues sometimes make a joke. They ask the retiree if they are going over to the dark side, a reference to Darth Vader in Star Wars, and work for the defense. It is important to consider who your clients are and who your friends will be. Larry E. Daniels, a computer forensics consultant, said that as a rule, "ex-law enforcement will not do defense work." Larry said that in 2006 he went to the CEIC conference where there were approximately 400 attendees and he was the only one doing defense work [16]. He said it that it can make one a bit of a pariah doing defense work [16]. This attitude is unfortunate because someone may be innocent and needs a good defense. Consider the person with an unsecured wireless network that is hacked and a wardriver comes by and exchanges child pornography with someone through that unsecured router. Where is that innocent person going to get a good defense? Perhaps Larry E. Daniels would take that case.

One of the things that young people can do is to get as many of the freeware and low-cost tools that he or she can. This book has discussed many of them that are for sale on eBay or are available from a link for download on websites. It is really important that one practices digital forensics on their own old laptops, desktops, and PDAs in order to get the necessary experience before one goes into the field to do that. A person may have to work for a private investigator or lawyer for a short time to gain sufficient capital for name brand cell phone forensic tools that include cable kits. The capital would also be used to get digital forensics insurance for at least one million dollars in case someone sues. Since society often seems to be on a trajectory of increasing regulations and putting more barriers to those starting a career in investigation, working for a private investigator for 5 years may be desirable. Such working experience would allow one to obtain one's own private investigator license after that. During that 5 year time, the employer may pay for the CCE, Encase certification, and other certificates that are expensive and could probably not be afforded by a person who has just started a business. During that 5 years working for a private investigator or lawyer, one could learn about the pitfalls and bad cases to stay away from.

Once someone has the CCE, Encase certification, some digital forensic tools, digital forensic investigator insurance, and a private investigator's license, then one is ready to start one's own business. Many former students of mine are private investigators who started out working for someone else and then went out on their own many years later. The students say that their former employers often do not resent their leaving and starting their own business. This lack of resentment seems odd, but the former employer often refers undesirable or less profitable work to these entrepreneurs. My former students say that they enjoy being their own boss, working when they feel like it, and having the flexibility to change hours to do activities with their families. Much of the private investigator work seems to center around insurance fraud investigation. Other work includes serving subpoenas and investigating a potentially wayward spouse. However, I was told that there is a new trend of work that could be considered a type of legal cyber-spying. Many people want to learn how to be a better parent and legally "monitor" their children's cell phone conversations, email, chat room activities, and web surfing activity. Cyber-parenting, for lack of a better word, seems to be an upcoming trend for these private investigators and that is why many are learning digital forensics. The parents are the customers and they purchase the PDAs, cell phones, and networks for their children. They would like to see a list of the people who their children talk to, what websites they visit, and what videos they look at. They want to make sure their young children are not looking at adult or child pornography and that they are not talking to child predators who wish to meet them at the local mall for illicit sex.

Susteen Secure View could be used by private investigators who are investigating the children's cell phones that are owned and brought in by the parents. There are add-on modules for Secure View that could be used to graph the children's activity times. It might be shown that they are communicating with others at times when it was agreed that they should be doing homework. Perhaps the features of SVProbe could be used to produce a graph of people that the child calls and how often. The list could be checked manually or perhaps against a list of registered sex offenders to make sure the child is safe. The gallery feature in the software also allows parents to see all the pictures taken with the phone. Some pictures may have been marked for deletion. Investigators may choose to use features in the Gallery to link photos with GPS coordinates to a street map. Some pictures might show that the child was in the red light district or an area that is known for drug dealing, even after agreeing with parents not to go there. The tool could be used as a verification tool. In the late twentieth century, President Ronald Reagan once spoke the immortal words, "Trust but Verify." Parents can use the digital forensic tools to make sure their children are telling the truth. If this seems a little extreme, ask a parent whose child was kidnapped, trafficked, and never found.

A private investigator could also give some advice about monitoring the children's whereabouts with tools such as "family locator" that come with the phone. These types of plans allow parents to see where the children are at all times. This way parents will never be unaware of their children's whereabouts and possibly be charged with failure to supervise. Parents may want to surprise their child with an unexpected call to make sure that the phone was not left with a friend while the child did activities in a place that the parent and child agreed would not be done. Law enforcement uses bracelets to keep track of adults for their own protection. The bracelet, made famous by Martha Stewart, makes sure parolees do not violate parole. Cell phones with GPS tracking ability might also be set up for adults with mental impairments from diseases such as Alzheimer's. This would assist in their being found if they wander off. I and a friend once saw a man with dementia of some sort wander off into the woods. The family asked me and my friend to help search and locate the man before nightfall and before the cold set in. The man only spoke Chinese and could not communicate with the police in Mountain Lakes, New Jersey when he was found. I asked the man in Chinese for his phone number and the Mountain Lakes, NJ police contacted the man's family. Everyone thought it would be a good idea for the man to at least carry a GPS-enabled cell phone so he could be called or located if lost. Limited English proficiency (LEP) coupled with dementia make GPS technology a must for some people.

Having your own business might also be good for being a cell phone consultant to schools. In New Jersey, every school in 2012 is supposed to institute a harassment intimidation and bullying (HIB) coordinator. This person has to file reports to the state and coordinate training for teachers. Since cell phones are used in cyberbullying, there may be occasions where the cell phone forensic investigator may be hired to make an image of the phone and recover SMS communications between the alleged bully and the victim. This type of work could be very lucrative given the frequency of cyberbullying in schools between children. The investigator may also make money creating and teaching workshops to teachers, students, and parents. Each workshop may need to discuss prevention, reporting methods, and consequences for certain behaviors. There may be an opportunity to be paid to create a data vault or storage area network to archive cases for school districts in case the action goes to court.

Parents may also wish to find out about legal issues, purchasing, and installing "phone spy" tools that would allow them to get a phone call at the same time as their children. The simultaneous notification also works for an IM or SMS message. This could allow the parents of cyberbullies to monitor and see the actions of their child's phone and mobile device use. Perhaps such reports could be provided to a school to show that a student was reformed, rehabilitated, and no longer offending others with cyberbullying behavior. A person who started his or her own mobile device forensics business would do well to attend some local board of education and parent–teachers association (PTA) meetings to tell them about the new business. The cell phone investigator should ask about the formats for reports for the state so that he or she is ready to assist the local HIB coordinator immediately with reports. The mobile forensic examiner should find out from parents and teachers about what type of mobile devices that the school children use and then make sure the investigator also has the tools and training to investigate those mobile devices. The person with a new mobile forensics business needs to find these new markets and then be ready and available for business. Perhaps being a sponsor for some local school sports events might help to get publicity for one's new business.

The mobile device investigator might also find that he or she needs to diversify his or her business and may do training in mobile device usage. Certain smartphones with cameras and Kurzweil software can be used by blind people to take pictures of signs and then verbally read the content aloud to the blind person. The investigation component of the business could be supplemented by a new branch that enables special needs populations. This would be a good business strategy. One could also do repairs and replace mini USB ports, LCD screens, and cases to put the Blackberry in. Having one's own business is good, the only concern seems to be the price of the tools, staying current with training, and having insurance in case there is a lawsuit. It seems that budgeting fifteen to twenty thousand U.S. dollars per year would be sufficient for the training, tools, hard drives, certifications, and professional associations and conferences to stay current in continuing education credits.

Consider a minimum of fifteen thousand dollars a year to be a mobile device forensics professional. If one charges two hundred dollars per hour, then one would have to be able to bill for at least 100 h just to break even. States vary with taxation amounts, but one must consider that one must pay taxes on money earned. There is often some part-time work with lawyers, but one would have to work hard to break even because of the cost of the tools. Who goes in business to break even? Perhaps someone who teaches full time for a university would not mind working to break even on the part-time business. The work would provide valuable experience for the classroom, as well as for material to write about in books. The academic with this type of experience might make good contacts and gain credibility in this field. The academic may later choose to change careers and apply for a job in the Department of Homeland Security since he or she would have the academic and real-life experience. In this case, the business would only be like a hobby to provide the training, certifications, tools, and experience to get a professor's job in a more prestigious university or with a job such as DHS.

If one bought all the good mobile device tools and could secure 1000 billable hours every year, then one's income would be two hundred thousand U.S. dollars. There would also be twenty thousand U.S. dollars to write off in tools, certification, and conferences. One might have to pay twenty thousand U.S. dollars for a health insurance plan. One is still at one hundred and sixty thousand U.S. dollars a year which is very good. That means one would have to get 83 billable hours per month or about 21 billable hours per week. If one works 5 days a week, it would mean 4 h and 15 min per day. If one has enough contacts with schools, lawyers, private investigators, and corporate investigation firms to make that, then one can do very well. A lot depends on one getting steady high-paying work and being able to collect the money.

The ideal setup for a new investigator would be to have Susteen Secure View, XRY Forensics, Mobil Edit, Paraben's Device Seizure, Cellebrite Ruggedized Cell Phone Forensics tools and the Chinese phone support, Logic Cube's CellDek, BitPim, and Encase. Then the Blackthorn Two software would be necessary for GPS devices. There are also a number of low-cost and freeware tools that one could use for both USB flash drives and digital cameras. Perhaps it is possible to have different services for different markets at different prices. The special needs services with GPS and with the blind might be fifty dollars an hour. Working with both of those previously discussed populations might lead to grant-fundable work with emergency managers who need locate and enable special needs populations. The key to success seems to be diversity. If one specialized only in mobile device investigation, there may not be enough work. However, a good living may be made if one supplements the investigation part of the business with additional work with special needs populations, repairs, selling accessories, and performing expert testimony.

Many people who have their own business in something such as computer forensics or mobile device forensics also wish to teach at a university and have a college connection. The college connection allows them to write academic papers in journals or present papers at a conference. A CCE can use a five page paper to get twelve credits of the forty that are needed every 2 years in order to keep the CCE certification. The papers also help keep one credible in the field of mobile device forensics and are a type of advertising too, since others will see it. A person who teaches at the university level part time might receive ten to twenty thousand U.S. dollars per year which might be enough to pay for the tools, insurance, and lower-cost certifications. The university is also a way for people to meet potential clients and advertise. Teaching at a university also gives people some instant respect if they have to testify in court as an expert witness. If they wrote papers as a coauthor with a professor who holds a doctorate, then that is even better. The university connection will also put them in touch with textbook publishers. Writing books may become an option for additional revenue streams.

One of the biggest dangers in having your own business is getting work that nobody else wants. A friend once told me that he had to testify as an expert witness in a case against a crime family. There was a period for at least a month that he said he worried every time he started his car. In court, the opposition never took their eyes off him and gave him a stone face look of constant disapproval. The other pitfall in having your own business is to be given work that looks easy, but is actually very time consuming. That may sound good, but one will never be able to collect for the hours it takes to do the job. Consider a new smartphone with a SIM card, high-capacity hard drive, and an 8 GB external storage card. Suppose that the sum of these three storage areas contain four thousand pictures, two thousand SMS messages, and five thousand emails. Many pictures are only vaguely about the case while many are directly related. Suppose it takes 500 h to seize the data, collate it, analyze it, create a report, and then testify in court. Who will pay the one hundred thousand U.S. dollars bill that it takes to do the work? Can you force them to pay? Will you take them to court to get your money? Is it worth the court costs, legal fees, and the stress? If one does not do a thorough job and it is discovered later, then some type of negligence lawsuit might occur. Does one hire and trust low-paid document readers, or does one try to do it all and not report the true hours? It is really a dilemma about what to do.

There are all kinds of new rules that must be complied with, if one works for financial corporations or health care providers. Who provides that training? HIPAA has to do with the health care industry; who provides the digital forensic examiner with the specialized training for HIPAA? Is the training free? Even if the training is free, the time engaged in training activities will take away from earning potentials that must be reached to break even or gain a profit. One may have to specialize in certain types of clients such as health care, and ask for some free yearly training that is specific to that particular industry. Many times when there are security positions open, people want security people who are specific to that industry. The casino industry is one example. ASIS International has a yearly conference for members who are security people specifically for the casino security. I once asked about doing some work for one of the casinos in Atlantic City and was politely told at a meeting that they already have a specialized pool of security people who are specific to the casino industries. That actually makes good sense. Casino owners want someone who knows the industry, knows the security specialty that they need, and is known to be trustable.

There is another factor of having your own business that is difficult to grasp unless one has lived through it. I was once an academic observer on a TSCM job with a friend who also does network security and computer forensic jobs. I and my friend drove to a job site. They parked around the corner out of sight from the job. Mr X took the magnetic signs for the business from the truck and put them in the vehicle. Mr X once told me that someone once stabbed the inside of his four tires with big knives. He did not know of the situation until he experienced four blowouts on the highway. He would have flipped the truck except that he was driving slowly due to heavy traffic. He said that if one finds an electronic bug, the person who planted it could get 3 years in prison. He said that there is a rumor that people will kill rather than do 3 years in prison. We carried the material in and Mr X conducted the job. We had to cover the windows, turn the blinds, and turn up the music to set off the bugs. Mr X used a large wand to try to find HF eavesdropping devices. We used hand signals to communicate since the suspect might have been listening to everything in the office. Mr X says that sometimes the person who places a bug will show up and may harm the TSCM. It is therefore important to be ready to call 911 and run. There is a certain amount of daily stress from facing real danger on a daily basis. Mr X had to pay for the four tires out of his own pocket. There is a certain amount of danger that many people do not really understand. It is one thing to see such things on TV and it is another thing to live it. I have taken cybercrime classes to visit the Bergen County, New Jersey jail. It is one thing to see a movie on television about prison and it is another thing to visit a jail, see prisoners up close, and even go into a cell, and hear the door shut behind you.

Investigation of any type is going to take people into a world that is dangerous and expose them to people who most people only see on television. The content on the mobile devices may bring us into contact with a world of sleaze and potential discomfort at times as we may see sexting on mobile devices and then may have to testify in court. The evidence we find may force us to testify in cases against organized crime and child traffickers. Some of these people may think no more of killing a person than most people would think of killing a ground hog that was eating vegetables in their prized garden. If one is going to go into investigation, expert testimony, and related careers, one has to ask if they have the nerves to deal with daily dangers. Does one have the stomach to see the vile pictures and video of children exploitation? As was stated earlier in a previous chapter, a U.S. government-funded study stated that these images have caused some mental health issues for some investigators. Before you go into any career, are you ready for what you will see on mobile devices? Are you ready to see the exploitive and ruthless side of humanity that will do anything to make money? Are you willing to testify in court against dangerous people?

Recently, a law enforcement officer asked me about a cell phone used in a homicide. I did not have any information about the phone nor did my tools have the capability to investigate that phone. It was too new and not part of the American cell phone consumer market. Had I been able to assist, I might have had to testify in the homicide case. A few years earlier, a similar occurrence happened with a phone used by someone who was arrested with large quantities of drugs and might have been a member of a drug cartel. The phone was foreign and not something I could help with. If I was able to help, I might have had to go to court to testify about evidence used against a drug cartel. Are you ready for that? It is one thing to watch *NCIS* and *Criminal Mind*, but it is another thing to face it personally.

Are schools doing enough to prepare people for a career in investigation and dealing with evidence that includes violent images and videos or pictures of child exploitation? Is that a school or police academy's responsibility? A career is more having the technical skills to collect evidence, analyze it, preserve it, and report on it. Can you live with what you do? In the American justice system, everyone has a right to a speedy trial and a lawyer. If a person cannot afford a lawyer, the court will appoint one. A public defender might hire you to help defend a child predator. Perhaps you may find a technical loophole to get him free. Perhaps you can show the system was hacked and others may have put some evidence there. You might bring some reasonable doubt to the case and he is free. Can you live with that? There is a mental health side of careers that technical institutes, colleges, technical schools, and law enforcement agencies may or may not be dealing with. If a person flips out and has to be hospitalized, it is a great cost to society. There is the lack of productivity. Someone goes from being a taxpayer to someone who collects benefits from a mental disability. Perhaps private investigation, mobile device forensics, or being a lawyer is for you. Perhaps it is not for you. It seems that one should try to speak to people in that career first before making a decision. It may be advisable to first obtain an internship for a summer or work 1 day a week in that industry. Perhaps even an unpaid internship with a private investigator or mobile device investigator/certified computer investigator would give a glimpse into a world that you are going to devote your life too. Whatever you do is your decision. The United States of America's constitution, the highest law of the land, says that people have the right to the pursuit of happiness. That is your right to pursue a career that makes you happy, but are you truly making an informed decision?

This book was written with the motivation to discuss all of the technical tools used for each type of technical examination of cell phones, PDAs, netbooks, digital cameras, flash drives, and GPS devices. The history of each device was discussed. The latest tools, techniques, and operating systems were discussed. Each chapter also helped you to understand the forensic preparation of the media, and how the machine should have licensed software. It is sad if the fruit of one's investigation is discarded due to something as foolish as improperly licensed software. The disposition of the examination machine was also discussed. Each machine that is used to examine evidence should be free of viruses, spyware, and rootkits, and should not be accessible to anyone else by any means of connectivity. This book also discussed the ways of isolating evidence with Faraday bags and documenting the trail of evidence with the chain of custody. There have also been some discussions about having both the technical skills and the mental skills needed to have a successful long-term career in this business. It is hoped that the reader will now have a full spectrum view of an investigative career that involves mobile devices.

It is hoped that you are now enlightened to some careers that utilize mobile device forensics. Choosing a career is more than making a living. It is doing something that enables you to live and should be something that you can live with. Many people enter careers or marriages because they met Mr Right-Now instead of Mr Right. Please think about the tools, the costs of starting your own business, your lifestyle, and your personality. Try to visit some law offices, some Regional Computer Forensic Labs (RCFL), talk to a private investigator and some forensic tool vendors. Then talk to some academics, your family, and some friends who know you. In the end, choose the career that is right for you. Find a good fit for your financial needs and lifestyle. Do not be one of those people who open up the drawer and have a load of medicine for stress, upset stomachs, headaches, and backaches. If you choose a line of work that you love, then it is said that you will never work a day in your life.

YOUR OWN BUSINESS: CONTRACTING W/SELF-PUBLISHED MOBILE FORENSICS BOOK AUTHOR

One of the careers that is open to anyone who likes to write is being an author. Some websites pay people by the word while others pay people per article. I wrote a series of books that discuss investigation and mobile device forensics. *E-Forensics and Signal Intelligence for Everyone* is one such book. In this arrangement, one gambles by paying someone such as AuthorHouse a reasonable fee to publish your book and advertise it on hundreds of websites including Barnes & Noble and Amazon. Then, if your book sells large quantities, then you receive quarterly royalty checks for the book over a period of many years. If the royalties for the book are less than the price of publishing the book, then you lose money. There is a common saying, "If there is no risk, there is no reward." AuthorHouse is what is commonly called "a self-publishing" publisher. Johnathon Clifford coined the term "vanity publishing" to this type of arrangement where one pays to publish [17].

YOUR OWN BUSINESS: CONTRACTING W/TRADITIONAL PUBLISHING FIRM OF MOBILE FORENSICS

It is better to write textbooks for a recognized publishing firm such as CRC Press, McGraw-Hill, and many others who come to mind. Suppose that one gets one thousand dollars to write the book and then one gets three dollars as a royalty for a book. If one hundred thousand books sell, then the author gets three hundred and one thousand U.S. dollars. If an author can produce such a book every 2 years, then one would have an income of approximately $150,000 per year before taxes. This seems a more sensible option than being a self-published author. If one combines this with teaching for an online university, then one can get knowledge from the questions of students as well as additional income.

CORPORATION/IT SECURITY AND INVESTIGATIONS

Corporations often need investigators who are part of an incident response team to seize mobile devices and investigate the contents, and prepare a report. Some corporations also create the mobile forensic tools with their own programmers, testing crew, and then need sales people to sell these tools. These are specialized careers but are interesting and can be financially rewarding. Salespeople might do training or there may be professional trainers. Salespeople who are around this industry may also decide to use the products and get a job in a new career as an investigator. A corporate investigator may be downsized and seek a career in sales. There are also nonprofit universities, for-profit universities, online universities, and traditional bricks-and-mortar universities. Many of these universities need teachers, researchers, and professors to create classes, teach, and advance the science of mobile device forensics. In this new economy, there is one thing that is constant and that is change.

CORPORATE: PREPARING FOR AN INVESTIGATIVE CAREER

A person who is a mobile forensic expert may want to get a job at a cell phone store. There are many people who are very upset when they lost all the data in their phone.

They go to the store to get a new phone because the old one does not function properly. Many of these people will pay to get their data back but have no idea who to ask. Many of these people would also like the opportunity to transfer all their data, including SMS, pictures, call logs, contact books, and email to the new phone. Data transfer could be considered a component of mobile device forensics. Being a cell phone technician is a great way to get initial experience on a large variety of phones and learning about the industry. Some people who sold and configured phones went on to the cell phone forensics industry. Working in a cell phone store could be a good first job and a great way to get practical experience on a wide variety of phones.

CORPORATE: RETAIL SECURITY, LOSS PREVENTION, AND INVESTIGATION

There is one area of security that seems to be growing and that is loss prevention and retail security. In Germany, the EHI Retail Institute says that 2.7 billion dollars worth of goods are shoplifted each year [18]. The National Association for Shoplifting Prevention says that approximately 13 billion dollars worth of goods are shoplifted annually in the United States [19]. One of the recent trends in shoplifting involves using social media such as Twitter and Facebook and portable computing devices [20]. Groups organize and swarm on a retailer to shoplift. This is known as flash robbing and it happened in Silver Springs, Maryland with a group possibly as large as fifty people going into a 7–11 store to steal [20].

There are times when people drop a cell phone or PDA in a robbery. If the store is big enough and has a store detective, he or she will examine the digital device to collect all of the evidence from the device concerning the flash robbery. The communications could reveal all of the identities of the robbers and then it might be possible to prosecute. Retail security and loss prevention is a large industry that is very stable. In fact as the economy sours and theft increases, the need for security, investigation, mobile device forensics, and loss prevention will increase.

FOR PROFIT UNIVERSITIES/PRIVATE UNIVERSITIES: MOBILE FORENSICS

The other area of growth for mobile forensics is academia. As the need for mobile device investigators increases, universities, trade schools, and county colleges will respond with programs for people to learn these skills. That means that there will be faculty positions to teach mobile device forensics and perhaps other forms of digital forensics. The faculty will most likely have to have a doctorate in computer science or criminal justice and have certifications for the industry such as Guidance Software's Encase certification and the ISCFE's CCE certification. An academic job is interesting because one constantly learns, develops new classes, and teaches students the latest methodologies for mobile device forensics. One is usually on a variety of committees at the school where one meets law enforcement, private investigators, state police, and other entities who advise one on the equipment and tools to teach. These committees will often review class materials, examine student feedback, and help revise the classes.

Some academic institutions that get a government or private grant may have a mobile lab that allows the instructor to give private classes to a variety of organizations who need classes on mobile forensics. This allows the instructor the opportunity to travel, see other people's investigative labs and tools, and see what type of investigations that they do. This allows one to learn practical knowledge and then revise one's class to include new knowledge. It makes for a dynamic and exciting career.

Many times there are government grants available for digital forensics research and training. An academic may get to write a proposal to get funding for a certain area of research that he or she may want to do. Perhaps one may propose a new mobile forensic investigation technique for finding all the participants of a flash robbing incident. Then if one gets the grant, one gets to do experiments, publish results, offer new classes, go to conferences, and write books. This is a great career. It is exciting, dynamic, and one gets to go to a variety of places one normally does not get to go to, such as Las Vegas, Nevada for the World Comp conference or to the High Tech Crime Investigative Association Conference in Hershey, Pennsylvania.

Academia is also good because one may have flexible hours and 1 or 2 months off in the summer. However, during months off, one is often reading books, developing new classes, and doing the activities that professors love to do. The only difference is that the intensity is less in the summer time and there is still time for vacation. Academia also allows one to collaborate with various government projects such as DARPA or HSARPA. There is also a possibility for a person in academia with ambition to move to the defense contractor sector, the intelligence sector, or move to Washington, D.C. to be part of an institute or think tank that influences policy or testifies before Congress.

There may even be opportunities to travel to another country to teach a class. Fairleigh Dickinson University has a campus in Wroxton, England and one in Vancouver, Canada. Academics have the opportunity to teach hybrid courses with an online component and in-person component in the other country. Oversees classes usually have guest lecturers from that country teach a relevant component of the class. With a mobile forensic class, a mobile forensic examiner from Scotland Yards may guest lecture and then take students to a digital forensic lab field trip in England. The academic life can be great if one gets into the right institution and makes use of the opportunities that are available.

Another great benefit for working for academia is that one can teach a 4-h continuing education class for a group who wants to learn some theory and how to use a mobile device forensic tool quickly. Then one can create and teach an undergraduate class on mobile device forensics that uses a textbook. Such environments allow the professor to go in depth much more than continuing education classes with discussions and exercises on topics such as file systems, data volatility, and the American legal system. One may also teach a graduate class on mobile device forensics and include elements of research and academic papers from a variety of conferences. Academia lets one really delve into the subject at various levels and teach it to various audiences at various educational levels.

CORPORATE: CORPORATE TRAINER FOR MOBILE FORENSICS

There are a number of tools that mobile forensics examiners use. Susteen has Secure View and Paraben has Device Seizure and there are dozens of other products such as Blackthorn 2's GPS Forensics by the Berla Corporation. People in the military, private investigation, academia, loss prevention, retail security, RCFL, intelligence agencies, and municipal police departments all need to purchase and learn how to use these tools. Even if they can learn how to use them on their own, they may need some type of class instruction certificate in case they go to court.

Corporate training can be a good life because one gets to go on a variety of trips all around the country or around the world teaching others the mobile forensic tools. Many of the students in these locations will invite the teacher to see their lab, ask their opinions on mobile forensic issues, and then sometimes ask for new devices to become supported. The students may also ask for new features such as a geo-mapping or creating a directed graph of all the contacts in the call log. The opportunity to meet users, drive product design, and teach is phenomenal. If people like the teacher, they may suggest that the teacher goes sightseeing in that location for the duration of when the class ends until the time that the trainer must get ready to go to the airport. This is not goof-off time but a time when the trainer can find out more about possible new sales leads for the product. It is also a time for people to share a meal and enjoy some touring.

Corporate trainers also get to go to a variety of conferences and trade shows. They obviously cannot see many presentations because they must be in the vendor area to meet potential customers and demonstrate products. However, they usually get to go to all the keynote speeches and see the main presentations by people who lead the field of mobile forensics. The trainers can also talk to other vendors, see their products, and report on new features to their management. It is an exciting career. If one can take the wife and children, they may go sightseeing during the day while one has the conference and then everyone may spend some time together after the evening activities are over.

CORPORATION/IT SECURITY AND INVESTIGATIONS

I know a man from ASIS International who was recently downsized and now collects unemployment. The man's position was cell phone examiner and computer forensic examiner. He worked for a large telecommunication company. He did not investigate cell phones very often, perhaps one every 2 months. He frequently examined desktops and laptops. He said that it is very difficult for companies to justify employing someone with an extremely focused career such as cell phone forensic examiner unless there is a tremendous need for that or unless the company gets outsourced work in that area of digital forensics. It is more likely that a person is a digital forensic investigator who works on a wide range of devices such as cameras, PDAs, cell phones, IPADS, desktops, laptops, notebooks, and netbooks. He also said that the person trained in this area will also be on call for an incident response team since most companies cannot justify someone on the payroll for only that function. In a slow economy, a person may have many functions that are somewhat related, in order for the company to save money on benefits, reduce manpower, and consolidate infrequent tasks such as incident response and mobile device investigation.

CORPORATE: MOBILE DEVICES FORENSICS SALESPERSON

Every forensics tool that exists has salespeople whose job is to sell those tools to people who go to conferences, trade shows, symposiums, seminars, and those who contact the organization after reading about the product on the website. The salesmen track for forensic tools is a wonderful career option that most people overlook. Many of these same people have provided me with information about advanced features of products over the years and have helped convince the management at their organizations to provide 30 day trail downloads of products for students of continuing education. These salesmen must learn every aspect of the product and be able to demonstrate to potential customers in law enforcement, intelligence, consulting, private investigation, and the military.

If the salesman is very educated about the product and has a computer science degree, he may be a much better salesman because he can talk about the more technical aspects of the products with those who will use it. The salespeople also speak to customers who want new features such as extracting a copy of the GPS data from a picture and geo-tagging it on a map. The salesperson is really on the cutting edge of digital forensics because he or she gets to utilize the latest products as well as future beta versions. However, he or she is also talking to investigators in the field who need advanced new capabilities and serves an important role in the contextual design of new products. Beyer and Holzblatt describe this customer-driven process in detail and the book describes various successes with the process [21]. I used Contextual Inquiry and Design to create successful communications applications on the computer for nonverbal persons with limited physical capabilities in the 19990s and early twenty-first century [22].

CORPORATE: PROGRAMMER FOR DIGITAL DEVICE FORENSIC APPLICATIONS

One of the careers that is often overlooked is the creation of new tools. One can create one's own tools and sell them or work for an established company such as Susteen, Paraben, or Cellebrite. If you work for someone else as a programmer, you have their financial backing and have the luxury of knowing that a regular check will be paid to you and that there are health benefits and perhaps a retirement plan. An established company also has a customer base, distribution chain, and professional salespeople and marketing professionals. The programmer can rest easy by not having to worry about the finances and can work purely on the product. However, if there is no risk, there is no reward. The programmer will only have enough money to survive while the owner of the mobile forensic tool may grow rich or go bankrupt.

Some cell phone and mobile forensic tools use a program such as Python to program in. At the High Tech Crime Investigative Conference (2010) in Atlanta, Georgia, many vendors of forensic tools recommended that academics learn Python scripting to create their own applications or modify existing tools. BitPim is a cell phone forensic application that is the result of five hundred lines of C/C ++ programming and thirty thousand lines of Python programming code [23]. If one uses the Cocomo model of predicting effort [24,25]

$$\text{Effort} = a(\text{size})^b$$
$$b = 1.05,\ a = 2.4 \text{ (organic, data processing in house)}$$

Effort is measured in staff months, and size is measured in thousands of lines. If an application such as BitPim is created organically by in-house data processing specialists, the effort by the Cocomo model shows that it takes 85.35 person months. This time frame should be significantly reducible by adding more than one expert programmer and by dividing the program into modules and having each programmer work on different modules. There are books that have many models that are excellent predictors of how long it takes to do a project. If one uses the Cocomo model, it would take a programmer a period of 7 years to create an application such as BitPim. Most people do not have the capability to support themselves and their families for such periods of time and then incur more debt to start a company to market and sell such a product. It seems a more logical choice to work as a programmer for someone else who is already selling other related products and finance your effort as an employee/programmer to create such applications.

Those programmers who learned programming in the late twentieth century would benefit greatly by learning to use Microsoft visual programming tools such as Visual Basic .NET, Visual C++ .NET, Visual Studio, and then other languages such as JAVA and Python. A working knowledge of Unix, Linux, and scripting would be beneficial too. Programmers would also greatly benefit by taking a course in human–computer interfacing and program design. They would also greatly benefit by reading Stephen Hill's book *A Practical Introduction to the Human Computer Interface*. Programmers could also benefit by learning how to ask questions of users at trade shows and conferences to learn what type of interfaces they find effective. Then the programmer could use a process of contextual inquiry and design to create a new tool with an easy-to-use interface and test it with users. In any case, the tool needs to be usable for investigators who are technically savvy and those who are not.

CORPORATE: MOBILE DEVICE FORENSIC PROGRAM TESTING PROFESSIONAL

One of the few unsung heroes in the mobile forensics community is the person or people who test applications and create reports for the investigative community. The Computer Security Division in the Information Technology Laboratory of the National Institute of Standards and Technology often tests new tools and writes reports on their findings. This is a good career path for those who like to test the latest and greatest versions of programs as well as hardware and then compare as well as write reports on their capabilities.

CORPORATE INVESTIGATIONS: UNITED KINGDOM

A job was advertised for a company called Disklabs in the United Kingdom in December 2011. Disklabs does work across Europe for corporations and law enforcement. It is important to look at a real example of a job, examine the salary, the duties, and then analyze if this is for you. The title of the advertised position was Mobile Phone Forensic Analyst. The posted salary £22K–30K based on experience. The job was based in Tamworth, Staffordshire. It is important to look at the rate of exchange of dollars to pounds to see the salary.

The website said, "The successful candidate will benefit from a package that includes Private Healthcare after 1 year, 1 extra days holiday per year served, (up to 5 days), paid charity day, 37.5 h working week. The successful candidate will have opportunities to progress in a company that is expanding, be provided with relevant ongoing training" [26]. Disklabs also asked for the following attributes: "Experience of working with .XRY, Experience of working with .XACT, Experience of Oxygen Forensic, Experience of MobilEdit, Experience of Hex Dumping, (manual and automated), Excellence in Report Writing, and Excellent References" [26]. When we analyze this, we can understand that they are looking for a person who can use the latest tools for examining UK phones, examine high-level evidence collected by the tools, and low-level bytes from a hex dump, and then analyze it and organize into a report.

They also want the potential candidate to "Be fully conversant with latest ACPO Guidelines, Have at least 12 months experience in this field, Hold a full UK driving license, be an excellent communicator, Be able to work in a dynamic team, and Be Methodical" [26]. A mobile phone examiner needs to be able to pick up phones, examine them, and work with a team to prepare a report for law enforcement or a corporation. These are all good skills to have for any aspect of investigation.

PUBLIC SECTOR/IT SECURITY INVESTIGATIONS

The public sector is where many investigators start out. It is not uncommon for someone to make $60,000.00 a year plus benefits in New Jersey working for a county prosecutor's office doing investigations with computers and cell phones. It could be as high as $90,000.00 in some circumstances. I know someone who was making $60,000 and then became a professor and was then making about $75,000.00. Then the person went to a corporate investigation firm in New York City and received $250,000.00 per year. Then after a few years the man progressed to another rank within the same organization and made $750,000.00 per year. The man switched jobs and the last I heard was making over 1 million dollars per year. That sounds great but there are tremendous hours, pressures, and lots of travel in the high-paying jobs. However, one good year as a top manager in corporate investigation/digital forensics (could include phones, computers, anything) could equal 20 years in the county prosecutor's office. One could live spartanly as a public employee for 25 years or one could retire after really good 5 years in the corporate life. Then one could move and pursue another line of work, teach, or retire.

Some people look at public services as a calling such as one would view the ministry or monastic life. Some people are not motivated by money and want to bring others to justice and devote their life to the law enforcement organization such as the CIA, DIA, FBI, county prosecutor's computer crime unit, or a municipal police department. For those who want a secure life where layoffs are infrequent, then public service might be the way to go. There is also in most cases a pension, lifelong medical benefits, and a sense of teamwork in the public sector that is not present in most corporations today. One has to evaluate their personality, willingness to travel, and willing to risk downsizing before making a decision. Changing jobs is not a problem in good economies but what about bad ones? Who wants to change jobs at age fifty with a wife and kids facing college? There is a lot to consider and one's family may have a lot to say about it for better or worse. In this part of the chapter, the reader would be advised to consider what was just discussed as he or she reads about careers in the public sector.

PUBLIC SECTOR: DEPARTMENT OF DEFENSE DoD CRIME LAB

There are sometimes civilian positions within the United States Department of Defense as a digital evidence investigator in their high-security lab in the greater Washington D.C. area. This is a fascinating place to work because there are tremendous amounts of devices that come in daily such as phones, computers, hard drives, and varied types of computer media. Some items have been damaged which adds to the challenge [27]. Such a place has pallet loads of incoming digital devices coming in daily which most likely offers some degree of job security.

PUBLIC SECTOR: UNITED STATES CIA OR DIA AS A CAREER

Sometimes there are positions advertised on the CIA website for mobile forensic examiners. These jobs usually require a degree in computer science, some foreign language skills, patriotism, and the willingness to travel as much as half of your time. This could be an exciting career for many and see events that most may only see in a movie or magazine many years later. One also gets to work with the top minds in his or her field and might have a large budget to work with. The website for the CIA is www.cia.gov and there is a section for jobs. The background investigation is lengthy so one may need other temporary work until all the clearances can be obtained. The CIA may be a challenging job and one may be killed in another country and some people know it as an anonymous star placed on the wall in Langley, Virginia.

The Defense Intelligence Agency (DIA) is one of sixteen intelligence agencies in the United States and could be an agency to make a career in. The DIA needs intelligence analysts, mobile device forensics specialists, and linguists. One has to go through a lengthy clearance process and each stage of your life has to be investigated. The DIA is a fascinating agency that uses a variety of computer-driven simulations with a variety of scenarios to train people. One can usually meet representatives of all the intelligence agencies at the yearly National Military Intelligence Conference if one is interested in a career as a mobile device forensic examiner in one of those agencies.

PUBLIC SECTOR: ONI AND NCIS

The acronym ONI means Office of Naval Intelligence and NCIS is an acronym for Naval Criminal Investigative Service. Both organizations are part of the United States Navy. The NCIS has a need to examine mobile devices and computers within the context of criminal investigations. It appears that more people carry mobile devices and therefore in a murder case, for example, they may have evidence about the deceased person's activities and communications. The NCIS being a federal government job should have a good pension plan.

The ONI has a counterpiracy, ship-watching, and maritime intelligence watch units. The ONI is tasked to watch ships in the Gulf of Aden and investigate cases of piracy. Many pirates are from Somalia and use small boats that may have pirates equipped with both cell phones and GPS navigation units. A book called *The Pirates of Somalia: Inside Their Hidden World* by Jay Bahadur discusses the motivations, equipment, and operations of the pirates of Somalia [28]. The ONI may need mobile device forensic examiners to examine both cell phones and GPS navigation devices from pirates who were captured in the Gulf of Aden. This could prove to be an exciting career with good health and retirement benefits.

PUBLIC SECTOR: COUNTY PROSECUTOR'S OFFICE COMPUTER CRIME UNIT

Each state has many counties. Even a small state such as New Jersey has 21 counties. Each county has a prosecutor's office and a computer crime unit of some sort. Many prosecutors' offices have some extra funding and equipment that is based on what they seize from illegal operations that they break up and arrest those running them. That can leave a lot of money for training, tools, and equipment that would not be possible in a university and available only in large corporations. For many who work in computer crimes units in their county, they have the benefit of living close to home, job security, benefits, somewhat regular hours, and often comp time. Comp time is if you work 5 h extra, you can take 5 h in the future when you need it and things are not too busy. For those who have children contemplating medical school, the computer crimes unit is a good place to start out, make contacts, and then go to the corporate computer or mobile forensics investigation world.

DEFENSE CONTRACTOR IN IRAQ OR AFGHANISTAN

I redacted some data from an email for security reasons, but did share it with my class regarding someone who was a defense contractor and working in Iraq. The email was from a man who we will call Victor, who discussed his pay and experiences with cell phone forensics with his fellow CCEs. Victor was stationed at a forward operating base in Iraq with a few other people and worked for 14 hours every day examining phones. Victor said that he received approximately ninety thousand dollars for 6 months work. He said that he slept in a cement crypt every night. His colleagues did not tuck him into bed but rather pulled over a large 1-inch thick sheet of copper over the crypt and then placed sand bags above the sheet. Victor said that many mortar rounds often hit close to his work area. He further stated that the work area and long hours "sucks" but the pay is great and the work is interesting. The class read the email and found that there was a lot of stress from extremely long workdays. Victor worked everyday with no time off and recreation.

Victor said that he also examines SD cards and a variety of external media cards for both existing and deleted files. He uses a variety of commercially available tools for recovering files from FAT file systems and cell phones. My class said that they felt that they could do the technical part of the job but were not sure if they could endure the conditions of the work environment. One student who was out of work and was at the end of his unemployment said that he would take the job immediately if he had the opportunity. The student told the class that he is now looking for such a job and that there is a website called "Danger Zone Jobs, Defense Contractors and Private Military" [29]. This website requires a small fee for joining but describes work conditions, pay, clearances, and many of the policies and procedures that one will have to abide by in such work environments. The website consolidates much information from public sources and allows the job searcher to learn about jobs and conditions that would otherwise require many hours of research.

Alison Doyle has a website about jobs in Iraq and says, "Some jobs in Iraq start at about $80,000 a year. Others pay more. In most cases, housing and meals are provided. If you work overseas long enough (over 330 days) that income is excluded from United States taxes" [30]. There are often many companies that hire defense contractors in places such as Afghanistan or Somalia to support the United States or its allies. It is important to go to websites such as www.glassdoor.com and look at jobs, salaries, and work conditions. It is good to be educated about the conditions and pay so that one is not taken advantage of.

DEFENSE CONTRACTOR IN IRAQ OR AFGHANISTAN AND MOONLIGHTING ONLINE

Online professors should have the same education and experience as the in-person professor. That means that a doctorate degree is the admission to the club of professorship.

There is a discussion about online professors getting tenure that is very interesting. Most of the discussion stems from full-time professors who taught in-person classes and that those classes migrated to online. The conversations also state that those who teach online are still expected to go to the universities for meetings, committees, and keep in-person office hours for online students. There is nothing written online about anyone who is teaching 100% online getting tenure [31].

Many people have opted for teaching for purely online universities such as the American Military University and a host of other schools. This seems to be a good option because one can live in a part of the United States where the cost of living and taxation is low. Then one can maximize the income earned. One can also focus on one's discipline and write books and teach in one's field. This maximizes one's efforts in one's interests, teaching, research, and writing. The online professor is not unproductive by serving on a plethora of committees that have no remote association with one's teaching expertise. A person can focus on their discipline and many online universities will pay for a reasonably priced conference if the professor spends one third of their time at the conference engaged in marketing activities and staffing a table with promotional literature for the online university.

Many professors at regular bricks-and-mortar universities complain that as much as 80% of their time is used up on activities that have nothing to do with their research and field of expertise. This can include commuting, marketing activities such as open houses, faculty meetings, various committees, and organizational speeches. Online professors often say that they can go to their in home office and write, correct papers, interact with students online, and then carry out research in a home lab. They can go to conferences, symposiums, trade shows, and other activities in their field and lose perhaps one third of that time to marketing. However, that one third is not lost time because the online professor is gathering information about students' desires for new classes, and perhaps gathering intelligence for new programs or research to perform. Many online professorships offer the same pay as in-person professorships and as was stated, the commuting costs are virtually nonexistent. For the online professor, location is no longer an issue, thus allowing one to minimize living costs.

STUDENTS AND GETTING JOBS

If a student can go to an ASIS International monthly meeting, he or she should go. The student gets to meet people, get advice, and hear about job opportunities. The student would also be advised to bring his or her resume to the meetings in case a job opening is available. The student should also go to conferences such as the High Tech Crime Investigative Association Conference. Even if one cannot afford the conference price, there are often opportunities for free or reduced fees if one helps give out bags, helps with registration, and does all the work associated with running a conference. While the student is there, he or she can network with vendors, presenters, attendees, and sometimes someone is looking to hire a student for a vacancy in their organization. It has been said that about 90% of success was showing up.

FUTURE OF DIGITAL EVIDENCE CAREERS

Gail Thackery, a U.S. attorney for the state of Arizona, says, "every case in America has a computer involved in it. The legal system is hungry for experts at digital evidence" [27].

She also said, "So computer forensics training and careers are going to be hot for a long time" [27]. We also know that in 2008, at least one third of humanity's six billion people had a cell phone [32]. Even if a subset of that is involved in illegal activities, a tremendous amount of digital evidence investigators will be needed. Many private investigators who take my continuing education class at FDU called "Introduction to Cell Phone Forensics" have stated that they are encountering more cell phones in investigations and need to know how to extract and examine the data. The same group of people have said that they also need to learn how to extract the data and examine GPS navigation devices on cars, and especially trucks that do interstate commerce.

SUMMARIZATION AND SOME CAREER ADVICE

Whatever path that you choose, you should make sure that the job, environment, and pay suit your likes, personality, life ambitions, and financial needs. However, if the economy is bad and you cannot afford to wait for a job, you may have to take the available job regardless of its suitability. Then work in that for as long as you must until a better job opportunity arises. One can always take new classes, volunteer one's time to advance technology tools such as BitPim, and then seek new jobs. If one has the time and education, a person can reinvent himself or herself and travel down a new path somewhere within the road of mobile forensics. It is often said that the people of Generation Y will have three or four different careers in their lifetime. As one contemplates this chapter, one can see it is true. A person could be a private investigator, go to law school and become a lawyer specializing in mobile device forensics, later become a professor, and lastly become a policy maker for the federal government. Keep learning, volunteering, and adapting.

REFERENCES

1. Williams, B. (October 21, 2011). Target's state-of-the-art forensics lab catches more than just shoplifters. *MPRnews*. URL accessed December 18, 2011. Retrieved from http://minnesota.publicradio.org/display/web/2011/10/21/target-forensics-lab/.
2. Gruber, M. Spying on your spouse, significant other, or domestic partner—Legal consequences. *Gruberlaw.biz*. URL accessed January 2, 2012. Retrieved from http://www.gruberlaw.biz/pdf/spying-on-spouse.pdf.
3. *Urban Dictionary* (April 22, 2010). Butt call (3). Posted by 12q23w34e45r. URL accessed December 18, 2011. Retrieved from http://www.urbandictionary.com/define.php?term = butt%20call.
4. Brown, S. (2003). *The Complete Idiot's Guide to Private Investigating*, Marie Butler-Knight, USA, p. 8, ISBN 0-02-864399-2.
5. Philip, V. (November 14, 2008). Virtual affair ends in real-life divorce. *ABC News*. URL accessed December 17, 2011. Retrieved from http://abcnews.go.com/International/SmallBiz/story?id = 6255277&page = 1.
6. Renaldo, C. (March 5, 2011). Stay away from fake social networking profiles. *Security World News*. URL accessed December 18, 2011. Retrieved from http://www.securityworldnews.com/2011/03/05/stay-away-from-fake-social-networking-profiles/.
7. Willard, R. (1997). *PI, A Self Study Guide on Becoming a Private Detective*. Paladin Press, Boulder, Colorado, ISBN 0-87364-954-0.
8. Beveridge, W. (1957). *The Art of Scientific Investigation*, Vintage Books, New York, NY.

9. Microsoft Support (Article ID: 982577, September 20, 2011, Revision: 4.0). The file size limits of .pst and .ost files are larger in Outlook 2010. URL accessed January 18, 2012. Retrieved from http://support.microsoft.com/kb/982577.
10. DataNumenInc. Advanced Outlook Repair. URL accessed January 17, 2012. Retrieved from http://www.repair-outlook.com.
11. PST Walker Software—PST Walker. URL accessed January 18, 2012. Retrieved from http://www.pstwalker.com/pstwalker.
12. Intella email investigation and ediscovery software. *Vound Software*. URL accessed January 19, 2012. Retrieved from http://www.vound-software.com/products.
13. Xerox. Xerox Litigation Services. URL accessed January 19, 2012. Retrieved from http://www.xerox-xls.com/pdf/XLS-email-analytics.pdf?PHPSESSID=3aed45ce87f02529534d9daf388a00cf.
14. Aid4Mail. URL accessed January 20, 2012. Retrieved from http://www.aid4mail.com/email.ediscovery.forensic.software.php.
15. *Boomerang Freight Solutions Corporation for International Business*. URL accessed December 19, 2011. Retrieved from http://www.atacarnet.com/.
16. Posted by The Hanged Juror (February 5, 2009). Computer forensics for the defense: A juror interviews Larry E. Daniel [Web log comment]. URL accessed December 19, 2011. Retrieved from http://blog.thejurorinvestigates.com/2009/02/05/computer-forensics-for-the-defense-a-juror-interviews-larry-e-daniel.aspx.
17. *Vanity Publishing*. URL accessed December 19, 2011. Retrieved from http://www.vanitypublishing.info/.
18. Walker, T. (September 6, 2011). Shoplifting costs German retailers 5 million euros a day. *DW*. URL accessed December 18, 2011. Retrieved from http://www.dw-world.de/dw/article/0,15142262,00.html.
19. *Loss Prevention Systems, Inc.* US shoplifting statistics—Atlanta. URL accessed December 18, 2011. Retrieved from http://www.preventshopliftingloss.com/u-s-shoplifting-statistics-atlanta.html.
20. *CIO*. 4 Steps retailers can take to combat flash robs. URL accessed December 18, 2011. Retrieved from http://www.cio.de/news/cio_worldnews/2011/2297171/index.html.
21. Beyer, H., Holtzblatt, K. (1997). *Contextual Design: Defining Customer-Centered Systems*. Morgan Kaufman Publishers, New York, NY, ISBN-13:978-1558604117.
22. Doherty, E. (2001). An investigation of bio-electric interfaces for computer users with disabilities, pp. 140–150, Doctoral thesis, University of Sunderland, Sunderland, England.
23. Binns, R. (n.d.). BitPim, an application in Python. *Bitpim.org*. URL accessed December 20, 2011. Retrieved from http://www.bitpim.org/papers/baypiggies/bitpim-piggies.pdf.
24. Faghih, F. (n.d.). Software effort and schedule estimation. URL accessed December 20, 2011. Retrieved from http://enel.ucalgary.ca/People/Smith/619.94/prev689/1997.94/reports/farshad.htm.
25. Montclair University (1996). Notes written from the blackboard from computer science Professor Eli Weiessman's class on software metrics.
26. Disklabs mobile phone forensics solutions. *Mobile Phone Forensics Careers*. URL accessed December 19, 2011. Retrieved from http://www.mobilephoneforensics.com/careers.php.
27. Radcliff, D. (March 8, 2004). Inside the DoD's Crime Lab. *Network World*. URL accessed December 18, 2011. Retrieved from http://www.networkworld.com/research/2004/0308dod.html?page=2.
28. Bahadur, J. (2011). *The Pirates of Somalia: Inside Their Hidden World*, Pantheon Books, New York, NY, ISBN 030737906X.

29. Danger Zone Jobs. URL accessed December 19, 2011. Retrieved from http://www.dangerzonejobs.com/artman/publish/index.shtml.
30. Doyle, A. (n.d.). Jobs in Iraq. *About.com Job searching*. URL accessed December 19, 2011. Retrieved from http://jobsearch.about.com/od/internationaljobs/a/iraqjobs.htm.
31. Chronicle Forums. *The Chronicle of Higher Education*. URL accessed December 20, 2011. Retrieved from http://chronicle.com/forums/index.php?topic = 64683.0.
32. Doherty, E. (2008). Cell Phone Forensics Class, p. 2. Fairleigh Dickinson University, supported by DOJ grant award 2005-DD-BX-1151.

Index

CPSIA information can be obtained
at www.ICGtesting.com
Printed in the USA
BVHW081521201218
535663BV00012B/178/P